NATURAL PRODUCTS CHEMISTRY

Biomedical and Pharmaceutical Phytochemistry

NATURAL PRODUCTS CHEMISTRY

Biomedical and Pharmaceutical Phytochemistry

Edited by

Tatiana G. Volova, DSc
Debarshi Kar Mahapatra, PhD
Sonia Khanna, PhD
A. K. Haghi, PhD

APPLE
ACADEMIC
PRESS

First edition published [2021]

Apple Academic Press Inc.
1265 Goldenrod Circle, NE,
Palm Bay, FL 32905 USA
4164 Lakeshore Road, Burlington,
ON, L7L 1A4 Canada

CRC Press
6000 Broken Sound Parkway NW,
Suite 300, Boca Raton, FL 33487-2742 USA
2 Park Square, Milton Park,
Abingdon, Oxon, OX14 4RN UK

First issued in paperback 2021

Library and Archives Canada Cataloguing in Publication

Title: Natural products chemistry : biomedical and pharmaceutical phytochemistry / edited by Tatiana G. Volova, DSc, Debarshi Kar Mahapatra, PhD, Sonia Khanna, PhD, A.K. Haghi, PhD.

Names: Volova, Tatiana G., editor. | Mahapatra, Debarshi Kar, editor. | Khanna, Sonia, editor. | Haghi, A. K., editor.

Description: Includes bibliographical references and index.

Identifiers: Canadiana (print) 20200279858 | Canadiana (ebook) 2020028035X | ISBN 9781771888769 (hardcover) | ISBN 9781003000693 (ebook)

Subjects: LCSH: Natural products—Biotechnology. | LCSH: Natural products—Therapeutic use. | LCSH: Biochemistry. | LCSH: Botanical chemistry. | LCSH: Pharmacology.

Classification: LCC QD415 .N38 2020 | DDC 572—dc23

Library of Congress Cataloging-in-Publication Data

Names: Volova, Tatiana G., editor. | Mahapatra, Debarshi Kar, editor. | Khanna, Sonia, editor. | Haghi, A. K., editor.

Title: Natural products chemistry : biomedical and pharmaceutical phytochemistry / edited by Tatiana G. Volova, Debarshi Kar Mahapatra, Sonia Khanna, A.K. Haghi.

Other titles: Natural products chemistry (Volova)

Description: Burlington, ON ; Palm Bay, Florida : Apple Academic Press, [2021] | Includes bibliographical references and index. | Summary: "Natural Products Chemistry: Biomedical and Pharmaceutical Phytochemistry focuses on the development of biochemical, biomedical and their applications. It highlights the importance of accomplishing an integration of engineering with biology and medicine to understand and manage the scientific, industrial, and clinical aspects. It also explains both the basic science and the applications of biotechnology-derived pharmaceuticals, with special emphasis on their clinical use. The biological background provided enables readers to comprehend the major problems in biochemical engineering and formulate effective solutions. This title also expands upon current concepts with the latest research and applications, providing both the breadth and depth researchers need. The book also introduces the topic of natural products chemistry with an overview of key concepts. This book is aimed at professionals from industry, academicians engaged in chemical science or natural product chemistry research, and graduate-level students"-- Provided by publisher.

Identifiers: LCCN 2020028857 (print) | LCCN 2020028858 (ebook) | ISBN 9781771888769 (hardcover) | ISBN 9781003000693 (ebook)

Subjects: MESH: Biological Products--chemistry | Biological Products--pharmacology | Nanotechnology | Drug Design | Phytotherapy

Classification: LCC RM301.25 (print) | LCC RM301.25 (ebook) | NLM QV 241 | DDC 615.1/9--dc23

LC record available at https://lccn.loc.gov/2020028857
LC ebook record available at https://lccn.loc.gov/2020028858

ISBN: 978-1-77188-876-9 (hbk)
ISBN: 978-1-77463-911-5 (pbk)
ISBN: 978-1-00300-069-3 (ebk)

About the Editors

Tatiana G. Volova, DSc

Professor and Head, Department of Biotechnology,
Siberian Federal University, Krasnoyarsk, Russia

Tatiana G. Volova, DSc, is a Professor and Head of the Department of Biotechnology, Siberian Federal University, Krasnoyarsk, Russia. She is the creator and head of the Laboratory of Chemoautotrophic Biosynthesis at the Institute of Biophysics, Siberian Branch of the Russian Academy of Sciences. She is researching the field of physicochemical biology and biotechnology and is a well-known expert in the fields of microbial physiology and biotechnology. She has created and developed a new and original branch in chemoautotrophic biosynthesis, in which the two main directions of the 21st century technologies are conjugated, hydrogen energy and biotechnology. The obtained fundamental results provided significant outputs and were developed by the unique biotechnical producing systems, based on hydrogen biosynthesis for single-cell proteins, amino acids, and enzymes. Under her guidance, the pilot production facility of single-cell protein, utilizing hydrogen, had been constructed and put into operation. The possibility of involvement of man-made sources of hydrogen into biotechnological processes as a substrate, including synthesis gas from brown coals and vegetable wastes, was demonstrated in her research. She had initiated and deployed, in Russia, the comprehensive research on microbial degradable bioplastics; the results of this research cover various aspects of biosynthesis, metabolism, physiological role, structure, and properties of these biopolymers and polyhydroxyalkanoates and have made a scientific basis for their biomedical applications and allowed them to be used in biomedical research. She is the author of more than 300 scientific works, including 13 monographs, 16 inventions, and a series of textbooks for universities.

Debarshi Kar Mahapatra, PhD

Assistant Professor, Department of Pharmaceutical Chemistry,
Dadasaheb Balpande College of Pharmacy, Rashtrasant Tukadoji Maharaj
Nagpur University, Nagpur, Maharashtra, India

Debarshi Kar Mahapatra, PhD, is currently an Assistant Professor in the Department of Pharmaceutical Chemistry, Dadasaheb Balpande College of Pharmacy, Rashtrasant Tukadoji Maharaj Nagpur University, Nagpur, Maharashtra, India. He was formerly an Assistant Professor in the Department of Pharmaceutical Chemistry, Kamla Nehru College of Pharmacy, RTM Nagpur University, Nagpur, India. He has taught medicinal and computational chemistry at both the undergraduate and postgraduate levels and has mentored students in their various research projects. His area of interest includes computer-assisted rational designing and synthesis of drug delivery systems and low-molecular-weight ligands against druggable targets and the optimization of unconventional formulations. He has published research as book chapters, reviews, and case studies in various reputed journals and has presented his work at several international platforms, for which he has received several awards by a number of bodies. He has also authored a book titled *Drug Design*. Presently, he is serving as a reviewer and an editorial board member for several journals of international repute. He is a Member of a number of professional and scientific societies, such as the International Society for Infectious Diseases, the International Science Congress Association, and the International Society of Exercise and Immunology.

Sonia Khanna, PhD

Assistant Professor, Department of Chemistry and Biochemistry,
Sharda University, Greater Noida, India

Sonia Khanna, PhD, is an Assistant Professor working at Sharda University, Greater Noida, Delhi NCR, India, with a teaching experience of more than 10 years. She has published 13 research articles in international journals, including *Coordination Chemistry Reviews, Dalton Transactions,* and publications of the American Chemical Society. She has also authored three book chapters. Her area of research includes synthesis and characterization of complexes and their anticancer properties. She has supervised many MSc and PhD students and is presently associated with many research projects. She is also a member of the review board for a national journal. She earned her PhD in Inorganic Chemistry from Guru Nanak Dev University, Amritsar, India, and was a topper in

BSc and MSc degree programs. She qualified in many national-level competitive exams (NET-JRF, GATE). She is a recipient of a gold medal and many merit certificates for her best performance in chemistry.

A. K. Haghi, PhD

Professor Emeritus of Engineering Sciences, Former Editor-in-Chief, International Journal of Chemoinformatics and Chemical Engineering and Polymers Research Journal; Member, Canadian Research and Development Center of Sciences and Culture

A. K. Haghi, PhD, is the author and editor of 165 books, as well as 1000 published papers in various journals and conference proceedings. He has received several grants, consulted for a number of major corporations, and is a frequent speaker to national and international audiences. Since 1983, he has been serving as a professor at several universities. He is the former Editor-in-Chief of the *International Journal of Chemoinformatics and Chemical Engineering* and *Polymers Research Journal* and is on the editorial boards of many international journals. He is also a member of the Canadian Research and Development Center of Sciences and Cultures, Montreal, Quebec, Canada.

Contents

Contributors

A. Andhare Aishwarya
Department of Microbiology, Biotechnology and Chemistry, Dayanand Science College, Latur 413512, Maharashtra, India

Vivek Asati
Department of Pharmaceutical Chemistry, NRI Institute of Pharmacy, Bhopal 462021, Madhya Pradesh, India

Ahlam Abdul Aziz
Department of Biotechnology, University of Calicut, Malappuram 673635, Kerala, India

Sonia Yesenia Silva Belmares
Research groups of Chemist-Pharmacist-Biologist and Nano-Bioscience

Sanjay Kumar Bharti
Institute of Pharmaceutical Sciences, Guru Ghasidas Vishwavidyalaya
(A Central University), Bilaspur 495009, Chhattisgarh, India

Elda Patricia Segura Ceniceros
School of Chemistry, Autonomous University of Coahuila, Blvd. Venustiano Carranza, Col. República Oriente, Saltillo 25280, Coahuila, Mexico

Amol J. Deshmukh
Department of Plant Pathology, College of Agriculture, Navsari Agricultural University, Waghai 394730, Gujarat, India

Anna Iliná
School of Chemistry, Autonomous University of Coahuila, Blvd. Venustiano Carranza, Col. República Oriente, Saltillo 25280, Coahuila, Mexico

P. R. Jayasree
Department of Biotechnology, University of Calicut, Malappuram 673635, Kerala, India

Jobish Joseph
Department of Biotechnology, University of Calicut, Malappuram 673635, Kerala, India

Hema Joshi
Hema Joshi, Department of Botany, Hindu College, Sonipat, Hryana

Moayad Naeem Khalaf
Chemistry Department, College of Science, University of Basrah, Basrah, Iraq

Sonia Khanna
Department of Chemistry and Biochemistry, Sharda University, Greater Noida, India

Poonam Khullar
Department of Chemistry, BBK DAV College for Women, Amritsar 143001, Punjab, India

P. R. Manish Kumar
Department of Biotechnology, University of Calicut, Malappuram 673635, Kerala, India

Debarshi Kar Mahapatra
Department of Pharmaceutical Chemistry, Dadasaheb Balpande College of Pharmacy, Nagpur 440037, Maharashtra, India

Debasish Kar Mahapatra
Medical and Occupational Health Department, Western Coalfields Limited, Nagpur 440037, Maharashtra, India

Bindu Panampilly
School of Chemical Sciences, Mahatma Gandhi University, Priyadarshini Hills P.O., Kottayam 686560, Kerala, India

S.V. Prudnikova
Siberian Federal University, 79 Svobodnyi Av., Krasnoyarsk 660041, Russia

Luis Enrique Cobos Puc
Research groups of Chemist-Pharmacist-Biologist and Nano-Bioscience

Crystel Aleyvick Sierra Rivera
Research groups of Chemist-Pharmacist-Biologist and Nano-Bioscience

V. A. Rudenok
Izhevsk State Agricultural Academy

Laura María Solís Salas
Research groups of Chemist-Pharmacist-Biologist and Nano-Bioscience

Maria Del Carmen Rodríguez Salazar
Research groups of Chemist-Pharmacist-Biologist and Nano-Bioscience

V. S. Shaniba
Department of Biotechnology, University of Calicut, Malappuram 673635, Kerala, India

Ravindra S. Shinde
Department of Microbiology, Biotechnology and Chemistry, Dayanand Science College, Latur 413512, Maharashtra, India

E.I. Shishatskaya
Siberian Federal University, 79 Svobodnyi Av., Krasnoyarsk 660041, Russia
Institute of Biophysics SB RAS, Federal Research Center "Krasnoyarsk Science Center SB RAS"50/50 Akademgorodok, Krasnoyarsk 660036, Russia

Hussein Ali Shnawa
Polymer Research Center, University of Basrah, Basrah, Iraq

Anamika Singh
Department of Botany, Maitreyi College, University of Delhi, Delhi, India

Rajeev Singh
Department of Environment Studies, Satayawati College, University of Delhi, Delhi, India

A.A. Shumiliva
Siberian Federal University, 79 Svobodnyi Av., Krasnoyarsk 660041, Russia

Therakathinal Thankappan Sreelekha
Regional Cancer Centre, Thiruvananthapuram, Kerala, India

Maya Sreeranganathan
Regional Cancer Centre, Thiruvananthapuram, Kerala, India

Lavnaya Tandon
Department of Chemistry, BBK DAV College for Women, Amritsar 143001, Punjab, India

Abed Alamer Hussein Taobi
Chemistry Department, College of Science, University of Basrah, Basrah, Iraq

Sabu Thomas
School of Chemical Sciences, Mahatma Gandhi University, Priyadarshini Hills P.O.,
Kottayam 686560, Kerala, India

Babukuttan Sheela Unnikrishnan
Regional Cancer Centre, Thiruvananthapuram, Kerala, India

Preethi Gopalakrishnan Usha
Regional Cancer Centre, Thiruvananthapuram, Kerala, India

Tatiana G. Volova
Siberian Federal University, 79 Svobodnyi Av., Krasnoyarsk 660041, Russia
Institute of Biophysics SB RAS, Federal Research Center "Krasnoyarsk Science Center SB RAS" 50/50
Akademgorodok, Krasnoyarsk 660036, Russia

Abbreviations

AG	arabinogalactan
ALT	alanine aminotransferase
AML	acute myeloid leukemia
AST	aspartate aminotransferase
BBB	blood–brain barrier
BBTB	blood–brain tumor barrier
BC	bacterial cellulose
BCE	brain capillary endothelium
CD	cyclodextrin
CMC	carboxymethyl chitosan CMC
CTs	condensed tannins
DAPI	4′,6-diamidino-2-phenylindole
DDS	drug delivery systems
DECO	direct electrochemical oxidation
DL	Dalton lymphoma
DLS	dynamic light scattering
DMA	dynamic mechanical analysis
DMSO	dimethyl sulfoxide
DNA	deoxyribonucleic acid
DOX	doxorubicin
DPBS	Dulbecco's phosphate-buffered saline
DSC	differential scanning calorimetry
DTA	differential thermal analysis
ECs	endothelial cells
EDX	energy dispersive X-ray spectroscopy
EPR	enhanced permeation and retention
ERK	extracellular signal-regulated kinase
FESEM	field-emission scanning electron microscopy
FGF	fibroblast growth factor
FTIR	Fourier transform infrared
GC-MS	gas chromatography coupled to mass spectrometry
GG	guar gum
GI	gastrointestinal
GMP	good manufacturing process

Gox	glucose oxidase
HAI	hospital-acquired infection
HAI	hyaluronic acid
HMEC	human microvascular endothelial cells
HPLC-MS	high-performance liquid chromatography–mass spectrometry
HS	Hestrin–Schramm
HTs	hydrolyzable tannins
HUVECs	human umbilical vein endothelial cells
IAP	index of atmospheric purity IAP
IR	infrared IR
KC	kappa carrageenan
LB	Luria–Bertani
MBC	minimum bactericidal concentration
MCS	mannosylated chitosan
MDR	multidrug resistant
MIC	minimum inhibitory concentration
MMPs	matrix metalloproteinases
mRNA	messenger ribonucleic acid
MRPs	multidrug-resistance-related proteins
MSCs	mesenchymal stem cells
MUF	melamine-urea-formaldehyde
NA	nutrient agar
NGO-HA	nanographene oxide HA conjugate
NIR	near-infrared
NMR	nuclear magnetic resonance
NPs	nanoparticles
OD	optical density
PBS	phosphate-buffered saline
PC	pericyte
PDI	polydispersity index
PEG	polyethylene glycol
PF	phenol-formaldehyde
PHA	polyhydroxyalkanoate
PLA	polylactic acid
PPE	personal protective equipment
PSA	prostate-specific antigen
PTX	paclitaxel
PUF	phenol-urea-formaldehyde
ROS	reactive oxygen species

RT-PCR	real-time reverse transcription polymerase chain reaction RT-PCR
SAMs	self-assembled monolayers
SC	stratum corneum
SEM	scanning electron microscopy
SERS	surface-enhanced Raman scattering
TAM	tamoxifen citrate
TCS	thiolated chitosan
TEM	transmission electron microscopy
TfR	transferrin receptor
TJs	tight junctions
TLC	thin layer chromatography
TPA	tons per annum
TPP	tripolyphosphate
UF	urea-formaldehyde
UV–vis	ultraviolet–visible
VCR	vincristine sulfate
VEGF	vascular endothelial growth factor
VEGFRs	vascular endothelial growth factor receptors
XG	xanthan gum
XRD	X-ray diffraction
ZOI	zone of inhibition

Preface

This book focuses on the development of biochemical and biomedical products and their applications. It highlights the importance of accomplishing an integration of engineering with biology and medicine to understand and manage the scientific, industrial, and clinical aspects.

It also explains both the basic science and the applications of biotechnology-derived pharmaceuticals, with special emphasis on their clinical use.

The biological background provided enables readers to comprehend the major problems of biochemical engineering and to formulate effective solutions.

Each chapter also includes detailed references and a list of recommended books for additional study, making this outstanding treatise a useful resource for teachers of chemistry and researchers working in universities, research institutes, and industries.

In the first chapter of this volume, on strain of acetic acid bacteria, *Komagataeibacter xylinus* B-12068, was studied as a source of bacterial cellulose (BC) production. The effects of cultivation conditions (carbon sources, temperature, and pH) on BC production and properties were studied in surface and submerged cultures. Glucose was found to be the best substrate for BC production among the sugars tested; an ethanol concentration of 3% (w/v) enhanced the productivity of BC. The highest BC yield (up to 17.0 g/L) was obtained under surface static cultivation conditions in the modified HS medium supplemented with ethanol, at pH 3.9, after 7 days of cultivation in the thinnest layer of the medium. Results suggest that BC composites constructed in the present study hold promise as dressings for managing wounds, including contaminated ones.

In the second chapter, we presented a detailed overview of nanoparticles, their types, synthesis, properties, and applications in fields of supramolecular chemistry, green reactions, and biomedical measurements. Nanoparticles have sizes ranging from a few nanometers to 500 nm and have a large surface area, which makes them suitable candidates for various applications. In addition to this, the optical properties of nanoparticles also increase their importance in photocatalytic applications. In this chapter, various methods of synthesis of nanoparticles by green routes are discussed. The applications of nanomaterials in biomedical conditions are also discussed. This information can be used to design new methodologies in nanotechnology for the benefit of the environment and humanity.

The biosynthesis and therapeutic potentials of silver nanoparticles are discussed in Chapter 3 in detail.

Xanthomonas axonopodis pv. punicae causes bacterial blight disease in pomegranate. A complete range of symptoms of bacterial blight caused by *Xanthomonas axonopodis* pv. punicae appear on various pomegranate plant parts, except roots. Chapter 4 was initiated to find a suitable alternative to synthetic antibiotics for the management of plant diseases caused by bacteria. The study aimed to use wild plant species, viz., *Abutilon indicum, Prosopis juliflora*, and *Acacia arabica* as antibacterial agent against *Xanthomonas axonopodis* pv. punicae. Aqueous extracts of *Abutilon indicum, Prosopis juliflora*, and *Acacia arabica* plants have antibacterial activity against *Xanthomonas axonopodis* pv. punicae. The antibacterial activity was tested by a well diffusion assay, minimum inhibitory concentration, and minimum bactericidal concentration.

The main objective of Chapter 5 is to review how biological hazards cause health issues to humans and animals. These hazards are very dangerous and have to be taken seriously by employers and employees. There are several ways to reduce contaminants. Engineering, administrative, workplace hazardous material information, personal protective equipment, standard precautions, and ISO standards are necessary for avoiding risks at all levels of biohazardous materials.

General and chemical perspectives and studies on tannins as natural phenolic compounds for some ecoefficient applications are discussed in Chapter 6.

Bryophyte's life structure is the simplest form, having recurring arrangements of the photosynthetic tissues, which provide maximum primary production and minimum water loss. Different life forms are found, which is a speciality of bryophytes. Bryophytes are mainly known as amphibians of the plant kingdom, as they successfully survive both on land and in water. Bryophytes are very economically important. Bryophytes are potential bioindicators of air pollution as they shows a wide range of habitat diversity, structural simplicity, totipotency, and fast multiplication rate. One most important features of bryophytes is that it shows a high metal accumulation capacity. Byometers are used for the measurement of phytotoxic air pollution. Bryophyta alone or along with lichens gives valuable information about air pollutants by an index of atmospheric purity, which is based on the number, frequency coverage, and resistance factor of species. In Chapter 7, we will discuss the role of bryophytes as bioindicators.

Applied techniques for extraction, purification, and characterization of medicinal plants' active compounds are investigated in Chapter 8.

Chapter 9 summarizes the studies reported on polysaccharide nanoparticles formulated for the transport across biological barricades with the aim of improving the anticancer potential.

Chapter 10 attempts to give a brief update on biocompatible polysaccharide-based nanodrug delivery systems for the administration of anticancer drugs through different routes, highlighting the potential to cross various biological barricades and reach the diseased target site, eliciting its anticancer therapeutic potentials.

Chapter 11 focuses on the budding perspective of chalcone (prop-2-ene-1-one)-based natural inhibitors (isoliquiretigenin, butein, garcinol, hydroxysafflor yellow A, broussochalcone A, 2,4-dihydroxy-6-methoxy-3,5-dimethylchalcone, 4′-hydroxy chalcone, and parasiticin-A, -B, and -C) and synthetic inhibitors (4-(*p*-toluenesulfonylamino)-4′-hydroxy chalcone, 4-maleamide peptidyl chalcone, and quinolyl-thienyl chalcone) that will prevent angiogenic switching (fibroblast growth factor angiogenin, TGF-β) by directly inhibiting the three vital therapeutic targets: vascular endothelial growth factor, vascular endothelial growth factor receptor-2, and matrix metalloproteinases-2/9, which will block neovascularization, vascular formation, and network formation by completely depriving the cells of the required nutrients, fluids, signaling molecules, and oxygen.

Direct electrochemical oxidation of blood is discussed in Chapter 12. The highlighted studies will positively motivate young minds, medicinal chemists, future researchers, and allied scientists for developing or exploring potential angiogenic inhibitors for the treatment of cancer with better pharmacodynamics attributes and less side or adverse effects.

CHAPTER 1

Bacterial Nanocellulose: Synthesis, Properties, Hybrids with Nano-Silver for Tissue Engineering

E. I. SHISHATSKAYA[1,2], S. V. PRUDNIKOVA[1], A. A. SHUMILIVA[1], and T. G. VOLOVA[1,2*]

[1]*Siberian Federal University, 79 Svobodnyi Av., Krasnoyarsk 660041, Russia*

[2]*Institute of Biophysics SB RAS, Federal Research Center "Krasnoyarsk Science Center SB RAS" 50/50 Akademgorodok, Krasnoyarsk 660036, Russia*

Corresponding author. E-mail: volova45@mail.ru

ABSTRACT

A strain of acetic acid bacteria, *Komagataeibacter xylinus* B-12068, was studied as a source for bacterial cellulose (BC) production. The effects of cultivation conditions (carbon sources, temperature, and pH) on BC production and properties were studied in surface and submerged cultures. Glucose was found to be the best substrate for BC production among the sugars tested; ethanol concentration of 3% (w/v) enhanced the productivity of BC. The highest BC yield (up to 17.0–23.2 g/L) was obtained under surface static cultivation conditions, in the modified Hestrin–Schramm medium supplemented with ethanol, at pH 3.9, after seven days of cultivation in the thinnest layer of the medium. C/N elemental analysis, emission spectrometry, scanning electron microscopy (SEM), differential thermal analysis (DTA), and X-ray were used to investigate the structural, physical, and mechanical properties of the BC produced under different conditions. The MTT assay and SEM showed that the native cellulose membrane did not cause cytotoxicity upon direct contact with NIH 3T3 mouse fibroblast cells and was highly biocompatible.

BC composites synthesized in the culture of the strain of acetic acid bacterium *K. xylinus* with silver nanoparticles, BC/AgNPs, were produced hydrothermally, under different $AgNO_3$ concentrations (0.0001, 0.001, and 0.01 M) in the reaction medium. The presence of silver in the BC/AgNP composites was confirmed by the elemental analysis conducted using scanning electron microscopy with a system of X-ray spectral analysis: silver content in the composites increased from 0.044 to 0.37 mg/cm^2. The surface structure, properties, and physicochemical characteristics of composites were investigated. The disk-diffusion method and the shake-flask culture method used in this study showed that all experimental BC/AgNP composites had pronounced antibacterial activity against *Escherichia coli*, *Pseudomonas eruginosa*, *Klebsiella pneumoniae*, and *Staphylococcus aureus*. No potential cytotoxicity was detected in any of the BC/AgNP composites in the NIH 3T3 mouse fibroblast cell culture, in contrast to the BC/antibiotic composites. These results suggest that BC composites constructed in the present study hold promise as dressings for managing wounds, including contaminated ones.

Hybrid wound dressings have been constructed using two biomaterials: BC and copolymer of 3-hydroxybutyric and 4-hydroxybutyric acids (P(3HB-co-4HB))—a biodegradable polymer. Some of the experimental membranes were loaded with drugs promoting wound healing and epidermal cells differentiated from multipotent adipose-derived mesenchymal stem cells. A study has been carried out to investigate the structural, physical and mechanical properties of the membranes. The *in vitro* study showed that the most effective scaffolds for growing fibroblasts were composite BC/P(3HB-co-4HB) films loaded with actovegin. Two types of experimental biotechnological wound dressings—BC/P(3HB/4HB)/ actovegin and BC/P(3HB-co-4HB)/fibroblasts—were tested *in vivo*, on laboratory animals with model third-degree skin burns. Wound planimetry; histological examination; and biochemical and molecular methods of detecting factors of angiogenesis, inflammation, type-I collagen, keratin 10, and keratin 14 were used to monitor wound healing. Experimental wound dressings promoted healing more effectively than VoskoPran—a commercial wound dressing.

1.1 INTRODUCTION

Cellulose is extracellular polysaccharide synthesized by higher plants, lower phototrophs, and prokaryotes belonging to various taxa (Ullah et al., 2016).

Although bacterial cellulose (BC) is produced in laboratories on a small scale for research, there are some commercial outlets for BC. In addition, traditional nata de coco (Iguchi et al., 2000), Fzmb GmbH, a German company, is considered one of the largest producers of BC for cosmetics and biomedical applications (Keshk et al., 2014a). In addition, Xylos Co., USA, is a producer of Prima CelTM, a type of BC used for wound dressing. Other brands of BC include Gengiplex® and Biofill® (Keshk et al., 2014a), which are used as a physical barrier for tissue regeneration. BC is also produced and used by many food industries in Asian countries (Budhiono et al., 1999; Ng and Shyu, 2004). Sony Corporation, Japan, in association with Ajinomoto, Japan, and other firms fabricated the first BC-based diaphragm for an audio speaker. Ajinomoto, Japan, also sells wet BC (Chawla et al., 2009; Czaja et al., 2006).

A promising material for biomedical application is BC—a biopolymer synthesized by microorganisms. The chemical structure of BC is similar to that of plant-derived cellulose, but it has unique physical, mechanical, and chemical properties, such as high strength, elasticity, gas permeability, good water-holding capacity, porosity, etc. This material shows high biocompatibility, without being cytotoxic or causing any allergic reactions. Studies on BC suggest that this natural polymer can be useful for cellular and tissue engineering as a material for constructing scaffolds and for reconstructive surgery as a material for skin defect reconstruction and as a matrix for drug delivery (Ma et al., 2010; Saska et al., 2011). Cellulose is used in a variety of applications in food and paper industries, medicine, and pharmaceutics. Gel pellicles of BC have an ordered structure: they are three-dimensional (3D) networks consisting of ribbon-like randomly oriented cellulose microfibrils. This structural arrangement of BC and its high compatibility with biological tissues make it an attractive material for reconstructive surgery; skin tissue repair; target tissue regeneration in dentistry, general surgery, and maxillofacial surgery; and cell and tissue engineering—as a carrier for drugs.

The physical and mechanical properties of BC can be enhanced by preparing BC composites with various materials: chitosan (Lin et al., 2013), collagen (Culebras et al., 2015), sodium alginate, gelatin, and polyethylene glycol (Shah et al., 2013). BC is not inherently antibacterial, but BC composites with chitosan and alginate inhibit the growth of pathogenic microorganisms such as *Escherichia coli*, *Candida albicans*, and *Staphylococcus aureus* (Lin et al., 2013; Kwak et al., 2015; Chang et al., 2016). Therefore, BC composite films can be considered for treating infected skin wounds. Owing to its 3D and porous structure, BC can be hybridized with metallic silver particles to produce an antibacterial and wound-healing formulation. Metallic silver and compounds thereof have a strong bactericidal effect, inhibiting the growth of a wide range of pathogenic microorganisms. Silver ions react with cell membrane protein thiol groups, affecting bacterial respiration and transportation of substances through the cell membrane (Percival et al., 2005). Several authors have described different techniques for preparing BC composites with silver nanoparticles (BC/AgNPs) and demonstrated thir high antibacterial activities (Sureshkumar et al., 2010; Feng et al., 2014; Wen et al., 2015; Wu et al., 2014; Sadanand et al., 2017). Another approach to imparting antibacterial activity to BC against pathogenic microflora is to prepare BC composites with antibiotics. While this approach has been described in a few studies (Shao et al., 2016; Wijaya et al., 2017), it remains poorly developed.

The purpose of the present study is to investigate the strain *Komagataeibacter xylinus* B-12068 as a new producer of the influence of culture conditions on the structure and properties of BC; to prepare BC/AgNPs and to investigate their antibacterial activity; to construct and investigate composites based on BC and polyhydroxyalkanoate (PHA) as biotechnological wound dressings; to evaluate their efficacy in managing model skin burns in experiments with laboratory animals.

1.2   MATERIALS AND METHODS

1.2.1   CHARACTERIZATION OF THE BACTERIAL STRAIN

A new bacterial strain *K. xylinus* B-12068 was isolated from the fermented tea (kombucha) *Medusomyces gisevii* J. Lindauon Hestrin–Schramm (HS) medium (Hestrin and Schramm, 1954). The strain was identified based on its morphological, biochemical, genetic, and growth parameters. The strain *K. xylinus* was deposited in the Russian National Collection of Industrial Microorganisms (VKPM) with registration number VKPM B-12068. The morphology of bacterial cells was studied in Gram-stained preparations. The phenotypic properties were studied using conventional microbiological techniques. The morphology of vegetative cells; spore formation; motility; reaction to Gram staining; requirement of growth factors; the presence of nitrate reductase; catalase, oxidase, amylase, and proteinase activities; antibiotic sensitivity; and NaCl sensitivity were determined. Growth on carbon sources such as glucose, sucrose, maltose, galactose, and mannitol was tested in the basal HS medium supplemented with 2% (w/v) of each carbohydrate. The strain produced BC on the surface and in submerged cultures (Prudnikova et al., 2014).

1.2.2 CULTIVATION OF THE STRAIN AND PRODUCTION OF BACTERIAL CELLULOSE

The collection culture of *K. xylinus* B-12068 was maintained on the HS agar medium. The standard HS medium contained (%, w/v) glucose, 2; peptone, 0.5; yeast extract, 0.5; Na_2HPO_4, 0.27; and citric acid, 0.115 (Hestrin and Schramm, 1954). Preculturing was performed on HS agar. Then, the colonies were transferred into a flask containing liquid HS medium and cultivated for seven days at a temperature of 30 °C under static conditions. To investigate the influence of culture conditions on BC biosynthesis and to find the conditions maximizing the cellulose yield, we modified the standard medium by replacing glucose with other carbon sources (sucrose, galactose, or glycerol), varied the initial pH values (3.2–4.8) by adding acetate or citrate, varied the temperature of the medium (20–37 °C), and added various concentrations of ethanol (0.5–3.0%) to the medium, based on the data, indicating that ethanol oxidized to acetate was a growth substrate and energy source for *K. xylinus* (Yamada et al., 2012). BC production by the strain was investigated under different culture conditions. Static cultivation was performed in the surface mode in Petri dishes, 1500 mL glass trays, or 250–500 mL glass flasks, which contained different volumes of the medium. Submerged cultivation was conducted in 500 mL glass flasks using a JeioTech SL-600 incubated shaker (JeioTech, Korea) at 100 rpm. Glucose concentration was determined using the "Glucose—FKD" kit, which contained a chromogenic enzyme substrate and a calibrator (a glucose solution of a known concentration). The optical densities of the study sample and calibration sample were compared photometrically with the optical density of the blank, with an optical path length of 10 mm at a wavelength of 490 nm.

BC yields in different modes of cultivation were compared by measuring the weight of the cellulose dried to constant weight, the pellicle thickness, and the amount of carbon substrate consumed. The total BC yield (g/L) and biosynthesis productivity for different fermentation processes were calculated using conventional methods. The BC yield was evaluated as the weight of dry cellulose per liter of medium (g/L). The dried BC pellicles were weighed using an Adventurer OH-AR2140 analytical balance (Ohaus, Switzerland).

1.2.3   STUDY OF PHYSICOCHEMICAL PROPERTIES OF BC

The synthesized BC was separated from the culture fluid and purified in a 0.5% NaOH solution for 24 h at 25–27 °C. Then, cellulose was placed in a 0.5% solution of hydrochloric acid for 24 h for neutralization and afterward rinsed with distilled water until pH 7. The BC pellicles were stored in a sterile solution or air-dried until they reached a stable weight.

The microstructure of the surface of BC pellicles was analyzed using scanning electron microscopy (S-5500, Hitachi, Japan). Prior to the analysis, the pellicles were freeze-dried using an ALPHA 1-2/LD freeze dryer (Martin Christ GmbH, Germany) for 24 h. The samples (5×5 mm) were placed onto the sample stage and sputter-coated with gold using an Emitech K575X sputter coater (10 mA, 2 × 40 s). Fiber diameters were measured by analyzing the SEM images with image analysis program ImageJ—Image Processing and Data Analysis in Java. The diameters of 50 individual ultrafine fibers were then measured in each SEM micrograph. The diameters were analyzed in 10 fields of the SEM images in triplicate.

1.2.4   PRODUCTION OF BC COMPOSITES WITH SILVER NANOPARTICLES, BC/AGNPS

BC/AgNPs were produced by a hydrothermal method without utilizing any catalyst, using the disks of the BC layer as a reducing and stabilizing agent (Yang et al., 2012). Purified raw BC films were cut into disks of 1 cm in diameter; placed in flasks with 0.0001, 0.001, and 0.01 M of $AgNO_3$; and heated for 60 min at a temperature of 90 °C. Composite BC films with silver nanoparticles were lyophilized at a temperature of −40 °C and a pressure of 0.12 mbar for 24 h in a vacuum drying unit ALPHA 1-2/LD (Martin Christ GmbH, Germany) or kept at room temperature in a laminar flow cabinet for 24 h. The parameters of the produced silver nanoparticles were investigated using a Zetasizer Nano ZS particle analyzer (Malvern, UK), by employing dynamic light scattering, electrophoresis, and laser Doppler anemometry.

1.2.5   IN VITRO ANTIBACTERIAL TESTS

The direct inhibitory effect of BC/AgNP and BC/antibiotic composites was tested on cultures of reference strains—ATCC 25922, *Pseudomonas aeruginosa* ATCC 27853, *Klebsiella pneumoniae* 204, and *S. aureus* ATCC 25923—using a disk-diffusion method in agar (20 mL) on Petri dishes. The dishes were allowed to stay at room temperature to solidify. The bacterial suspension was standard inoculum whose density corresponded to the 0.5 McFarland standard, which contained approximately 1.5×10^8 CFU/mL. The Petri dishes were placed upside-down into an incubator and kept at a constant temperature of 35 °C for 18–24 h (depending on the microorganism tested). The diameter of the growth retardation zones and the distance from the edge of the film to the end of the absence zone were measured by the photographs of dishes, using the ImageJ program. The results were processed using the Microsoft Excel application package. The arithmetic mean and standard deviation were calculated.

1.2.6   CYTOTOXICITY ASSAYS

The ability of BC films to facilitate cell attachment was studied using NIH 3T3 mouse fibroblast cells. The films were placed into 24-well cell culture plates (Greiner Bio-One, USA) and sterilized using a Sterrad NX medical sterilizer (Johnson&Johnson, USA). Cells were seeded at 1×10^3 cells/mL per well. Cells were cultured in the Dulbecco's modified Eagle's medium (DMEM) supplemented with 10% fetal bovine serum and a solution of antibiotics (streptomycin 100 µg/mL, penicillin 100 IU/mL) (Sigma) in a CO_2 incubator with the CO_2 level maintained at 5% at a temperature of 37°C. The medium was replaced every three days. The number of cells attached to the film surface was determined using DAPI. Cell viability was evaluated using the MTT assay at day 7 after cell seeding onto films. The reagents were purchased from Sigma-Aldrich. A 5% MTT solution (50 µL) and complete nutrient medium (950 µL) were added to each well of the culture plate. After 3.5 h incubation, the medium and MTT were replaced by dimethyl sulfoxide (DMSO) to dissolve the MTT-formazan crystals. After 30 min, the supernatant was transferred to the 96-well plate, and the optical density of the samples was measured at a wavelength of 540 nm using a Bio-Rad 680 microplate reader (Bio-Rad Laboratories Inc., USA). The measurements were performed in triplicate. The number of viable cells was determined from the standard curve.

1.2.7   CONSTRUCTING BC/P(3HB-CO-4HB) COMPOSITES AS WOUND DRESSINGS

The composites were prepared from BC films or powder and solutions of poly(3-hydroxybutyrate-co-4-hydroxybutyrate) (P(3HB-co-4HB)) (85 3HB/

15 mol.% 4HB). BC/P(3HB-co-4HB) hybrids were constructed using different methods: (1) dried BC pellicles and wet (chloroform-impregnated) BC pellicles were soaked in 2% and 5% P(3HB-co-4HB) solutions in chloroform; (2) dried BC pellicles were soaked in a 2% P(3HB-co-4HB) solution in chloroform, kept for 24 h, and dried in a dust-free cabinet until the solvent had completely evaporated; (3) nonwoven membranes were placed into the *K. xylinus* B-12068 culture, and BC synthesized in it grew on the surface of the membranes under static conditions; (4) powdered cellulose (particle size of 120 μm) was added to a 3% P(3HB-co-4HB) solution in chloroform and mixed ultrasonically to homogeneity; then, the films were produced by the solvent evaporation technique, at polymer-to-cellulose ratios of 2:1 and 1:1; (5) P(3HB-co-4HB) powder was added to the *K. xylinus* B-12068 culture; and (6) a film of BC grew on nonwoven membranes in the submerged culture of *K. xylinus* B-12068.

1.2.8 LOADING OF CELLS AND WOUND-HEALING DRUGS INTO BC/P(3HB-CO-4HB) COMPOSITES

To prepare composites loaded with cells, 10-mm-diameter BC/P(3HB-co-4HB) disks were placed into 24-well culture plates (TPP Techno Plastic Products AG, Switzerland) and sterilized using a Sterrad NX medical sterilizer (Johnson & Johnson, USA). Fibroblasts were derived from the adipose tissue mesenchymal stem cells (MSCs) of Wistar rats. The adipose tissue cells were isolated enzymatically. The adipose tissues were rinsed in the Dulbecco's phosphate-buffered saline (DPBS) solution with antibiotics, ground, and incubated in a type-I collagenase solution (100 units/mL) at 37 °C until the tissue particles were dissolved, but for no more than 1 h. Then, collagenase was inactivated with an albumin solution, centrifuged, and rinsed in the DMEM medium several times, until the lipid layer had been completely removed. The settled cells were suspended and seeded onto Petri dishes that contained the DMEM medium (Gibco) with 10% fetal bovine serum (HyClone) and a solution of antibiotics: penicillin, streptomycin, and amphotericin B (Gibco). On the DMEM medium, adipose tissue cells showed fibroblast-like morphology. After 3–4 passages, the cells were removed from the Petri dishes with a trypsin solution (Gibco) and seeded onto sterile BC/P(3HB/4HB) disks: 10^5 cells per disk. Loading of BC/P(3HB/4HB) composites with drugs promoting wound healing was performed using solutions of actovegin (Takeda Pharmaceuticals, Russia) and solcoseryl (MEDA Pharma, Switzerland). Sterile composite samples were exposed to PBS to investigate drug release.

Under aseptic conditions, the samples were placed into vials, each containing 50 mL of saline at pH 6. The vials were incubated in a thermostat for 72 h at a temperature of 37 °C. The experiment was done with samples of 1 cm in diameter prepared with different percentages of drugs (1–5%).

1.2.9 STUDY OF THE BIOLOGICAL COMPATIBILITY OF BC/P(3HB-CO-4HB) COMPOSITES IN CELL CULTURE

BC/P(3HB-co-4HB) samples loaded with drugs were shaped as 10 mm-diameter disks. They were placed into 24-well culture plates (TPP Techno Plastic Products AG, Switzerland) and sterilized using a Sterrad NX medical sterilizer (Johnson & Johnson, USA). Scaffolds were seeded with cells at 10^5 cells per scaffold for 7 days. Cells were cultured on the DMEM medium with 10% fetal bovine serum (HyClone) and a solution of antibiotics (Gibco reagents) in a 5% CO_2 atmosphere at 37 °C, with the medium replaced by the fresh one every three days. For the MTT assay, 5% MTT-bromide solution (50 µL) and complete nutrient medium (950 µL) were added to each well of the culture plate. After 4 h cultivation, the medium and the MTT solution were replaced by DMSO to dissolve MTT-formazan crystals. After the dissolution of MTT-formazan crystals, 100 µL of supernatant was transferred to the 96-well plate, and the optical density of the samples was measured at a wavelength of 540 nm using an iMark microplate reader (Bio-Rad Laboratories Inc., USA). The number of viable cells was determined from the calibration graph.

1.2.10 IN VIVO STUDY OF BC/P(3HB-CO-4HB) COMPOSITES AND HYBRIDS AS EXPERIMENTAL WOUND DRESSINGS

Experiments were conducted in accordance with the Russian State Standard (GOST R ISO) 10993-1-2009 "Medical products. Estimating biological effects of medical products" and the international regulations "International Guiding Principles for Biomedical Research Involving Animals (CIOMS, 1985), WMA Declaration of Helsinki on treating research animals with respect (2000), and European Convention for the Protection of Vertebrate Animals used for Experimental and Other Scientific Purposes (Strasburg, 18.03.1986, ratification date 01.01.1991, CETS No. 123)." The protocol of the experiments was approved by the Local Ethics Committee at the Siberian Federal University. The research animals were maintained and used in accordance with the rules accepted in the Russian Federation. The efficacy of the wound dressings based on BC and P(3HB/4HB) with and without fibroblasts differentiated from the adipose-derived MSCs was evaluated in experiments on research animals with model skin defects (burns) and compared to the efficacy of commercial wound dressings VoskoPran (Biotekfarm, Russia).

Sexually mature female Wistar rats were kept in the animal facility and fed their usual diet. The rats were anesthetized with 0.5–0.7 mL of a 1% solution of sodium thiopental administered intraperitoneally. Then, the animals were placed on their backs, and a 1 cm^2 section of the skin was shaved. The burn was done with a 16-mm-diameter steel plate preheated in boiling water for 10 min at 100 °C. The exposure lasted 8 s to create a second-degree skin burn (in accordance with ICD-10). After 5–10 min, the epidermis was separated from the underlying dermis with a sterile gauze swab. The animals were divided into four groups, six rats per group. In the treatment groups, three types of experimental wound dressings were used: BC/P(3HB/4HB), BC/P(3HB/4HB)/actovegin, and P(3HB/4HB)/BC/fibroblasts. The dressings were fixed with single surgical sutures around wound edges. In the control group, the wounds were covered with polyamide mesh VoskoPran.

After the wounds were closed, each rat was placed in a separate cage for 14 days. The animals were euthanized with a lethal dose of Zoletil. The evaluation of the efficacy of the wound dressings was based on the intensity of the inflammatory response in the wound and surrounding soft tissues and the rate and completeness of skin repair. The changes in the area of the wound surface were monitored by Popova's method (Popova et al., 1942) to determine the wound healing rate. At days 3, 7, and 14 of the treatment, the animals were sacrificed and the skin samples were removed and used in histology. Tissue layers with edges from the wound site were sectioned and fixed in a 10% formalin solution. The samples were processed by conventional procedures and embedded in paraffin. The histological sections were stained with hematoxylin and eosin.

To get insights into the mechanism of the wound healing process, using molecular methods, we detected the factors that characterized the degree of the inflammatory process, vascularization, and formation of the new connective tissue at the defect site: angiogenesis (VEGF), inflammation (TNF-a), type-I collagen (Col-1) (an indicator of connective tissue formation), keratin 10 (K10), and keratin 14 (K14). Gene expression of type-I collagen (Col-1) was determined by real-time reverse transcription polymerase chain reaction (RT-PCR) to confirm the fibroblast phenotype of the cells and their viability. The cells were lysed, and messenger ribonucleic acid (mRNA) was extracted using an RNA-Extran reagent kit (Sintol, Russia). The complementary deoxy-ribonucleic acid (cDNA) was synthesized by reverse transcription reaction using MMLV reverse transcriptase (Sintol, Russia). RT-PCR was performed with the corresponding primers using a Real-time CFX96 Touch detection system (Bio-Rad Laboratories Inc., USA), following the manufacturer's instructions. The housekeeping gene was β-actin. The comparative threshold cycle method ($2^{-\Delta\Delta Ct}$) was used to determine the relative level of expression of the marker genes. The total RNA was isolated from the granulation tissue of the treatment and control groups by extraction with a guanidinium thio-cyanate–phenol–chloroform mixture from the RNA-Extran reagent kit. Then, from the RNA template, using reverse transcriptase, we synthesized cDNA with several types of oligonucleotide primers: with a mixture of random primers—hexa primers—and oligo-dT primer, following the procedure recommended by the manufacturer (Sintol). Negative controls of the reverse transcription reaction were prepared for all samples to confirm the absence of DNA contamination in the initial RNA. Real-time PCR amplification for the quantification of cDNA fragments of rat VEGF, rat TNF-a, rat Col-1, rat K10, and rat K14 was performed by using a CFX-96 thermocycler, according to the manufacturer's protocol, using the color channel of HEX.

1.2.11 STATISTICAL ANALYSIS

Statistical analysis of the results was performed using electronic tables in Microsoft Excel 2010 and the Statistica 6.0 program for the Windows XP operating system. A comparison of two related groups for quantitative attributes was performed using the parametric method with the Wilcoxon matched-pairs test. A comparison of two independent groups for quantitative attributes was performed using the nonparametric method with the Mann–Whitney *U* test. If the probability was greater than 95%, the differences were considered statistically significant.

1.3 RESULTS AND DISCUSSION

1.3.1 SYNTHESIS AND PROPERTIES OF BC OBTAINED SYNTHESIZED IN THE CULTURE OF ACETIC ACID BACTERIA K. XYLINUS B-12068

The production of BC in the culture of *K. xylinus* B-12068 at various carbon sources and in the modification of culture parameters was studied. Figure 1.1

shows the results of the evaluation of the ability of the strain *K. xylinus* B-12068 to synthesize cellulose under static conditions in glass flasks at a temperature of 30 °C and an initial pH of 6.0 in the HS medium with varied carbon sources. The highest BC yield (about 2.2 g/L) was obtained in the experiment with the *K. xylinus* B-12068 cells cultivated for 7 days in the HS medium containing glucose. Similar was the production of BC on glycerol reaching up to 2.1 g/L. The BC yield in the media with sucrose and galactose was somewhat lower—1.6 and 1.4 g/L, respectively. The physiological glucose range for this strain is rather wide; *K. xylinus* B-12068 growth and BC production were inhibited at glucose concentrations in the culture medium higher than 25% (w/v).

As the BC production is influenced not only by carbon sources but also by the agitation conditions, the medium volume-to-surface area ratio, pH of the medium, and addition of the secondary carbon substrates to the medium (Pokalwar et al., 2010; Ruka et al., 2012; Fu et al., 2013; Keshk, 2014; Li et al., 2015), we investigated the cellulose production under varied conditions of *K. xylinus* B-12068 cultivation. Figure 1.1 shows how the initial pH value influenced the cellulose production by the strain. The physiological pH range for *K. xylinus* B-12068 is between 3.2 and 5.0. The trend in the graph indicates the optimal pH range for BC biosynthesis, which is very narrow— about 3.6. The highest BC yield from the standard HS medium (2.73 g/L) was obtained at pH 3.6. When *K. xylinus* B-12068 cells were cultivated in the glucose enriched medium, the initial pH value of 6.0 dropped to 3.4 after 7 days of cultivation due to the accumulation of products of glucose oxidation, mainly, gluconic acid. As *K. xylinus* B-12068 was found to be acid tolerant, the pH of the medium, which decreased during cultivation, did not need to be adjusted. The ability of bacteria to synthesize BC at low pH values can be used to reduce the risk of contamination of the commercial strain by foreign microflora.

The effect of the temperature of the *K. xylinus* B-12068 culture medium on BC production is shown in Figure 1.1, suggesting that the optimal temperature range for cell growth and BC production is rather narrow about 30 °C, while the physiological temperature range is wide (20–35 °C). Based on the data suggesting that acetic acid bacteria are capable of oxidizing ethanol and that in some cases, ethanol enhances the productivity of BC, for example, in *Gluconacetobacter hansenii* culture (Park et al., 2003), we investigated the effect of ethanol as a secondary substrate on BC production by *K. xylinus* B-12068 cultivated in the HS medium enriched with glucose. The increase in ethanol concentration from 0.5 to 3.0% (v/v) resulted in an increase in cellulose production, from 1.38 to 3.06 g/L. The highest BC yields were obtained at an ethanol concentration of 3%. As glucose was determined as the best substrate for *K. xylinus* B-12068, it was used as a carbon substrate in the subsequent experiments.

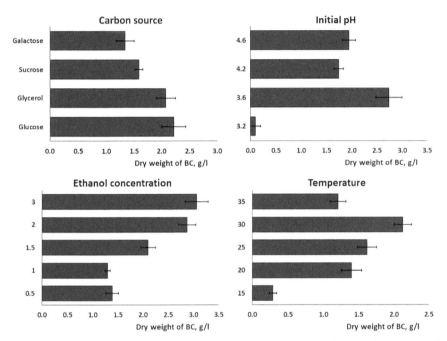

FIGURE 1.1 Effect of cultivation conditions on cellulose yield in *K. xylinus* B-12068 culture (g/L).

TABLE 1.1 Bacterial Cellulose Production on Standard and Modified Hestrin–Schramm Media in Different Volumes

Culture Conditions	BC Yield (g/L)	Productivity (g/L/day)
Static conditions (carbon source, medium volume)		
Flask, HS standard medium (glucose), 100 mL	1.7	0.24
Flask, HS modified medium (glucose + ethanol), 100 mL	7.9	1.13
Petri dishes, HS standard medium (glucose), 25 mL	4.3	0.61
Petri dishes, HS modified medium (glucose + ethanol), 25 mL	17.0	2.43
Glass tray, HS modified medium (glucose + ethanol), 500 mL	6.8	0.97
Petri dishes, HS-modified medium (glycerol 5% + ethanol), 25 mL	23.2	3.30
Petri dishes, HS modified medium (glycerol 10% + ethanol), 25 mL	16.9	2.29
Agitated conditions (carbon source, medium volume)		
Flask, standard medium (glucose), 150 mL	2.8	0.41
Flask, modified medium (glucose + ethanol), 150 mL	6.2	0.89

When the strain was grown on the surface of glucose enriched HS medium (100 mL, the layer of the medium is 2.0 ± 0.1 cm high) under static conditions, the BC productivity was the lowest. Under agitated conditions, BC yield was 1.6 times higher than under static ones. The modification of HS medium by adding ethanol concentration of 3% (v/v) and decreasing the initial pH value to 3.9 enhanced the BC productivity of K. *xylinus* B-12068 to 0.97–1.13 g/L/day under static conditions and to 0.89 g/L/day under shaking conditions.

The highest productivity of the culture on glucose was obtained in a static culture on HS modified medium (glucose + ethanol) at a layer height of 25 mL; for seven days, the yield of the BC was 17 g/L; the productivity of the process was 2.43 g/L/day (Figure 1.2). Similar high yields of BC (2.05–15.0 g/L) were also obtained in the culture of *K. hansenii* as a result of extensive research by the authors (Lima et al., 2017).

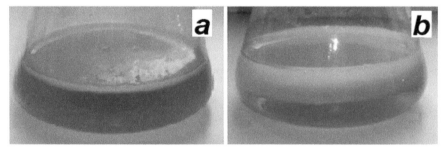

FIGURE 1.2 Bacterial cellulose films synthesized by *K. xylinus* B-12068 on standard medium (a) and supplemented with ethanol (b) medium.

An important result was obtained by replacing glucose as the main C substrate with glycerol. When cultured by the surface method in 250 mL flasks per 100 mL of standard HS medium, the BC output after seven days was 2.08 ± 0.17 g/L, relative to the previously obtained 1.77 ± 0.33 g/L on glucose (Figure 1.3). With a decrease in the height of the medium layer to 50 ± 1 mm, the output of the BC for 7 days increased by 23.2 g/L; the productivity was up to 3.3 g/L/day (Table 1.1).

In the present study, the highest BC yield 17.0–23.2 or 2.4–3.3 g/L/day was observed in the experiment with *K. xylinus* B-12068 grown in Petri dishes containing 25 mL of medium (the medium layer being 0.5 ± 0.1 cm thick). In Petri dishes, the volume of the medium was four times less than that in the flasks, the surface area was the same as in the flasks (60 ± 1 cm^2), and the amount of the substrate (glucose) was 2.2–2.6 times smaller (Table 1.1). These results are in good agreement with the data in the study of Ruka et al. (2012), showing that the lower volume of the nutrient medium in the vessel did not decrease BC production but increased the efficiency of utilization of carbon substrate.

FIGURE 1.3 Bacterial cellulose films synthesized by *K. xylinus* B-12068 on HS medium with glycerol: (a)10 % and (b) 5%.

The use of larger bulbs (3 L) and fermentation vessels in the form of trays allowed us to scale the process and gave us the opportunity to produce BC films of various sizes (Figure 1.4). Under agitated conditions, however, in contrast to the static culture, cellulose did not form a pellicle at the air/culture medium interface but exhibited the shapes of spheres, filaments, and fibers distributed over the volume of the culture medium.

Thus, in the present study, high yields of BC were obtained from the culture of *K. xylinus* B-12068—a new strain of acetic acid bacteria.

1.3.2 PRODUCTION OF BC COMPOSITES WITH SILVER NANOPARTICLES, BC/AGNPS

The dried BC films synthesized in the *K. xylinus* B-12068 culture had similar thickness (1.8 ± 0.2 mm) and density (0.15 ± 0.01 cm³), while their surface properties differed considerably, as determined by the water contact angle, which varied between 36° and 57°. The ultrastructure and size of fibrils in BC is a critical factor that determines the unique properties of BC films. SEM images of the microstructure of BC films show that BC films were layered nets of different densities composed of randomly oriented microfibrils (Figure 1.5). Whatever the drying method used, the average diameter of BC film microfibrils was 110 nm, with the smallest and the largest diameters being 52 nm and 173 nm, respectively. However, in freeze-dried films, the fibrils were positioned more loosely, and the distance between them was 2.5–3.0 times greater than that in the films dried at room temperature, reaching about 1.6 μm.

FIGURE 1.4 Scaling process of the bacterial cellulose synthesis in agitated culture in the fermenter and surface static culture.

The average crystallinity of the series of BC films was $60 \pm 15\%$; the thermal decomposition temperature of the films was 260 ± 30 °C. The films had the following parameters of mechanical strength: Young's modulus, 4.6–5.3 MPa; elongation at the break, 12–13%.

During the hydrothermal synthesis of silver nanoparticles, variations in AgNO₃ concentration in the reaction medium at 90 °C changed the number of the synthesized nanoparticles (Figure 1.6A) without considerably influencing their size (Figure 1.6B). For instance, the average sizes of Ag nanoparticles at AgNO₃ concentrations of 0.0001, 0.001, and 0.01 M were 13, 23, and 12 nm, respectively (Figure 1.7).

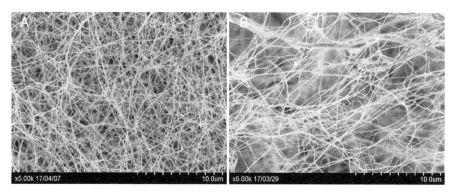

FIGURE 1.5   SEM images of initial BC films dried at room temperature (A) and freeze-dried ones (B). Scale bar =10 μm.

The increase in the number of silver nanoparticles in BC films is illustrated in Figures 1.7 and 1.8. The SEM images show that as the AgNO₃ concentration of the reaction medium was increased, the number of silver nanoparticles adhering to BC fibrils and between them increased (Figure 1.8). Ag nanoparticles showed different aggregation behaviors in freeze-dried BC films compared to those in the films dried at room temperature. On the films dried at room temperature, the size of Ag particles was 25–60 nm, the size of their aggregates was 85–350 nm, and the number of aggregates reached 15 per 1 μm². On freeze-dried films, particle aggregates were larger, between 350 and 780 nm, and their number reached 19 per 1 μm². The presence of silver in the BC/AgNPs was confirmed by the elemental analysis performed using scanning electron microscopy with a system of X-ray spectral analysis. The analysis showed that the average atomic number of silver particles in composite samples depended on the concentration of AgNO₃; as the AgNO₃ concentration in the reaction solution was increased from 0.0001 to 0.01 M, the silver content in the composites increased from 1.08 to 5.32.

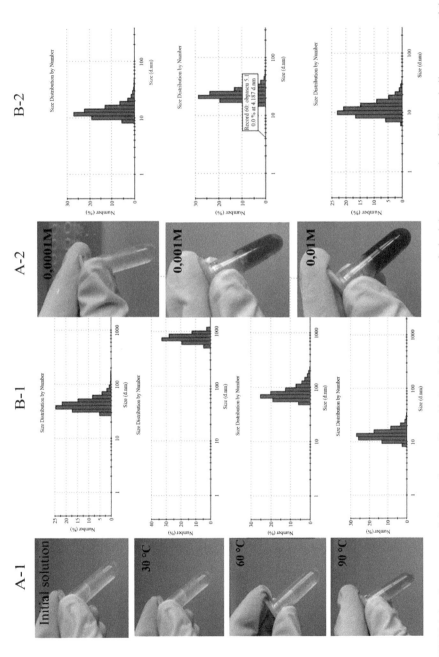

FIGURE 1.6 Appearance of the reaction medium (A-1 and A-2) in the process of obtaining composites of BC and silver nanoparticles at different temperatures and concentrations of $AgNO_3$; size distribution of silver nanoparticles (B-1 and B-2).

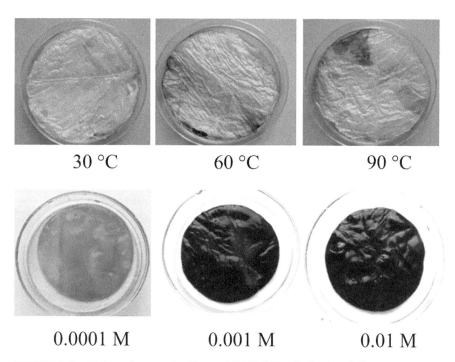

30 °C   60 °C   90 °C

0.0001 M   0.001 M   0.01 M

FIGURE 1.7 Photos of composite films of BC/AgNPs obtained at different temperature conditions and AgNO$_3$ concentrations in the reaction solution.

One of the factors determining the biocompatibility of implants is the physicochemical reactivity of their surface. The main factors that influence surface interaction with blood and tissue cells and components of biological fluids are surface topography, microstructure, and adhesive properties. An indirect indicator of surface hydrophilicity is the liquid contact angle. The properties of BC composites are listed in Table 1.2. BC films are inherently quite hydrophilic, with the water contact angle below 50° (45.5° ± 17.6°). The surfaces of BC/AgNP composites tend to become less hydrophilic (more hydrophobic) as their silver content increases. The dispersive and polar components increase, too. Such changes reduce the fouling of the surfaces by proteins and cells, as described in a study by Jia et al. (2012). The authors of that study observed an increase in the water contact angle on BC/CuNPs composite films to 110 ± 0.4°—the value that was twice higher than that on the initial BC (50.1 ± 1.8°).

All samples of BC composites exhibited bactericidal activity against test microbial cultures—the most common representatives of nosocomial infection and pathogenic microflora of contaminated wounds. However, the

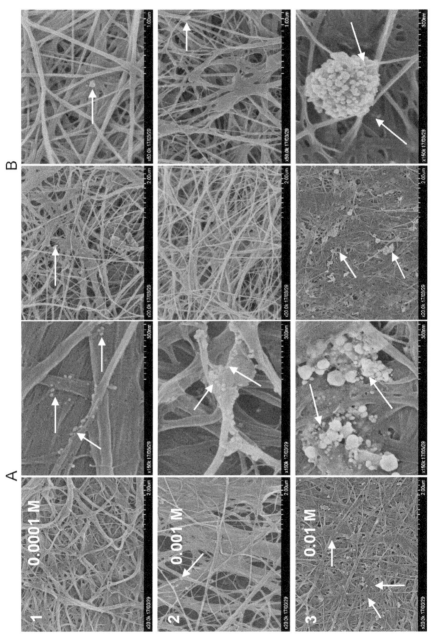

FIGURE 1.8 SEM images of the composites films of BC/AgNPs produced under different AgNO$_3$ concentrations in the medium: the films were dried at room temperature. Arrows denote aggregates of Ag nanoparticles. Scale bars = 200 and 300 µm.

levels of antibacterial activity of the composites were different (Table 1.3, Figure 1.9). BC/antibiotic composites had a stronger inhibitory effect on the growth of pathogenic microorganisms.

TABLE 1.2 Surface Properties of BC/AgNP Composites

Samples	Water Contact angles	Dispersive Component (mN/m)	Polar Component (mN/m)
Pristine BC	45.5 ± 17.6	28 ± 9.24	17.5 ± 8.37
BC/AgNPs			
0.0001 M BC/AgNPs	50.3 ± 4.61	42.6 ± 1.01	27.7 ± 3.6
0.001 M BC/AgNPs	68.8 ± 1.76	46.2 ± 0.99	22.6 ± 0.77
0.01 M BC/AgNPs	69 ± 2.71	44.9 ± 1.85	25.1 ± 0.86

TABLE 1.3 Inhibition of Pathogenic Bacteria on the Solid Medium by BC/AgNP Composites

Samples	Diameter of Inhibition Zones (mm)		
	P. aeruginosa	*E. coli*	*S. aureus*
Pristine BC	–	–	–
BC/AgNPs			
0.0001 M	12 ± 0.44	11 ± 0.20	15 ± 0.73
0.001 M	13 ± 1.15	13 ± 0.50	14 ± 0.28
0.01 M	14 ± 2.11	14 ± 0.61	15 ± 1.58

In experiments with BC/AgNPs, the largest zone of inhibition (15 ± 1.58 mm) was observed for *S. aureus*, at the highest concentration of AgNPs (0.01 M); the smallest zone of inhibition (11 ± 0.20 mm) was created by 0.0001 M BC/AgNPs for *E. coli*. These differences are caused by the different susceptibility of bacteria to BC composites with Ag particles, their cell structure and physiology. Gram-positive bacteria are generally more susceptible to the bactericidal effect of AgNPs (Ruparelia et al., 2007). Similar data were reported by (Shao et al. (2016): in their study, at a concentration of BC–Ag of 0.01 M, the diameters of the zones of inhibition of *E. coli* and *S. aureus* growth were 11.7 ± 0.1 and 11.6 ± 0.1 mm, respectively. Feng et al. (2014) reported a study of BC composites with silver reduced using $NaBH_4$, which created zones of inhibition of diameter about 16.1 mm for *E. coli* and 17.7 mm for *S. aureus*; the BC composites with silver reduced with sodium citrate created zones of inhibition with smaller diameters: 13.7 and 13.2 mm, respectively. The diameters of the zones of inhibition of *B. subtilis* and *E. coli*

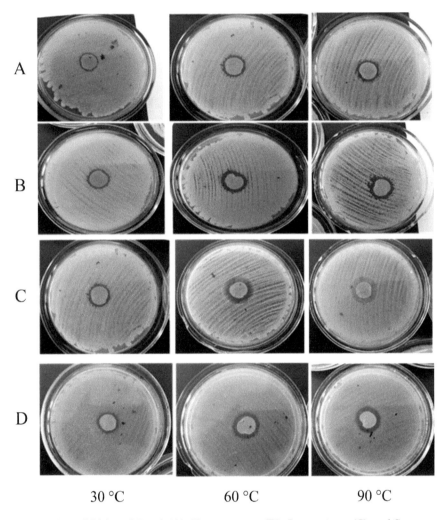

30 °C   60 °C   90 °C

FIGURE 1.9   Inhibition of *E. coli* (A), *K. pneumoniae* (B), *P. aeruginosa* (C), and *S. aureus* (D) with BC/AgNP composites obtained at different temperatures of the reaction medium.

by composites based on polydopamine magnetic BC and Ag were 20 mm and 17 mm, respectively (Sureshkumar et al., 2010). However, a study by Barud et al. (2011) showed that the composite membranes of 1 mol/L^{-1} BC/Ag/TEA created inhibition zones for *P. aeruginosa* ATCC-27853, *E. coli* ATCC 25922, and *S. aureus* ATCC 25923 reaching 20 mm. The differences between results may be caused by the different methods employed to produce composites, sizes of silver nanoparticles, and their concentrations in the BC films. It has been assumed that the lowest inhibitory concentration of Ag nanoparticles is about 0.05–0.1 mg/mL (Sureshkumar et al., 2010). In the present study, the strongest inhibition of pathogenic microflora by BC/AgNP composites was observed under the highest silver concentration in the reaction solution during composite production, when the silver content was 5.32 (or 0.024 mg/mL), that is, these results are consistent with the data reported by other authors.

The zones of inhibition by BC/antibiotic composites were generally larger than those produced by BC/AgNP composites (Table 1.3). It is well known that antibiotics penetrate through the cell membrane and irreversibly bind to specific receptor proteins of the bacterial cell, thus effectively suppressing the synthesis of bacterial membranes.

As the experimental BC composites were investigated as possible candidates for wound dressings on skin defects and injuries, including those contaminated by pathogenic microflora, it was important to study the effect of silver and/or antibiotics impregnated into BC films on dermal cells. The BC composites were investigated for their potential cytotoxicity in fibroblast cell culture using 4′,6-diamidino-2-phenylindole (DAPI) stain—a marker of nuclear DNA—and MTT assay, which determines the number of viable cells. The evaluation of the effect of BC composites on fibroblasts using DAPI staining was in good agreement with the results of the MTT assay (Figures 1.10 and 1.11). The results obtained in the present study are consistent with the published data, suggesting that BC/AgNPs produce a strong inhibitory effect on pathogenic and opportunistic pathogenic microflora, without inhibiting the growth of epidermal cells (Wu et al., 2014; Zhang et al., 2015).

No potential cytotoxicity was detected in all BC/AgNP composites in the NIH 3T3 mouse fibroblast cell culture, in contrast to the BC/antibiotic composites. These results suggest that BC composites constructed in the present study hold promise as dressings for managing wounds, including infected ones.

1.3.3 BIOTECHNOLOGICAL WOUND DRESSINGS BASED ON BACTERIAL CELLULOSE

Facilitating skin wound healing by using new technologies and materials is one of the main challenges faced by science and clinical practice. Hundreds of surgical and therapeutic materials and tools can be used to reconstruct skin defects. The properties of the material used to fabricate wound dressings must correspond to the type of injury and the development and phase of the healing process. The fabrication of biotechnological dermal equivalents carrying cells is a rapidly developing line of research aimed at reconstructing skin defects. The main challenge facing researchers is the choice of materials, which should have a number of special properties and impart characteristics of living tissues to the engineered constructs (grafts). These characteristics include (1) the ability to regenerate themselves, (2) the ability to maintain the blood supply, and (3) the ability to alter their structure and properties in response to effects of the external factors, including mechanical loads (Artyukhov et al., 2011).

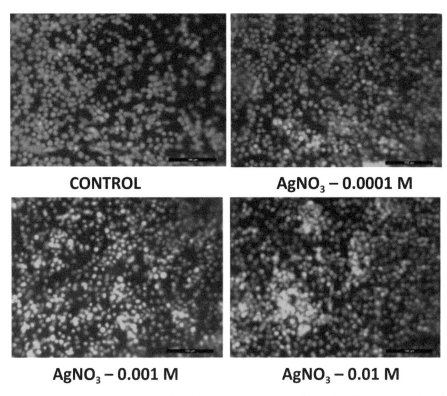

FIGURE 1.10 NIH 3T3 mouse fibroblast culture grown on films of BC/AgNPs, produced under different AgNO$_3$ concentrations in the medium. DAPI staining at day 3 of the culture. Scale bar = 100 μm.

BC and bioresorbable PHAs are promising materials for reconstructive biomedical technologies including tissue engineering. BC shows high biocompatibility, without being cytotoxic or causing any allergic reactions *in vivo*; it has a unique structure and physical/mechanical properties, such as high mechanical strength, elasticity, gas permeability, good water holding capacity, porosity, etc. The BC structure and high biocompatibility suggest that this natural polymer can be useful for reconstructive surgery as a material for skin defect reconstruction and as a matrix for drug delivery (Lin et al., 2013; Kwak et al., 2015; Chang et al., 2016; Shao et al., 2015; Kirdponpattara et al., 2015; Brassolatti et al., 2018).

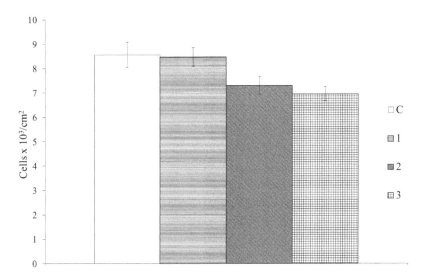

FIGURE 1.11 Number of viable cells (MTT assay) of NIH 3T3 mouse on BC/AgNPs after three days: C, control; 1, 0.0001 M BC/AgNPs; 2, 0.001 M BC/AgNPs; 3, 0.01 M BC/AgNPs.

A number of studies have shown the effectiveness of using BC alone and BC composites with various materials in reconstructing skin defects. Cellulose impregnated with lidocaine considerably facilitated the healing of third-degree burns (Shao et al., 2015). Full-thickness skin wounds in rats healed effectively under a film of BC nanocrystals/chitosan incorporating silver and curcumin particles (Bajpai et al., 2015). BC scaffolds with curcumin particles encapsulated in chitosan and composition with gelatin and fibrin were used to manage skin wounds in mice and rats (Shefa et al., 2017). Singla et al. (2017) investigated hydrogels prepared from bamboo cellulose nanocrystals loaded with silver particles, which successfully healed skin wounds in diabetic mice. Other authors (Kwak et al., 2015; Lin et al., 2013) tested gels based on cellulose and acrylate loaded with fibroblasts and keratinocytes, which considerably accelerated the healing of model burns and synthesis of type-1 collagen.

P(3HB-co-4HB) is one of the most attractive representatives of PHAs. PHAs are biopolymers of microbial origin, which are biodegradable, highly biocompatible, and effective in various biomedical applications. PHAs have been successfully tested as resorbable surgical sutures (Shishatskaya et al., 2004), artificial pericardium (Protopopov et al., 2005), wound dressings (Shishatskaya et al., 2016), implants for repairing cranial defects (Shumilova et al., 2017), biocompatible coatings for metallic stents (Protopopov et al., 2008), fully resorbable stents (Myltygashev et al., 2017), biliary stents (Markelova et al., 2008), etc. (Murueva et al., 2013; Eke et al., 2014; Chen et al., 2018). Thus, each of these biomaterials (BC and PHA) can be used to regenerate tissues, including the skin. One of the new lines of research is constructing composite materials and implants based on BC and PHA to repair tissue defects. The preparation of the BC/PHA composite materials and their characterization are described in a number of studies (Barud et al., 2011; Zhijianga et al., 2011; Akaraonye et al., 2016; Chiulan et al., 2016; Zhijiang et al., 2016; Abdalkarim et al., 2017).

The purpose of the present study was to construct and investigate composites based on BC and PHA as biotechnological wound dressings and to evaluate their efficacy in managing model skin burns in experiments with laboratory animals.

Composites based on BC and P(3HB-co-4HB) were constructed using different methods. The photographs of the samples are shown in Figure 1.12A, and the SEM images are in Figure 1.12B.

The BC/P(3HB-co-4HB) composites prepared by submerging nonwoven membranes into the *K. xylinus* B-12068 culture had the highest mechanical strength (Young's modulus of 272.3 ± 27.2 MPa). The most likely reason for this is the high mechanical strength of the nonwoven membranes (113.54 ± 9.85 MPa). The values of Young's modulus of the composites prepared by soaking BC films in the 2% P(3HB-xo-4HB) solution were 115 and 146 MPa, and they were higher than Young's modulus of the initial P(3HB-co-4HB) film. Although Young's modulus of the composites prepared by mixing the polymer solution and BC powder was low, it was higher than that of the corresponding parameter of pure BC (47.60 ± 6.32 MPa), but the elongation at break of the P(3HB-co-4HB) + BC powder (1:1) (2:1) composites ($3.45 \pm 0.34\%$) was insignificantly lower than the elongation at break of the initial BC film ($4.35 \pm 0.82\%$). The data reported in a number of studies suggested that BC/P(3HB-co-4HB) composites prepared using different methods had better mechanical properties than pure BC. In a study by Zhijiang et al. (2016), the mechanical properties of the nonwoven fiber based on P(3HB) and cellulose

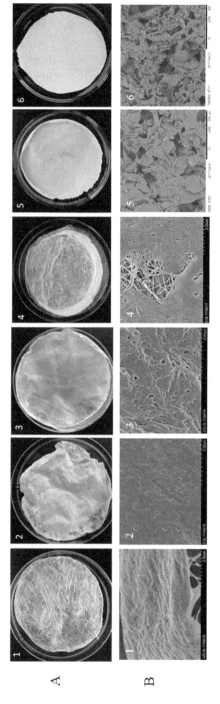

FIGURE 1.12 Photographs (A) and SEM images (B) of composite BC/P(3HB-co-4HB) films prepared using different methods: 1, BC+2% P(3HB-co-4HB) solution; 2, BC+5% P(3HB-co-4HB) solution; 3, BC+2% P(3HB-co-4HB) solution, soaked for 24 h; 4, wet BC+2% P(3HB-co-4HB) solution; 5, P(3HB-co-4HB) + BC powder (2:1); and 6, P(3HB-co-4HB) + BC powder (1:1).

acetate produced by electrospinning were better than the mechanical properties of the nonwoven fiber based on cellulose acetate. With P(3HB) content
increasing from 60% to 90% in the blend nanofiber, the tensile strength,
elongation at break, and Young's modulus increased from 4.52 ± 0.34 MPa,
6.53 ± 0.48%, and 806.9 ± 168.2MPa to 7.86 ± 0.67MPa, 16.7 ± 1.52%, and
854.2 ± 187.6MPa, respectively. In a study by Akaraonye et al. (2016), the
compressive modulus of porous composites based on P3HB was 0.08±0.01
MPa, but the addition of microfibrillated cellulose (10, 20, 30, and 40 wt%)
increased it by 35%, 37%, 64%, and 124%, respectively().

The BC film is rather hydrophilic, with a water contact angle of 45.5° ±
17.6°, while the water contact angle of the P(3HB-co-4HB) film is 66.5° ±
4.4° (Table 1.4).

TABLE 1.4   Surface Properties of BC/P(3HB-co-4HB) Hybrids

Sample	Water Contact Angle (degree)	Dispersive Component (mN/m)	Polar Component (mN/m)	Water Vapor Transmission Rate (g/m²/d)
Initial BC	45.5±17.6	28±9.24	17.5±8.37	2655±21
P(3HB-co-4HB) film	66.5± 4.4	22.90± 17.6	42.2± 17.6	404±9.1
P(3HB-co-4HB) nonwoven membrane	60.5±4.4	22.40±15.0	12.2±10.5	1000±7
BC + 2% P(3HB-co-4HB) solution	46.2±13.77	45.2±11.74	21±2.03	4473±10
P(3HB-co-4HB) + BC powder (2:1)	43.9±17.55	22±9.39	21.9±8.17	5014±20
P(3HB-co-4HB) + BC powder (1:1)	36.1±0.66	35.5±0.58	29.6±0.08	1732±9
BC + nonwoven membrane	65.3±2.72	45.6±1.29	19.7±1.42	2005±15

Soaking of the BC film in a 2% P(3HB-co-4HB) solution produced an
insignificant effect on the surface properties of BC. The composites prepared
by mixing the P(3HB-co-4HB) solution and BC powder at ratios of 2:1 and
1:1 had water contact angles of 43.9 ± 17.55° and 36.1 ± 0.66°, respectively;
the dispersive and polar components were larger than those of pure BC,
suggesting higher hydrophilicity of the surface of composites. The reason
for the high values of the water contact angle of the BC/P(3HB-co-4HB)
nonwoven membrane composites, 65.3 ± 2.72°, was that the nonwoven
membranes used to grow cellulose were initially dense, with a water contact
angle of about 60.5 ± 4.4°. After drying, the films impregnated with a 5%
P(3HB-co-4HB) solution had a very rough surface and a nonuniform coating,
making it impossible to examine their surface properties.

An important parameter of wound dressings is the water vapor transmission rate—a measure of the passage of water vapor through the material. The experimental composites showed different water vapor transmission rates: for the P(3HB-co-4HB) + BC powder (2:1), it was the highest (5014 ± 20 g/m²/d), while for the P(3HB/4HB) film, it was the lowest (404 ± 9.1 g/m²/d). It is important to increase this parameter to control the build-up of fluid underneath the dressing, make it more breathable, and prevent the establishment of conditions favorable for the development of pathogenic microorganisms. The models of third-degree skin burns were created on anesthetized rats. The animals were divided into four groups, six rats in each group. In the control group, the defect was closed with a VaskoPran dressing and covered with a dry sterile gauze bandage. In treatment groups, three types of experimental wound dressings were used: BC/P(3HB-co-4HB), BC/P(3HB-co-4HB)/actovegin, and BC/P(3HB-co-4HB)/fibroblasts differentiated from MSC wound dressing (Figure 1.13).

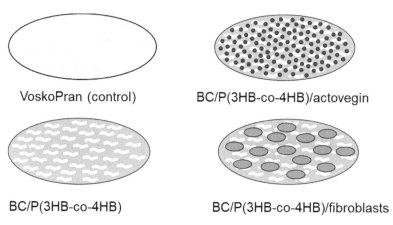

VoskoPran (control) BC/P(3HB-co-4HB)/actovegin

BC/P(3HB-co-4HB) BC/P(3HB-co-4HB)/fibroblasts

FIGURE 1.13 Types of experimental wound dressings used.

After surgical intervention, all animals were healthy and active, displayed normal eating behavior, and moved on their own. After the treatment group animals awakened from anesthesia, they did not show any signs of pain. The wound healing process was monitored using the following approaches: planimetric measurements of the wounds and the determination of wound surface area reduction and healing rate; histological examination; and biochemical and molecular techniques of detecting angiogenesis factor, inflammation, type-I collagen, keratin 10, and keratin 14. Measurements of the wound surface area showed that wound healing occurred at a faster rate in the treatment groups (Figure 1.14).

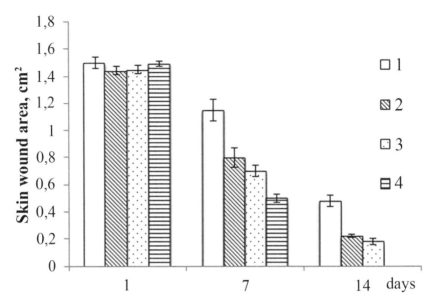

FIGURE 1.14 Dynamics of reduction in the skin wound area in the groups of animals with different wound dressings: 1, VoskoPran (control); 2, BC/P(3HB-co-4HB); 3, BC/P(3HB-co-4HB)/actovegin; and 4, BC/P(3HB-co-4HB)/fibroblasts.

The surface area of the wound closed with the BC/P(3HB-co-4HB)/actovegin wound dressing was reduced to 1 cm^2 (67% of the initial area) at day 3, to 33% at day 7, and to 7.3% at day 14. A similar healing process was observed in the group with the BC/P(3HB-co-4HB)/fibroblasts wound dressings: the wound area was reduced to 69.3% at day 3, to 40% at day 7, and to 9% at day 14. The healing rates in these groups were similar too, about 0.4 cm^2/d. In the group with the wounds closed with the BC/P(3HB-co-4HB) films containing no cells or actovegin, the healing rate was somewhat slower, with the wound area reduced to about 12% of the initial area by the end of the experiment. In the control group, a much poorer result was obtained: at day 14, the wound area was reduced to 30.6% of its initial area and the average healing rate was 0.19 cm^2/d.

Histological examination showed that at day 3, the necrotic zones in all groups were characterized by considerable destructive changes of the epidermis, dermis, and subcutaneous tissue, with the total necrosis of skin appendages. The surface of the wound was uneven and covered by a layer of necrotic masses with residual epidermis. Under the necrotic zone, in the layer adjacent to the undamaged tissue, there was edema pronounced to different extents and infiltration dominated by band neutrophils. In the control group, the thickness of epithelial injury (scab) was 28.45 ± 8.3 µm, and the number of band neutrophils reached 10 ± 2 cells (in 10 fields of view). In the BC/P(3HB-co-4HB) group, the scab was 20.29 ± 2.3 µm thick and the number of band neutrophils was 6 ± 2 cells (in 10 fields of view). In the BC/P(3HB-co-4HB)/ actovegin and BC/P(3HB-co-4HB)/fibroblasts groups, the thickness of the scab was 18.1 ± 1 µm and the number of band neutrophils was 6 ± 2 cells (in 10 fields of view). In the center of the burn, there were no hair follicles and sebaceous glands or they showed pronounced necrotic changes. There were no signs of regeneration in the center and at the edges of the wounds.

At day 14, in all groups, the thickness of the layer of necrotic masses covering the wound was reduced (Figure 1.15). In all treatment groups, the wound bed was represented by fibrous connective tissue with nonuniformly distributed infiltration by lymphocytes, macrophages, and segmented leukocytes. The rapid proliferation of fibroblasts was observed at the wound edges beneath the epidermis. It was accompanied by the development of relatively few small blood vessels. In groups with membranes loaded with actovegin and fibroblasts, either there was no leukocyte infiltration in the wound or it was represented by the minimal number of segmented leukocytes (2 ± 1 cells in 10 fields of view). In the control group, leukocyte infiltration on the wound bed, in the layer adjacent to viable tissues, was still observed, but it was weaker than in the previous phases of the experiment. Rapid epidermization was observed in all groups, but in the treatment groups, it was more effective, especially in the groups with membranes loaded with agents promoting wound healing (actovegin, fibroblasts). In those groups, the wounds were almost completely covered with a layer of the epidermis.

To study the mechanism of the healing process, real-time PCR was performed to determine the factors characterizing the degree of inflammation, vascularization, and formation of the new connective tissue at the defect site. Relative expression of VEGF (vascular endothelial growth factor, an indicator of vasculogenesis and angiogenesis) was higher in the groups with hybrid membranes loaded with fibroblasts or actovegin (Figure 1.16). In the BC/P(3HB-co-4HB)/actovegin group, ΔCt was 20% higher than in the VoskoPran group and 30% higher than in the BC/P(3HB-co-4HB) group. In the group with the membranes loaded with fibroblasts, relative expression was 5–6% lower but it was also higher than in the control group.

Relative TNF-α (tumor necrosis factor, a protein involved in acute inflammation) gene expression decreased considerably by the end of the experiment, and in the groups with membranes loaded with fibroblasts and actovegin, it was lower by a factor of two than in the BC/P(3HB-co-4HB) and VoskoPran groups, suggesting alleviation of inflammation (Figure 1.16).

The PCR method was used to determine the expression of type-I collagen (Col-1) (a signal of connective tissue formation), keratin 10 (K10) and keratin 14 (K14) (signals of the formation of the spinous and granular layers in the epidermis during tissue regeneration) genes. The number of Col-1, K10, and K14 cDNA copies was higher than that in the control by a factor of 1.5–2.0, also suggesting more effective wound healing beneath the experimental hybrid wound dressings (Figure 1.16).

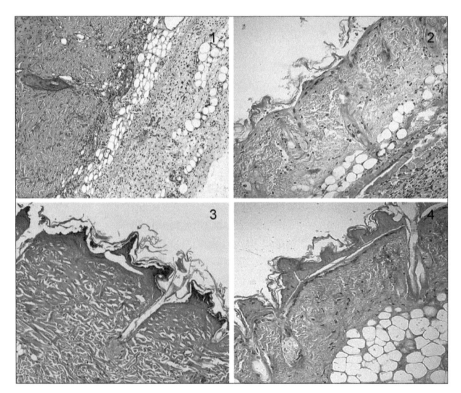

FIGURE 1.15 Histological sections collected at the defect site. Wound healing (day 14 post defect creation) under different wound dressings: 1, commercial material VoskoPran; 2, BC/P(3HB-co-4HB); 3, the BC/P(3HB-co-4HB)/actovegin hybrid membrane; and 4, the BC/P(3HB-co-4HB)/fibroblasts hybrid membrane.

The results of experiments in which the BC-based composites were tested as wound dressings are consistent with the literature data suggesting BC effectiveness in healing skin wounds. However, other studies described using BC in combination with the materials other than PHAs. Bajpai et al. (2015) demonstrated 97.2% healing of full-thickness skin wounds in Albino

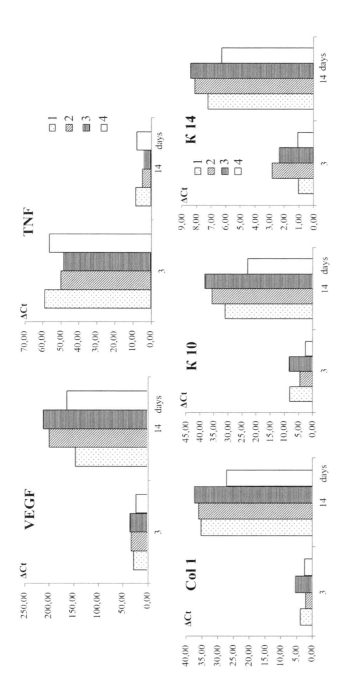

FIGURE 1.16 Relative expression of VEGF (vascular endothelial growth factor, an indicator of vasculogenesis and angiogenesis), TNF-α genes (tumor necrosis factor, a protein involved in acute inflammation), type-I collagen (Col-1), keratin 10 (K10), and keratin 14 (K14) (results normalized relative to the housekeeping gene—β-actin)—indicators of the regeneration dynamics of skin wounds in Wistar rats with hybrid wound dressings: 1, BC/P(3HB-co-4HB); 2, BC/P(3HB-co-4HB)/fibroblasts; 3, BC/P(3HB-co-4HB)/actovegin; and 4, VoskoPran (control).

Wistar rats under a film of BC nanocrystals/chitosan incorporating silver and curcumin particles. The potential use of BC modified with carboxymethyl groups and loaded with chitosan particles with encapsulated curcumin as material for healing skin wounds was shown in a study by Rojewska et al. (2017). Gelatin/BC scaffolds accelerated healing of skin wounds in mice (Goma et al., 2017); BC composites with silk fibroin effectively healed model skin wounds in rats (Shefa et al., 2017), and that was confirmed by measuring the levels of expression of wound healing markers. Singla et al. (2017) used hydrogels prepared from bamboo cellulose nanocrystals loaded with silver particles to heal skin wounds in streptozotocin-induced diabetic mice. Wounds were 98–100% recovered within 18 days as confirmed by detecting molecular markers of tissue genesis and a significant increase in the expression of collagen and growth factors (FCF, PDGF, and VEGF), which caused enhanced re-epithelization, vasculogenesis, and collagen deposition.

In a study by Kwak et al. (2015), the second-degree 1 cm^2 burn injury in a rat model was closed with a BC membrane. The expression of VEGF and angiopoietin-1 (Ang-1) was considerably decreased at days 10 and 15. Moreover, the level of collagen expression was much higher in the BC group than in the group of rats with the wounds closed with gauze. Although the topical application of BC membranes for two weeks accelerated wound healing, including tissue regeneration, connective formation, and angiogenesis, no complete skin healing occurred. In a study by Lin et al. (2013), membranes based on BC and chitosan and cellulose alone were used to close 1.2-cm × 1.2-cm degloving wounds in rats. The wounds were 90 and 88% healed under the composite material and cellulose, respectively, after 21 days of the experiment. The authors concluded that skin regeneration should be accelerated by adding agents promoting wound healing (drugs or cells).

1.4   CONCLUSION

A strain of acetic acid bacteria, *K. xylinus* B-12068, was studied as a source for BC production. The effects of cultivation conditions (carbon sources, temperature, and pH) on BC production and properties were studied in surface and submerged cultures. Glucose was found to be the best substrate for BC production among the sugars tested; ethanol concentration of 3% (w/v) enhanced the productivity of BC. The highest BC yield (up to 17.0 g/L) was obtained under surface static cultivation conditions, in the modified HS medium supplemented with ethanol, at pH 3.9, after seven days of cultivation in the thinnest layer of the medium.

BC with silver nanoparticles, BC/AgNPs, were produced hydrothermally, under different $AgNO_3$ concentrations in the reaction medium. The disk-diffusion method used in this study showed that all experimental BC/AgNP composites had pronounced antibacterial activity against *E. coli*, *P. eruginosa*, *K. pneumoniae*, and *S. aureus*. No potential cytotoxicity was detected in any of the BC/AgNP composites in the NIH 3T3 mouse fibroblast cell culture, in contrast to the BC/antibiotic composites. These results suggest that BC composites constructed in the present study hold promise as dressings for managing wounds, including contaminated ones.

Hybrid wound dressings have been constructed using two biomaterials: BC and the copolymer of 3-hydroxybutyric and 4-hydroxybutyric acids (P(3HB-co-4HB))—a biodegradable polymer. Two types of experimental biotechnological wound dressings—BC/P(3HB-co-4HB)/actovegin and BC/P(3HB-co-4HB)/fibroblasts—were tested *in vivo* on laboratory animals with model third-degree skin burns. Wound planimetry; histological exami-nation; and biochemical and molecular methods of detecting factors of angiogenesis, inflammation, type-I collagen, keratin, 10 and keratin 14 were used to monitor wound healing. Experimental wound dressings promoted healing more effectively than VoskoPran—a commercial wound dressing.

ACKNOWLEDGMENTS

This study was financially supported by the State assignment of the Ministry of Science and Higher Education of the Russian Federation No. FSRZ-2020-0006.

KEYWORDS

- **bacterial cellulose**
- **growth conditions**
- **Komagataeibacter xylinus**
- **AgNP composites**
- **bioresorbable PHA**
- **properties**
- **hybrids**
- **wound dressings**

REFERENCES

Abdalkarim S.Y.H., Yu H.Y., Song M.L., Zhou Y., Yao J., Ni Q.Q., In vitro degradation and possible hydrolytic mechanism of PHBV nanocomposites by incorporating cellulose ZnO nanohybrids, Carbohydr. Polym. 15 (2017) 38–49.

Akaraonye E., Filip J., Safarikova M., Salih V., Keshavarz T., Knowles J.C., Roy I., Composite scaffolds for cartilage tissue engineering based on natural polymers of bacterial origin, thermoplastic poly(3-hydroxybutyrate) and micro-fibrillated bacterial cellulose, Polym. Int. 65 (2016) 780–791.

Artyukhov A.A., Shtilman M.I., Kuskov A.N., Fomina A.P., Lisovy D.E, Golunova A.S., Tsatsakis A.M., Macroporous polymeric hydrogels formed from acrylate modified polyvinyl alcohol macromers, J. Polym. Res. 18 (2011) 667–673.

Bajpai S.K., Chand N., Ahuja S., Investigation of curcumin release from hitosan/cellulose micro crystals (CMC) antimicrobial films, Int. J. Biol. Macromol. 79 (2015) 440–448.

Barud S.H., Regiani T., Marques R.F.C., Lustry W.R., Messaddeq Y., Ribeiro S.J.L., Antimicrobial bacterial cellulose-silver nanoparticles composite membranes, J. Nanomater. (2011) 1–8.

Barud H.S., Souza J.L., Santos D.B., Crespi M.S., Ribeiro C.A., Messaddeq Y., Ribeiro S.J.L., Bacterial cellulose/poly(3-hydroxybutyrate) composite membranes, Carbohydr. Polym. 83 (2011) 1279–1284.

Budhiono A., Rosidi B., Taher H., Iguchi M., Kinetic aspects of bacterial cellulose formation in nata-de-coco culture system, Carbohydr. Polym. 40 (1999) 137–143.

Brassolatti P., Kido H.W., Bossini P.S., Gabbai-Armelin P.R, Otterço A.N, Almeida-Lopes L., Zanardi L.M., Napolitano M.A., L.R.D.S. de Avó, Forato L.A., Araújo-Moreira F.M., Parizotto N.A., Bacterial cellulose membrane used as biological dressings on third-degree burns in rats, Biomed. Mater. Eng. 29 (2018) 29–42.

Chang W.S., Chen H.H., Physical properties of bacterial cellulose composites for wound dressings, Food Hydrocoll. 53 (2016) 75–83.

Chawla P.R., Bajaj I. B., Survase S.A., Singhal R.S. Microbial cellulose: fermentative production and applications, 47 (2009) 107–124.

Chen G.Q., Zhang J., Microbial polyhydroxyalkanoates as medical implant biomaterials, Artif Cells Nanomed. Biotechnol. 46 (2018) 1–18.

Chiulan I., Mihaela Panaitescu D., Nicoleta Frone A., Teodorescu M., Andi Nicolae C., Căşărică A., Tofan V., Sălăgeanu A., Biocompatible polyhydroxyalkanoates/bacterial cellulose composites: preparation, characterization, and in vitro evaluation, J. Biomed. Mater. Res. A 104 (2016) 2576–2584.

Culebras M., Grande C.J., Torres F.G., Troncoso O.P., Gomez C.M., Bañó M.C., Optimization of cell growth on bacterial cellulose by adsorption of collagen and poly-L-lysine, Int. J. Polym. Mater. Polym. Biomater. 64 (8) (2015) 411–415.

Czaja W., Krystynowicz A., Bielecki S., Brown Jr R. M. Microbial cellulose—the natural power to heal wounds, Biomaterials 27 (2006) 145–151.

Ullah H., Wahid F., Santos H.A., Khan T. Advances in biomedical and pharmaceutical applications of functional bacterial cellulose-based nanocomposites, Carbohydr. Polym. 150 (2016) 330–352.

Iguchi M., Yamanaka S., Budhiono A. Bacterial cellulose—a masterpiece of nature's arts, J. Mater. Sci. 35 (2000) 261–270.

Jia B., Mei Y., Cheng L., Zhou J., Zhang L., Preparation of copper nanoparticles coated cellulose films with antibacterial properties through one-step reduction, ACS Appl. Mater. Interfaces 4 (6) (2012) 2897–2902.

Feng J., Shi Q., Li W., Shu X., Chen A., Xie X., Huang X., Antimicrobial activity of silver nanoparticles in situ growth on TEMPO-mediated oxidized bacterial cellulose, Cellulose 21 (2014) 4557–4567.

Fu L., Zhang J., Yang G. Present status and applications of bacterial cellulose-based materials for skin tissue repair, Carbohydr. Polym. 92 (2013) 1432–1442.

Goma S.F., Madkour T.M., Moghannem S., El-Sherbinyb I.M., New polylactic acid/cellulose acetate-based antimicrobial interactive single dose nanofibrous wound dressing mats, Int. J. Biol. Macromol. 105 (2017) 1148–1160.

Eke G., Kuzmina A.M., Goreva A.V., Shishatskaya E.I., Hasirci N., Hasirci V., In vitro and transdermal penetration of PHBV micro/nanoparticles, J. Mater. Sci. Mater. Med. 25 (2014) 1471–1481.

Hestrin S., Schramm M. Synthesis of cellulose by *Acetobacter xylinum*. 2. Preparation of freeze-dried cells capable of polymerizing glucose to cellulose, J. 58 (1954) 345–352.

Keshk S.M. Bacterial cellulose production and its industrial applications, J. Bioprocess. Biotechnol. 4 (2014) 1–10.

Kirdponpattara S., Khamkeaw A., Sanchavanakit N., Pavasant P., Phisalaphong M., Structural modification and characterization of bacterial cellulose–alginate composite scaffolds for tissue engineering, Carbohydr. Polym. 132 (2015) 146–155.

Kwak H., Kim J.E., Go J., Koh E.K., Song S.H., Son H.J., Kim H.S., Yun Y.H., Jung Y.J., Hwang D.Y., Bacterial cellulose membrane produced by *Acetobacter* sp. A10 for burn wound dressing applications, Carbohydr. Polym. 122 (2015) 387–398.

Li Z., Wang L., Hua J., Jia S., Zhang J., Liu H. Production of nano bacterial cellulose from waste water of candied jujube-processing industry using *Acetobacter xylinum,* Carbohydr. Polym. 120 (2014)115–119.

Lin W.C., Lien C.C., Yeh H.J., Yu C.M., Hsu S., Bacterial cellulose and bacterial cellulose-chitosan membranes for wound dressing applications, Carbohydr. Polym. 94 (2013) 603–611.

Lima H.L.S., Nascimento E.S., Andrade F.K., Brígida A.I.S., Borges M.D.F., Cassales A.R., Muniz C.R., Souza Filho M. de S.M., Morais J.P.S., Rosa, M.D.F. Bacterial cellulose production by *Komagataeibacter hansenii* ATCC 23769 using sisal juice—an agroindustry waste, Braz. J. Chem. Eng. 34 (2017) 671–680.

Lin W.C., Lien C.C., Yeh H.J., Yu C.M., Hsu S.H., Bacterial cellulose and bacterial cellulose chitosan membranes for wound dressing applications, Carbohydr. Polym. 94 (1) (2013) 603–611.

Ma X., Wang R.M., Guan F.M., Wang T.F., Artificial dura mater made from bacterial cellulose and polyvinyl alcohol, CN Patent ZL200710015537 (2010).

Markelova N.M., Shishatskaya E.I., Vinnik Y.S., Cherdantsev D.V., Beletskiy I.I., Kuznetsov M.N., Zykova L.D., In vitro justification of using endobiliary stents made of polyhydroxyalkanoates, Macromol. Symp. 269 (2008) 82–91.

Murueva A.V., Shishatskaya E.I., Kuzmina A.M., Volova T.G., Sinskey A.J., Microparticles prepared from biodegradable polyhydroxyalkanoates as matrix for encapsulation of cytostatic drug, J. Mater. Sci. Mater. Med. 24 (2013) 1905–1915.

Myltygashev M.P., Boyandin A.N., Shumilova A.A., Kapsargin F.P., Shishatskaya E.I., Kirichenko A.K., Volova T.G. Study of the effectiveness of the use of biodegradable stents on the basis of polyhydroxyalkanoates in the plasty of the pyeloureteral segment, Urology 1 (2017) 16–22.

Ng C.C., Shyu, Y.T. Development and production of cholesterol-lowering Monascus-nata complex, World J. Microb. Biotechnol. 20 (2004) 875–879.

Park J.K., Jung J.Y., Park Y.H. Cellulose production by *Gluconacetobacter hansenii* in a medium containing ethanol, Biotechnol. Lett. 25 (2003) 2055–2059.

Percival S.L., Bowler P.G., Russell D., Bacterial resistance to silver in wound care, J. Hosp. Infect. 60 (1) (2005) 1–7.

Pokalwar S.U., Mishra M.K., Manwar A.V. Production of cellulose by *Gluconacetobacter sp.*, Recent Res. Sci. Technol. 2 (2010) 14–19.

Popova L.N., How to Change Borders of Epidermis During Wound Healing, Ph.D. Thesis 1942 1–14.

Prudnikova S.V., Volova T.G., Shishatskaya E.I. A strain of bacterium *Komagataeibacter xylinus*—a producer of bacterial cellulose. RF Patent 2568605 (in Russian) (2015).

Protopopov A.V., Kochkina T.A., Konstantinov E.P., Shishatskaia E.I., Efremov S.N., Volova T.G., Gitelson I.I., Investigations of application of PHA coating to enhance biocompatibility of vascular stents, Dokl. Biol. Sci. 401 (2005) 85–87.

Protopopov A.V., Konstantinov E.P., Shishatskaya E.I., Efremov S.N., Volova T.G., Gitelson I.I., The use of resorbable polyesters to increase the biocompatibility of intravascular stents, Technol. Liv. Syst. 2 (2008) 25–33.

Rojewska A., Karewicz A., Boczkaja K., Wolski K., Kępczyński M., Zapotoczny S., Nowakowska M., Modified bionanocellulose for bioactive wound-healing dressing, Eur. Polym. J. 96 (2017) 200–209.

Ruka D.R., Simon G.P., Dean K.M. Altering the growth conditions of *Gluconacetobacter xylinus* to maximize the yield of bacterial cellulose, Carbohydr. Polym. 89 (2012) 613–622.

Ruparelia J.P., Strain specificity in antimicrobial activity of silver and copper nanoparticles, Acta Biomater. 4 (3) (2008) 707–716.

Sadanand V., Tian H., Varada Rajulu A., Satyanarayana B., Antibacterial cotton fabric with in situ generated silver nanoparticles by one-step hydrothermal method, Int. J. Polym. Anal. Charact. 22 (3) (2017) 275–279.

Saska S., Barud H.S., Gaspar A.M.M., Marchetto R., Ribeiro S.J.L., Messaddeq Y., Bacterial cellulose-hydroxyapatite nanocomposites for bone regeneration, Int. J. Biomater. 2011 (2011) 1–8.

Shah N., Ul-Islam M., Khattak W.A., Park J.K., Overview of bacterial cellulose composites: a multipurpose advanced material, Carbohydr. Polym. 98 (2) (2013) 1585–1598.

Shao W., Liu H., Wang S., Wu J., Huang M., Min H., Liu X., Controlled release and antibacterial activity of tetracyclinehydrochloride-loaded bacterial cellulose composite membranes, Carbohydr. Polym. 145 (2016) 114–120.

Shao W., Liu H., Liu X., Wang S., Wu J., Zhang R., Min H., Huang M., Development of silver sulfadiazine loaded bacterial cellulose/sodium alginate composite films with enhanced antibacterial property, Carbohydr. Polym. 132 (2015) 351–358.

Shefa A.A., Amirian J., Kang H.J., Bae S.H., Jung H., Choi H.J.,,. Lee S.Y, Lee B.T., In vitro and in vivo evaluation of effectiveness of a novel TEMPO-oxidized cellulose nanofiber-silk fibroin scaffold in wound healing, Carbohydr. Polym. 177 (2017) 284–296.

Singla R., Soni S., Patial V., Kulurkar P.M., Kumari A., Padwad Y.S., Yadav S.K., Cytocompatible anti-microbial dressings of syzygiumcumini cellulose nanocrystals decorated with silver nanoparticles accelerate acute and diabetic wound healing, Sci. Rep. 7 (2017) 1–13.

Shishatskaya E.I., Volova T.G., Puzyr A.P., Mogilnaya O.A., Efremov S.N., Tissue response to the implantation of biodegradable polyhydroxyalkanoate sutures, J. Mater. Sci. Mater. Med. 15 (2004) 719–728.

Shishatskaya E.I., Nikolaeva E.D., Vinogradova O.N., Volova T.G., Experimental wound dressings of degradable PHA for skin defect repair, J. Mater. Sci. Mater. Med. 27 (2016) 1–16.

Sureshkumar M., Siswanto D.Y., Lee C.K., Magnetic antimicrobial nanocomposite based on bacterial cellulose and silver nanoparticles, J. Mater. Chem. 20 (2010) 6948–6955.

Shumilova A.A., Myltygashev M.P., Kirichenko A.K., Nikolaeva E.D., Volova T.G., Shishatskaya E.I., Porous 3D implants of degradable poly-3-hydroxybutyrate used to enhance regeneration of rat cranial defect, J. Biomed. Mater. Res. A 105 (2017) 566–577.

Wen X., Zheng Y., Wu J., Yue L., Wang C., Luan J., Wu Z., Wang K., In vitro and in vivo investigation of bacterial cellulose dressing containing uniform silver sulfadiazine nanoparticles for burn wound healing, Prog. Nat. Sci. Mater. Int. 25 (2015) 197–203.

Wijaya C.J., Saputra S.N., Soetaredjo F.E., Putro J.N., Lin C.X., Kurniawan A., Ju Y.H., Ismadji S., Cellulose nanocrystals from passion fruit peels waste as antibiotic drug carrier, Carbohydr. Polym. 175 (2017) 370–376.

Wu J., Zheng Y., Song W., Luan J., Wen X., Wu Z., Chen X., Wang Q., Guo S., In situ synthesis of silver-nanoparticles/bacterial cellulose composites for slow-released antimicrobial wound dressing, Carbohydr. Polym. 102 (2014) 762–771.

Yamada Y., Yukphan P., Lan Vu H.T., Muramatsu Y., Ochaikul D., Tanasupawat S., Nakagawa Y. Description of *Komagataeibacter* gen. nov., with proposals of new combinations (*Acetobacteraceae*), J. Gen. Appl. Microbiol. 58 (2012) 397–404.

Yang G., Xie J., Deng Y., Bian Y., Hong F., Hydrothermal synthesis of bacterial cellulose/AgNPs composite: a «green» route for antibacterial application, Carbohydr. Polym. 87 (4) (2012) 2482–2487.

Zhang J., Chang P., Zhang C., Xiong G., Lio H., Zhu Y., Ren K., Yao F., Wan Y., Immobilization of lecithin on bacterial cellulose nanofibers for improved biological functions, React. Funct. Polym. 91 (2015) 100–107.

Zhijianga C., Guanga Y., Kim J., Biocompatible nanocomposites prepared by impregnating bacterial cellulose nanofibrils into poly(3-hydroxybutyrate), Curr. Appl. Phys. 11 (2011) 247–249.

Zhijiang C., Yi X., Haizheng Y., Jia J., Liu Y., Poly(hydroxybutyrate)/cellulose acetate blend nanofiber scaffolds: preparation, characterization and cytocompatibility, Mater. Sci. Eng. C: Mater. Biol. Appl. 58 (2016) 757–767.

CHAPTER 2

Applications of Nanotechnology in Diverse Fields of Supramolecules, Green Chemistry, and Biomedical Chemistry: A Very Comprehensive Review

SONIA KHANNA

Department of Chemistry and Biochemistry, Sharda University, Greater Noida, India. E-mail: sonia.khanna@sharda.ac.in

ABSTRACT

Nanotechnology has emerged as one of the most dynamic science and technology domains of physical sciences, molecular devices, green chemistry, biotechnology, and medicine. Nanomaterials (nanoparticles, nanowires, nanofibers, and nanotubes) have been explored in many biological applications (biosensing, biological separation, molecular imaging, and anticancer therapy). The unusual properties of nanoparticles, such as hardness, rigidity, high yield strength, flexibility, ductility, are attributed to the high surface-to-volume ratio. Supramolecular chemistry and green chemistry are also related to concepts of nanotechnology, depicted in their applications. The realization that the nanoscale has certain properties needed to solve important medical challenges and cater to unmet medical needs is driving nanomedical research. The present chapter explores the significance of nanoscience and the latest nanotechnologies for human health. The objective of this chapter is to describe the potential benefits and impacts of nanotechnology in different areas such as green chemistry, biotechnology, and supramolecular chemistry.

2.1 INTRODUCTION

Nanotechnology is a multidisciplinary area involved in the design, synthesis, characterization, and application of materials and devices having at least one dimension on the nanometer scale. It shows properties different from its bulk counterparts such as extreme hardness, high rigidity and yield strength, flexibility, ductility, quantum size effect, and macroquantum tunneling effect. Nanomaterials have a number of fascinating potential applications in a wide range of industrial sectors such as electronics, magnetic and optoelectronics, biomedical, pharmaceutical, cosmetics, energy, environmental, catalytic, space technology, and many others (Puzyn et al., 2009; Leszczynski, 2010). Supramolecular chemistry represents the chemistry beyond the molecule, noncovalent intermolecular interactions constituting the driving force for the preparation of molecular and supramolecular assemblies. Upon molecular recognition between discrete units having dimensions on the nanometer scale, chemical processes such as self-assembly and self-organisation start operating and are the leading processes to build up supramolecular aggregates and materials. The processes of self-assembly and molecular recognition of supramolecules are important in the functioning of many discrete numbers of assembled molecular subunits or components. The supramolecular recognition properties of the nanoparticles (NPs) are used to generate stable and ordered 3D functional nanostructures. The synthesis of NPs involves the use of toxic chemicals and harmful processes. Recent research is on the synthesis of NPs by green methods with no harmful effects and toxicity. Such a methodology is also discussed in this chapter. Within such nanoscale, we could include supramolecular biological systems, such as cell membranes, nucleic acids, proteins, as well as supramolecular artificial nanostructured materials; among them, carbon nanotubes, liquid crystals, self-assembled monolayers (SAMs), or supramolecular systems based on colloids or liposomes. Nowadays, the synthesis of nanomaterials has attracted increasing interest because of their unique properties and promising applications (Ehrlich, 1906; Fischer, 1894).

Nanotechnology has ventured into the field of biotechnology, now also known as nanobiotechnology, to study its applications to medicine and physiology. These nanomaterials and devices are designed to allow their interaction with cells and tissues at the molecular (i.e., subcellular) level with a high degree of functional specificity.

2.2   NANOTECHNOLOGY

Nanotechnology is the science of extremely small materials. It deals with the creation of devices and systems at different levels and explores their novel properties (physical, chemical, and biological) on the nanometer scale.

2.2.1 CLASSIFICATION OF NANOMATERIALS

Nanomaterials can be classified as follows:

1. Zero-dimensional (quantum dots) in which the movement of electrons is confined in all three dimensions.
2. One-dimensional (quantum wires) in which electrons can move in one direction.
3. Two-dimensional (thin films) in which the free electrons can move in the X–Y plane.
4. Three-dimensional nanostructures in which electrons can move in the X, Y, and Z directions.

Semiconductor nanocrystals are zero-dimensional quantum dots, in which the spatial distributions of the excited electron–hole pairs are confined within a small volume. Nanorods and nanowires have dimensions less than 100 nm; tubes, fibers, and platelets have dimensions less than 100 nm; and particles, quantum dots, and hollow spheres have 0 or 3 dimensions greater than 100 nm.

Nanomaterials in different phases can be classified as follows:

1. Single phase: Crystals, amorphous particles, and layers are included in this class.
2. Multiphase: Matrix composites and coated particles are included in multiphase solids.

Multiphase systems of nanomaterials include colloids, aerogels, ferro-fluids, etc.

2.2.2 SYNTHESIS OF NANOPARTICLES

The synthesis of NPs adopts two approaches—a top-down approach and a bottom-up approach.

In the *top-down approach*, bigger materials are broken down into smaller units using many physical, chemical, and thermal techniques. Top manufacturing involves the construction of parts through methods such as cutting, carving, and molding, and due to our limitations in these processes, highly advanced nanodevices are yet to be manufactured. Laser ablation, milling, nanolithography, hydrothermal techniques, physical vapor deposition, and electrochemical methods (electroplating) use the top-down approach for nanoscale material manufacturing. In a study, colloidal carbon spherical particles were synthesized by continuous chemical adsorption of polyoxometalates on the carbon interfacial surface. Adsorption made the carbon black to aggregate into relatively smaller spherical particles, with high dispersion capacity and narrow size distribution (Garrigue et al., 2004).

The *bottom-up approach* results in the generation of nanoparticles with uneven edges and cavities on the surface of NPs. The bottom-up approach results in the synthesis of good NPs with minimum wastage of chemicals. Examples of the bottom-up approach are sedimentation and reduction techniques (Iravani, 2011). The bottom-up method prepares nanomaterials with a very small size, whereas the top-down process cannot deal with very tiny objects. The bottom-up approach generally produces structures with fewer defects as compared to the nanostructures produced by the top-down approach. The main driving force behind the bottom-up approach is the reduction in Gibbs free energy. Therefore, the materials produced are close to their equilibrium state.

Nanotechnology in its broadest terms refers to devices with dimensions in the range of 1–100 nm, while nanofabrication involves the manipulation of matter on the nanoscopic length scale to design structures and patterns with desired functions. It occurs by bottom-up synthetic chemistry; self-assembly has provided a powerful way of making materials and organizing them into functional constructs designed for a specific purpose. Self-assembly teaches that matter of all kinds, exemplified by atoms and molecules, colloids, and polymers, can undergo spontaneous organization to a higher level of structural complexity, driven by a map of forces operating over multiple length scales.

The fabrication of NPs is done by shaping, positioning, and organizing at the nanoscale using two approaches, that is, top-down and bottom-up nanofabrication. The former involves the use of ion and electron, photon, and atom beams to sculpt matter from macroscopic to nanoscopic dimensions in a serial process to form functional constructs with purposeful utility, while the latter self-assembles these constructs in a parallel manner from nanoscale building blocks.

The stability and ordering of these NP structures are the most important features to achieve function for long-term applications. In nanotechnology, items are constructed from the bottom-up method using techniques and tools to make complete, high-performance products. Nanostructured materials are assembled by physical assembly and chemical assembly. The physical assembly techniques are based on the assembly of nonfunctionalized NPs on the surfaces by physical forces such as convective or capillary assembly (Denkov et al., 1993), spin coating (Ozin and Yang, 2001), and sedimentation (Wijnhoven and Vos, 1998). These techniques result in relatively simple close-packed 2D or 3D particle arrays with limited stabilities.

Self-assembly of NPs is an important feature of nanomaterials and refers to the autonomous organization of components into patterns or structures without human intervention. It is a fundamental principle that creates structural organization from the disordered components in a system. The design of a nanodevice lies in the self-assembly of a target structure from the spontaneous organization of building blocks like molecules or nanoscale clusters. The nanomaterials are functionalized to deposit NPs with surface functionalities onto a functionalized substrate. The nanosurface can be modified to cater to the requirements of the applications. The surface of nanotubes can be modified by covalent interactions including oxidation, cycloaddition reactions, esterification, and amidation. The nanostructures are designed by controlling the functional groups on the surface of NPs and tailoring the nanostructures in a predictable manner to generate functional, more complex nanostructured architectures on the surfaces to meet the needs for specific applications, such as molecular electronics and biosensing (Lahav et al., 2000). Various chemical interaction strategies, for example, covalent bonding (Paraschiv et al., 2002), electrostatic forces (Decher, 1997), and host–guest interactions (Crespo-Biel et al., 2005), have been employed to chemically govern the self-assembly of NPs onto surfaces. Cross-linking of the neighboring particles with chemical forces by selective binding can further enhance the stability of NP assemblies (Zirbs et al., 2005). These methods control the spatial distribution of NPs across a large area in more complex patterns when combined with nanopatterning schemes. The ability to attach NPs onto planar surfaces in a well-defined, controllable, and reliable manner is an important prerequisite for the fabrication of micro- or nanostructured devices suitable for the application in the field of (bio) nanotechnology. Nanofabrication allows control over functional groups on the surface of NPs to tailor the nanostructures in a predictable manner, resulting in the formation of functional, more complex nanostructured architectures on the surfaces to meet the needs for specific applications such as in molecular electronics and biosensing (Lahav et al., 2000).

Metallic nanomaterials have received peculiar interest in diverse fields of applied science such as materials science and biotechnology (Greque de Morais et al., 2014). The extremely small size and a high surface-to-volume ratio of nanomaterials have evolved nanomaterials as sensors (Collins et al., 2000). Research in nanotechnology reflects the improvement in the design and application of materials, devices, and models that exhibit sustainable future. All of these methods employ hazardous chemicals. In the case of NP synthesis, the bottom-up (or self-assembly) procedures involve a homogeneous system wherein catalysts (e.g., reducing agents and enzymes) are producing nanostructures affected by catalyst properties, reaction media, and reaction conditions (e.g., solvents, stabilizers, and temperature). For example, the chemical reduction method is the most commonly employed synthetic route for metallic particle synthesis. Silver NPs are increasingly used in various fields such as medical, food, health care, and industries due to their unique physical and chemical properties (Gurunanthan et al., 2015; Li et al., 2010). The chemical reduction technique is based on the reduction of metal salts like silver nitrate in an appropriate medium using reducing agents like citrate or branched polyethyleneimine. However, citrate produces negatively charged silver NPs, while branched polyethyleneimine produces positively charged ones (Mallick et al., 2004). Particularly, intermediates and byproducts are playing a crucial role in NP synthesis. Indeed, the physicochemical properties and surface, and morphological characteristics of NPs will influence their fate, activity, transport, and toxicity.

2.3   SUPRAMOLECULAR CHEMISTRY

Supramolecular chemistry, as defined by Lehn, "chemistry beyond the molecule," focuses on the development of functional complex architectures through noncovalent interactions. The domain of supramolecular chemistry came of age when Donald J. Cram, Jean-Marie Lehn, and Charles J. Pedersen were jointly awarded the Nobel Prize for chemistry in 1987 in recognition of their work on "host–guest" assemblies (in which a host molecule recognizes and selectively binds a certain guest). It established supramolecular chemistry as a discipline, which is now being explored in various areas such as drug development, sensors, catalysis, nanoscience, molecular devices, etc. Supramolecular chemistry is an interdisciplinary field of research, reaching across from chemistry to the physics and biology of chemical species molecules, and focuses on the chemical systems made up of a discrete number of assembled molecular subunits or components. It can be described as the study of systems that contain more than two molecules, having convergent binding sites such as donor atoms, sites for the formation of hydrogen bonds and sizable cavity, and another molecule (analyte or guest) with divergent binding sites such as hydrogen bond acceptor atoms. A supramolecule is a well-defined discrete system generated through interactions between molecules (receptors or hosts).

2.3.1 FORCES IN SUPRAMOLECULES

Just as molecular chemistry is based on covalent bonds, supramolecular chemistry is based on noncovalent interactions. These noncovalent interactions are as follows:

1. electrostatic interactions (ion–ion, dipole–dipole),
2. hydrogen bonding (4–120 kJ/mol),
3. π–π stacking (0–50 kJ/mol),
4. cation–p interactions (5–80 kJ/mol),
5. van der Waals forces (<50 kJ/mol), and
6. Hydrophobic effects.

However, the small stabilization gained by one weak interaction when added to all small stabilizations from the other interactions leads to the generation of a stable architecture or a stable host–guest complex. The formation of a complex via supramolecular interactions is fast, facile, and more stable. The nature of forces depends upon the degree of electronic coupling between the molecular units. Electrostatic interactions occur between charged molecules. An attractive force is observed between oppositely charged molecules, and a repulsive force between molecules with the same type of charge (both negative or both positive). The magnitude of this interaction is relatively large compared to other noncovalent interactions and cannot be ignored in the formation of a supramolecule. The strength of this interaction changes in inverse proportion to the dielectric constant of the surrounding medium, making it stronger in a hydrophobic environment. Dipole–dipole and dipole–ion interactions play important roles in neutral species compared to electrostatic interactions.

Hydrogen bonding also plays an important role in molecular recognition, as in deoxyribonucleic acid (DNA). Hydrogen bonding interaction is weaker (4–120 kJ/mol) than an electrostatic interaction (200–300 kJ/mol) and occurs only when the functional groups that are interacting are properly oriented. That is why hydrogen bonding is the key interaction during recognition in many cases. The double-helical structure of DNA is stabilized primarily by two forces: hydrogen bonds between nucleotides and base-stacking interactions among the aromatic nucleobases (Yakovchuk et al., 2006). In the aqueous environment of the cell, the conjugated π bonds of nucleotide bases align perpendicular to the axis of the DNA molecule, minimizing their interaction with the solvation shell and therefore, the Gibbs free energy.

The van der Waals interaction is weaker (<5 kJ/mol) and less specific than those described above, but it is undoubtedly important because this interaction generally applies to all kinds of molecules such as dipoles and make a significant contribution to molecular recognition.

π–π interactions play a crucial role in recognition of aromatic compounds. When the aromatic rings face each other, the overlap of π-electron orbitals results in an energetic gain. For example, the double-helical structure of DNA is partially stabilized through π–π interactions between neighboring base pairs.

2.3.2 HOST–GUEST THEORY

Host–guest chemistry plays an important role in supramolecular chemistry. Hosta are the molecules having binding sites facing inward or converged, and guest molecules have binding sites facing outward or diverged. Supramolecules are composed of two or more molecules or ions that are held together in a unique structural relationship than those of full covalent bonds. Noncovalent bonding is important in the study of the 3D structure of large molecules, such as proteins, and is involved in many biological processes in which large molecules bind specifically but transiently to one another. In host–guest chemistry, the host is the bigger molecule that binds a smaller molecule via the lock and key principle. Emil Fischer proposed that enzymes recognize substrates by a "lock and key" mechanism, where the structural fit between the recognizing molecule and the recognized molecule is important. Multivalent binding has added advantages of high binding strength and reversibility of the binding process, and due to the organization of binding sites in the space, the geometry of the supramolecular complex can be controlled. Moreover, the multivalency effect ensures the formation of discrete supramolecular systems rather than oligomers. Thus, the interactions between a receptor (host) and an analyte (guest) strongly depend on multiple binding. On the other hand, in the case of self-assembled polymeric supramolecular systems, cooperative interactions between multiple chemical components are needed. Due to these cooperative interactions (noncovalent in supramolecular systems) or cooperativity, the free energy change (ΔG) is either decreased or increased over the subsequent interaction steps in a reaction as compared to the first step. If the free energy vary interestingly, many supramolecular architectures are very stable than what would be expected from the presence of chelate/ positive cooperativity alone. If the change is decreased, it is called positive cooperativity, while an increase in the free energy change suggests negative cooperativity. The positive cooperativity of multiple binding sites present on the host molecule to bind a guest molecule, in a way, is the chelate effect.

There are two types of host molecules: (1) cation-binding hosts, such as crown ethers, spherands, podants, and cryptands, and (2) anion-binding hosts, such as guadinium and extended porphyrins (Figure 2.1).

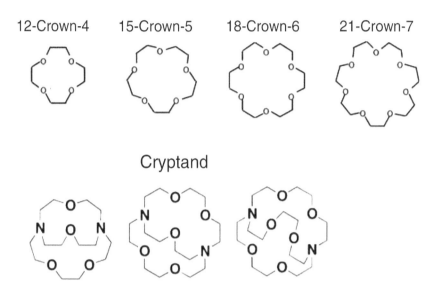

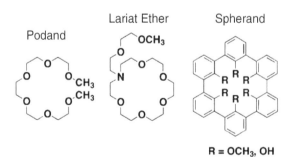

FIGURE 2.1 Host molecules in supramolecular chemistry.

Supramolecular chemistry is based on three concepts: binding, recognition, and coordination. In molecular recognition, a molecule selectively recognizes its partner through various molecular interactions. The complexation event takes place between a host and a guest. The host can be a large molecule, an aggregate having a sizeable cavity with convergent binding sites, while a guest could be a cation, anion, or a neutral molecular species with divergent binding sites (Steed and Atwood, 2009). Molecular self-assembly and molecular recognition are important processes in the generation of a supramolecule. Molecular recognition is a specific interaction between two molecules, which are complementary in their geometric and electronic features. Self-assembly refers to the recognition between molecules leading to well-defined aggregates, for example, DNA.

The self-organization may be referred to as an ordered self-assembly. Self-assembly and self-organization of a supramolecular architecture are both multistep processes. A sequence or hierarchy of assembly steps may be followed.

Like supramolecular interactions, host–guest binding relies on manifold noncovalent interactions, with the added requirement that the host possesses an interior cavity that is complementary in size and shape to the guest molecule (Hof et al., 2002; Pedersen, 1998). The "inner phase" of a synthetic host presents a different chemical environment to a bound guest than what it would experience in the surrounding bulk solvent. Molecular machines based on supramolecules are capable of showing controlled repetitive motion and function at the nanoscale. Thus, such materials have great potential in the field of nanotechnology.

2.3.3   APPLICATIONS OF NANOTECHNOLOGY IN SUPRAMOLECULAR CHEMISTRY

Supramolecular chemistry and nanochemistry are two strongly interrelated cutting-edge frontiers in research in the chemical sciences. Supramolecular chemistry is recognised by its process of self-assembly and molecular recognition, where the host recognises the guest molecule on the basis of the available cavity. This concept is extended for the formation of stable 2D and 3D nanostructures, while using molecular recognition for establishing stability and order as well as creating a functionality of the resulting structure. Supramolecular chemistry is also known by its applications in molecular devices. Molecular devices are functional materials that are structurally based on using the concepts of supramolecular chemistry. Supermolecules are capable of electron conduction and electrical switches (molecular electronic devices); supermolecules that respond to light and manipulate photonic information (molecular photonic devices), supermolecules that can be used for information processing and calculations (molecular computer), and supermolecules that move, rotate, and catch targets (molecular machines) are introduced as examples of molecular devices. Molecular wires and molecular switches have been developed using carbon nanotubes. Molecular recognition devices or photonics can control light emission in devices, photoelectronic devices. In particular, much of the progress made in the field of molecular electronics is based on the application of carbon nanotubes. Very interestingly, many supramolecular architectures are very stable than what would be expected from the presence of chelate/positive cooperativity alone. The coupling chemistries are being incorporated to direct and control the deposition of NPs with surface functionalities onto a functionalized supramolecular substrate. Various chemical interaction strategies, for example, covalent bonding, electrostatic forces, and host–guest interactions (Atwood et al., 1996; Beer et al., 1999; Stoddart; Jehn, 1995), have been employed to chemically govern the self-assembly of NPs onto surfaces.

The neighboring molecules can be cross-linked by selective binding and enhance the stability of NP assemblies. It can directly control the spatial distribution of NPs across a large area in more complex patterns when combined with nanopatterning schemes. The integration of particles into devices usually requires placing them in specific positions on surfaces. Hence, the combination of top-down patterning techniques, for example, microcontact printing (Park et al., 2005), transfer printing (Meitl et al., 2006), nanoimprint lithography (Maury et al., 2005), and photolithography (Yin et al., 2001), and bottom-up self-assembly is crucial in obtaining (submicron) patterned functional nanostructures on surfaces. The introduction of SAMs on localized areas of a substrate allows straightforward further functionalization and directed assembly of NPs. In lithography, specific binding can be introduced, allowing the control of NP assembly into desired lithographic patterns. Wet chemical self-assembly of NPs is used for the fabrication of NP-based nanostructures because of its compatibility with various kinds of substrates with complex shapes.

The host–guest complexation of cyclodextrin (CD) and its guest molecules (e.g., adamantane and ferrocene) are applied to assist the nanoparticle assembly. Some naturally occurring cyclic hosts that possess molecular recognition capabilities were known before crown ethers (the first artificial host molecules) were discovered. For example, the cyclic oligopeptide valinomycin and the cyclic oligosaccharide CD were found to bind specific guest molecules. CDs are the naturally occurring cyclic hosts possessing molecular recognition capabilities. They are obtained from the action of enzymes on polysaccharide starch (a) converting it into a cyclic oligomer with an appropriate number of glucopyanoside units.

Huang and co-workers (Huang et al., 2004) reported a new route for the formation of supramolecular isomers via molecular templating in isolation of two polygons of a higher number of sites constructed by simple bent imidazolate bridging ligands and two CuI ions by templating of circular organic molecules. Synthetic macrocyclic host compounds, like calixarenes, CDs, etc., can interact with suitable guest molecules via noncovalent interactions to form functional supramolecular systems. The response of molecules and the unique properties at the nanoscale can synergistically interact to functionalize NPs with the host–guest supramolecular systems. This has shown great potentials for a broad range of applications in the fields of nanoscience and nanotechnology. The applications of the NPs functionalized with supramolecular host–guest systems in nanomedicine and healthcare, including therapeutic delivery, imaging, sensing, and removal of harmful substances, have been reported (Zilong et al., 2015).

2.4   GREEN CHEMISTRY

Green chemistry, also called sustainable chemistry, focuses on the design and processes that minimize or eliminate hazardous chemicals. Chemicals should be designed such that, at the end of their function, they break down into innoc-uous degradation products and do not harm the environment. It is based on 12 principles that provide a framework for learning about designing processes and products to minimize hazards. These 12 principles are as follws:

1. Prevent waste—It is better to prevent waste than to treat or clean up after it has been created.
2. Atom economy—To maximize the incorporation of all materials used in the process into the final product.
3. Less hazardous chemical syntheses—To design such synthetic methods to use and generate substances that possess little or no toxicity to human health and environment.
4. Designing safer chemicals—Chemicals should be designed to prevent the efficacy of function while reducing toxicity.
5. Safer solvents and auxillaries—The use of auxiliary substances such as solvents, separation agents should be made unnecessary wherever possible.

6. Design for energy efficacy—Energy requirements should be recognized for their environmental and economic impacts. Synthetic methods should be conducted at ambient temperature and pressure.
7. Use of renewable feedstocks—A raw material or feedstock should be renewable rather than depleting wherever possible.
8. Reduce derivatives—Unnecessary derivatization of chemical processes should be minimized or avoided.
9. Catalysis—Catalytic reagents used should be as selective as possible.
10. Design for degradation—Chemicals should be so designed that they break down into innocuous products at the end of their function and do not persist.
11. Real-time analysis for pollution prevention—Analytical methodologies should be further developed to allow real-time in-process monitoring and control prior to the formation of hazardous substances.
12. Inherently safer chemistry for accident prevention—Substances used in the chemical processes should be such chosen to minimize the potential for chemical accidents, explosions, fire, etc.

2.4.1 APPROACH OF NANOTECHNOLOGY IN NANOTECHNOLOGY

Green and biogenic bottom-up synthesis attracting many researchers due to the feasibility and less toxic nature of processes. These processes are cost-effective and environmentally friendly, where the synthesis of NPs is accomplished via biological systems such as using plant extracts. Bacteria, yeast, fungi, aloe vera, tamarind, and even human cells are used for the synthesis of NPs. Au NPs have been synthesis from the biomass of wheat and oat (Parveen et al., 2016) and using microorganism and plant extracts as a reducing agent (Ahmed et al., 2016). NPs have been prepared by several physical and chemical approaches depending on the accessibility and viability of procedures to attain the essential applications (Mohanpuria et al., 2008; Sun and Zeng, 2002). The wet chemistry method is one among the methods for the preparation of nanomaterials, where the metal reduction is carried out in solution using several chemical reductants, such as hydrazine hydrate, sodium borohydride, and trisodium citrate (Wu and Chen, 2003; Qi et al., 1997). All of these chemicals are toxic. They are stabilized by applications, frequently hazardous stabilizers, to avoid aggregation. These approaches have been widely applied, but the stabilizers, reactants, reductants and several organic solvents employed in these approaches are toxic and hazardous to the environment (Tolaymat et al., 2010; Thomas et al., 2011). Various physical and chemical processes have been exploited in the synthesis of several NPs. However, these methodologies involve the use of toxic, hazardous chemicals that may pose potential environmental and biological risks. The process of agglomeration and sintering involves many toxic hydrocarbons and solvents.

Capping agents are widely used in the stabilization and controlling the shape morphology of NPs. Most of the stabilizers, usually adsorbing on the surface of the NPs, may significantly alter the surface properties of the nanomaterials, and increasing severe toxicity concerns strictly hinder the biological applications of such nanomaterials (De Jong and Borm, 2008). The synthesis of NPs by physical approaches involving laser ablation, arc discharge, ball milling regularly needs high-temperature and -pressure conditions, demanding a tremendous amount of energy (Thakkar et al., 2010). The concept of green chemistry becomes more significant in these circumstances (Sheldon, 2012). Heteroatom functionalized long-chain hydrocarbons, such as oleic acid, oelylamine, and trioctylphosphine, control the size and dispersability of NPs by binding strongly through heteroatoms (Park et al., 2007). Biochemical NP synthesis, beyond being environmentally friendly, is simple, cost-effective, more reproducible, and have defined physicochemical properties.

2.4.2 *PREPARATION OF NANOPARTICLES BY GREEN METHOD*

Various methods have been reported for the preparation of different NPs by green methods:

2.4.2.1 *FROM PLANT RESIDUES*

Various plant residues have been used for the synthesis of various NPs. These plant residues act as reducing agents and reduce metallic NPs to the zerovalent state.

Gold NPs: The production of gold NPs has been done by the controlled reduction of the Au^{3+} ion to Au^0. Green synthesis of gold NPs was reported by using a variety of plant materials, such as *Achillea wilhemsii* flowers (Andeani et al., 2011), ethanolic extract of the plant *Mirabilis jalapa* flowers (Vankar and Bajpai, 2010), *Coleus amboinicus* L. leaf extract (Narayanan and Sakthivel, 2010), aqueous extract of cypress leaves (Noruzi et al., 2012),

Barbated Skullcup herb extract (Wang et al., 2009), aqueous extract of rose petals (Noruzi et al., 2011), leaf extracts of two plants *Magnolia kobus* and *Diopyros kaki* (Song et al., 2004), oat (*Avena sativa*) biomass (Armendariz et al., 2004), flower extracts of *Ixora coccinea* (chetty flower) (Nagaraj et al., 2011), *Azadirachta indica* plant leaf extract (Thirumurugan et al., 2010), ethanol extract of black tea and its tannin free fraction (Banoee et al., 2010), alfalfa plants (Gardea-Torresdey et al., 2002), and coriander leaves (Narayanan and Sakthivel, 2008). Gold NPs were also synthesized using banana peel extract, and their antimicrobial activity toward most of the tested fungal and bacterial cultures was described (Bankar et al., 2010).

Silver NPs: The synthesis of NPs using plant extracts as reducing and capping agents has received special attention. The synthesis of plant-mediated silver NPs by *Solanum torvum* (Govindaraju et al., 2010), papaya fruit extract (Jain et al., 2009), *Nicotiana tobaccum* leaf extract (Prasad et al., 2011), *Elettaria cardamomom* (Jobitha et al., 2012), and onion (*Allium cepa*) (Saxena et al., 2010) and their promising antimicrobial activity are reported. Krishnaraj and co-workers reported the synthesis of silver NPs using *Acalypha indica* leaf extracts (Krishnaraj et al., 2010). Green synthesis of small silver NPs using Geraniol and their cytotoxicity against Fibrosarcoma-Wehi 164 were evaluated by Safaepour et al. (Safaepour et al., 2009). Ghosh et al. reported the synthesis of silver NPs using *Dioscorea bulbifera* tuber extract and evaluated their synergistic potential in combination with antimicrobial agents (Ghosh et al., 2012). Green synthesis, antimicrobial, and cytotoxic effects of silver NPs using *Eucalyptus Chapmaniana* leaves extract are also reported (Sulaiman et al., 2013).

Titanium dioxide NPs: The biosynthesis of titanium dioxide NPs by Nyctanthes leaves extract (Sundrarajan and Gowri, 2011), *Eclipta prostrata* leaf extract (Rajakumar et al., 2012), and *Annona squamosa* peel extract (Roopan et al., 2012) was reported. The synthesis of titanium dioxide NPs utilizing aqueous leaf extract of *Catharanthus roseus* against the adults of hematophagous fly, *Hippobosca maculata* leach (Diptera: Hippoboscidae), and sheep-biting louse, *Bovicola ovis* Schrank (Phthiraptera: Trichodectidae), was reported for their antiparasitic activities (Velayutham et al., 2012).

Zinc oxide NPs: Zinc oxide NPs have been synthesized using orange juice (Jha et al., 2011), aloe vera extract (Sangeetha et al., 2011), *Citrus aurantifolia* extracts (Samat and Nor, 2013), and milky latex of *Calotropis procera* (Singh et al., 2011).

Copper oxide NPs: Copper oxide NPs have been synthesized using plant leaf extract of *Magnolias* as a reducing agent, (Lee et al., 2011) latex of

Calotropis procera L. (Harne et al., 2012), soybean (Guajardo-Pacheco et al., 2010), and many more, and their antibacterial activity was assessed.

2.4.2.2   BACTERIA- AND FUNGUS-MEDIATED SYNTHESIS

Different bacteria species have been successfully used to reduce gold ions to a zerovalent metal. Gold particles of nanoscale dimensions may be readily precipitated within bacterial cells by incubation of the cells with Au^{3+} ions. Lower pH values and initial concentrations of Au(III) were individually responsible for reductions in nanogold particle size. Gold NPs were synthesized using *Magnetospirillum gryphiswaldense* MSR-1 (Cai et al., 2011), *Bacillus megatherium* D01 (Wen et al., 2009), mesophilic bacterium *Shewanella* (Ogi et al., 2010), *Rhodopseudomonas capsulata* (Shiying et al., 2007), *Pseudomonas aeruginosa* (Husseiny et al., 2007), *Stenotrophomonas maltophilia* (Nangia et al., 2009), *Escherichia coli* and *Desulfovibrio desulfuricans* (Deplanche and Macaskie, 2008), marine alga *Sargassumwightii Greville* (Singaravelu et al., 2007), marine microalga *Tetraselmis suecica* (Shakibaie et al., 2010), marine bacterial strain of *Marinobacterpelagius* (Sharma et al., 2012), aqueous extract of brown algae *Laminaria Japonica* (Ghodake and Lee, 2011), and alkalotolerant actinomycete (*Rhodococcus* sp.) (Ahmad et al., 2003). The synthesis of gold NPs using the cell-free filtrate of fungus, *Sclerotium rolfsii* (Narayanan and Sakthivel, 2011) fungus *Verticillium* sp. (Mukherjee et al., 2001), fungus *Trichoderma harzianum* (Singh and Raja, 2011) was reported in the literature. The synthesis of silver NP by eukaryotic organisms like *Verticillium* sp. is also reported (Mukherjee et al., 2012). Among fungi, *Aspergillus* sp. is the least expensive and economical biomaterial for the biosynthesis of silver NP. The use of filamentous fungus *Phoma* sp. 3.2883 (Chen et al., 2003) and white-rot fungus *Phanerochaete chrysosporium* in the synthesis of silver NP was reported (Vigneshwaran et al., 2006). *Penicillium fellutanum* isolated from coastal mangrove sediment (Kathiresan et al., 2009) for silver NP production was reported earlier. The biosynthesis of silver NPs using fungi *Penicillium diversum* and their antimicrobial activity against *E. coli*, *Salmonella typhi*, *Vibrio cholerae*, and the clinical isolate of *Paratyphia* were investigated (Ganachari et al., 2012). Titanium dioxide NP synthesis by using *Lactobacillus* sp., *Saccharomyces cerevisae* (Jha et al., 2009), fungus *Fusarium oxysporum*, and *Bacillus subtilis* (Bansal et al., 2005) was reported. In another study, the biosynthesis of titanium NPs was achieved by using *Aspergillus flavus* as a reducing and capping agent and assessed to be a good novel antibacterial material (Rajakumar et al., 2012). The manganese and zinc

NPs were successfully synthesized by using *Streptomyces* sp. HBUM171191 (Waghmare et al.). The biosynthesis of phase-pure metallic copper NPs using a silver-resistant bacterium *Morganella morganii* was investigated (Ramanathan et al., 2013). The biosynthesis of copper oxide NPs using *Penicillium aurantiogriseum, Penicillium citrinu,* and *Penicillium waksmanii* was also reported (Honary et al., 2012). Biological method for extracellular bacterium belonging to the genus *Serratia* is known (Hasan et al., 2008).

2.4.3 APPLICATION OF GREEN CHEMISTRY IN SUPRAMOLECULAR CHEMISTRY

The field of supramolecular chemistry concerns the design of molecular entities that are defined by reversible, noncovalent interactions. While each supramolecular interaction is quite weak individually, the effect of many such interactions working in combination can produce strongly associated and structurally well-defined molecular species. Such additive effects are responsible for the spectacular structural complexity found in biomacromolecules such as proteins. Supramolecular chemistry represents a complementary approach toward molecular construction and relies upon noncovalent interactions to provide the primary associative interaction between the catalyst and the substrate, a factor that is responsible for the spectacular selectivity and reactivity of enzymes. Supramolecular interactions can be involved in catalysis in several ways. Supramolecular encapsulation of one or more substrate molecules within a host (which itself is often self-assembled through supramolecular chemistry) can promote or modulate reactivity. Supramolecular binding can enforce substrate–catalyst interactions through molecular recognition processes that function independently of the reactive functional groups. Finally, it is possible to install catalytic moieties within the cavity of a molecular host, which can then bind substrate molecules. The synthesis of supramolecular devices utilize toxic chemicals posing a threat to the environment. Hence, the emphasis has now shifted to the synthesis of supramolecules using green solvents. Solvents account for a large fraction of waste generated in chemical reactions, and switching to environmentally innocuous solvents is one of the 12 principles of green chemistry (Anastas and Werner, 1998). Many studies have been reported for the same. A substantial fraction of self-assembled molecular hosts is soluble in water and possess hydrophobic interiors. When the reactants and/or catalyst of an organic reaction are encapsulated within a hydrophobic cavity, the molecular host can act as a nanometer-sized reaction flask, bringing together reactants that would otherwise be insoluble (Yoshizawa et al., 2009). While many examples of supramolecular catalysis in the water now exist, the most practical and synthetically useful strategy that has emerged is the use of micellar hosts that spontaneously self-assemble in water. The advantages of these systems over other water-soluble hosts include their reliable self-assembly under a wide range of conditions, commercial availability, low cost of many micelle-forming surfactants, and a wide range of hydrophobic molecules that are encapsulated.

Tamiaki and co-workers reported synthetic zinc and magnesium chlorine aggregates as a model for supramolecular antenna complexes of green photosynthetic bacteria (Tamiaki et al., 1996). A large-channel supramolecular structure composed entirely of organic molecules was synthesized using a green chemistry approach, combining mechanochemistry and solid-state photodimerization.

2.5   NANOTECHNOLOGY IN BIOMEDICAL MEASUREMENTS

The combination of nanotechnology and molecular biology has developed into an emerging research area, called nanobiotechnology. Nanomaterials such as NPs, nanofibers, nanowires, and nanotubes have novel properties and functionalities, which make them attractive to explore and modify biological processes, with potential applications in biomedicine. Nanotechnology frames the study, manipulation, and control of chemical and/or biological materials at the nanoscale, which corresponds to structures or systems with dimensions within the range of $1-100$ nm (1 nm $= 10^{-9}$ m) (Cademartiri and Ozin, 2009) that enable technology that deals with nanometer-sized objects. The nanomaterials are the most advanced area at present for developing scientific knowledge and commercial applications. Living organisms are built of very small cells, nearly 10 μm wide. Even smaller are the proteins with a typical size of just 5 nm, which is comparable with the dimensions of smallest manmade NPs. Such a small size of human cells gives an idea of using NPs as very small probes that would allow us to investigate at the human cellular level without causing too much interference (Taton, 2002). Understanding biological processes at the nanoscale level is a strong driving force behind the development of nanotechnology (Whitesides, 2003). The fact that NPs exist in the same size domain as proteins makes nanomaterials suitable for biotagging or biolabeling. However, size is just one of the many characteristics of NPs that itself is rarely sufficient if one is to use NPs as biological tags. To interact with the biological target, a biological or molecular coating or layer like antibodies, biopolymers acting as a bioinorganic interface should be attached to the NPs (Sinani et al., 2003) or monolayers of small molecules that make the NPs biocompatible (Zhang et al., 2002). In addition, as optical detection techniques are widespread in biological research, NPs should either fluoresce or change their optical properties.Nanomedicine has been applied in various areas such as imaging, tissue regeneration, diagnosis, detection, and drug delivery. Biomolecules such as proteins folding into precisely defined three-dimensional shapes, nucleic acids assembling, and antibodies with specificity for certain molecules can perform transport operations. One interesting application involves the use of nanoscale devices that may serve as vehicles for the delivery of therapeutic agents and act as detectors or guardians against early diseases. The metabolic and genetic defects can also be repaired at the nanolevel. They would seek out a target within the body such as a cancer cell and perform some functions to fix it by releasing a drug in the localized area. This has opened the doors of promising future developments intreatments such as gene replacement, tissue regeneration, or nanosurgery. Biological molecules and systems have many attributes that make them highly suitable for nanotechnology applications. Various polymers have been used in drug delivery research as they can effectively deliver the drugs to the target site, thus increasing the therapeutic benefit while minimizing side effects. The controlled release of pharmacologically active drugs at the precise action site and at the therapeutically optimum degree and dose regimen has been a major goal in designing such devices.

2.5.1 NANOPARTICLES AS BIOMEDICINE

A lot of research has focused on NPs because of their superior chemical, biological, and magnetic properties including chemical stability, nontoxicity, and biocompatibility (Sun et al., 2008; Karimi et al., 2013; Chen et al., 2008). NPs have received attraction for their ability to deliver drugs in the optimum dosage range, often resulting in increased therapeutic efficiency of the drugs, weakened side effects, and improved patient compliance (Alexis et al., 2008). Nanomedicine can provide protection to agents susceptible to degradation and denaturation in regions of harsh pH, prolonging the duration of exposure of a drug by increasing retention of the formulation through bioadhesion (Sahoo et al., 2007). NPs usually form the core of nanobiomaterials. They can be used as a convenient surface for molecular assembly and may be composed of inorganic or polymeric materials. They can also be in the form of a nanovesicle surrounded by a membrane or a layer. The spherical or plate-like shape and other shapes permit penetration through a pore structure of the cellular membrane. The control of size and size distribution of NPs allows creating very efficient fluorescent probes that emit narrow light in a very wide range of wavelengths and are used as nanodevices. A combination of magnetic and luminescent behaviors can both detect and manipulate the particles.

Metallic NPs have been used in drug delivery, especially in the treatment of cancer, bacteria, fungi, etc. Among various metals, silver (Huang et al., 2014; Panáček et al., 2009; Ragaseema et al., 2012) and gold NPs (Bhattacharya et al., 2007; Muddineti et al., 2015) are of prime interest. Other NPs include metal oxides including titanium dioxide (TiO_2) (Wu et al., 2011; Ren et al., 2013; Du et al., 2015), cerium dioxide (CeO_2) (Renu et al., 2012; Nesakumar et al., 2013), and zinc oxide (ZnO) (Wang et al., 2015). In the current world, various diseases such as diabetes, cancer, Parkinson's disease, Alzheimer's disease, cardiovascular diseases, and different kinds of serious inflammatory or infectious diseases constitute a high number of serious and complex illnesses. Nanomedicine makes use of nanomaterials and nanoelectronic biosensors for the detection and treatment of diseases. With the help of nanomedicine, early detection and prevention, improved diagnosis, proper treatment, and follow-up of diseases are possible. Certain nanoscale particles are used as tags and labelsin processes can be performed quickly, making the testing more flexible and sensitive. The use of nanotechnology, here exemplified as the use of NPs, has some advantages in gene delivery: the structure of the NPs protects the nucleic acids from degradation by nucleases and the environment; they also minimize the side effects by directing the nucleic acid to the specific location of action; they facilitate cell entry of nucleic acids; and normally, they sustain gene delivery for longer periods compared to other vehicles (Kompella et al., 2013).

Gold NPs have marked their presence in the nanomedicine field as agents for labeling and imaging (Kooyman, 2008), diagnostic, or carriers for delivery of biomolecules or small drugs since they have many features that make them suitable for such applications. The inert nature of gold particles and their nontoxic behavior make them attractive for biomedical applications. They can be synthesized by simple methods that allow obtaining NPs that are monodisperse, and their surface can be easily functionalized, mainly with thiols but also with other capping agents, such as amines (Aslam et al., 2004), carboxylates (Turkevich et al., 1951), or phosphines (Weare et al., 2000). Gold NPs have been reported to successfully detect early stage inflammatory processes through surface-enhanced Raman scattering (SERS). They can be conjugated with monoclonal antibodies specific for intercellular adhesion molecule-1 (ICAM-1) (McQueenie et al., 2012). The ICAM-1 expression in endothelial cells can be linked to the progression of a wide range of inflammatory, autoimmune, and infectious diseases. The Au NPs result in a dramatic increase in signal contrast compared to other antibody–fluorescent dye-targeting agents (Suh et al., 2009).

Aaron et al. (2007) have shown that 25 nm gold NPs when conjugated with antiepidermal growth factor receptor monoclonal antibodies can be efficiently used as *in vivo* targeting agents for imaging cancer markers, specifically epidermal growth factor receptors. Gold NPs have been used in a variety of optical and electrical assays. The electrical properties of the gold NPs were harnessed for the development of a piezoelectric biosensor, for real-time detection of a food-borne pathogen. Noble metal nanostructures are being utilized for biodiagnostics, biophysical studies, and medical therapy (Jain et al., 2008). For example, taking advantage of the strong localized surface plasmon resonance scattering of gold NPs conjugated with specific targeting molecules allows the molecule-specific imaging and diagnosis of diseases such as cancer (Huang et al., 2007).

Elemental silver and silver salts have been used as antimicrobial agents in curative and preventive health care. The antimicrobial activity of the silver salts and complexes (ionic silver) act by their interaction in various biomacromolecular components causing structural changes and deformations in bacterial cell walls, membranes, and nucleic acids (Abu-Youssef et al., 2010; Cavicchioli et al., 2010). The use of colloidal silver as an anticancer agent on human breast cancer cells was investigated by Franco-Molina et al. (Franco-Molina et al., 2010). MCF-7 human breast cancer cells were grown and different concentrations of colloidal silver NPs were used to determine whether a cytotoxic effect had occurred. Sanpui et al. (2011) developed a silver NP–chitosan nanocarrier and examined the cytotoxic effect this nano-carrier had on HT 29 colon cancer cell lines. At low concentrations, silver NPs–chitosan nanocarriers induced cell apoptosis in HT 29 colon cancer cells, indicating their potential use in cancer therapy. The use of modified nanosilver as a treatment for multidrug resistant (MDR) cancer was studied by Liu et al. (2012). MDR cancer is a major issue in the treatment of the disease as the cancer cells can survive the chemotherapy treatment. Both in vitro and *in vivo* studies were carried out to investigate the antitumour effect of both silver NPs and modified silver NPs. Guo et al. (2013) investigated the cytotoxic effect that silver NPs on acute myeloid leukemia (AML) (SHI-1, THP-1, DMAI, NB-4, HL-60, and HEL cell lines). The results showed that PVP-coated silver NPs had an antileukaemia effect against AML cell lines and isolates.

Silver NPs have been reported to be an effective bactericide against *E. coli* by accumulating in the bacterial membrane and increasing its permeability, causing the death of the cell (Sondi and Salopek-Sondi, 2004). Buttacavoli et al. (2018) investigated the activity of silver NPs embedded into a specific polysaccharide (EPS), biogenerated by Klebsiella oxytoca DSM 29614 under aerobic (AgNPs-EPS[aer]) and anaerobic conditions, and tested by the MTT assay for cytotoxic activity against human breast (SKBR3 and 8701-BC) and colon (HT-29, HCT 116 and Caco-2) cancer cell lines, revealing AgNPs-EPS[aer] as the most active, in terms of IC50, with a more pronounced efficacy against breast cancer cell lines (Buttacavoli et al., 2018).

The utilization of metal and semiconductor NPs in biomedical applications has been demonstrated by many research groups. The gold NPs result in a dramatic increase in signal contrast compared to other antibody–fluorescent dye-targeting agents (Suh et al., 2009). The advantages of nanobodies in developing therapeutics are their extreme stability, the ability to bind antigens with nanomolar affinity, high target specificity, low toxicity, the ability to combine the advantages of conventional antibodies with important features of small-molecule drugs, and the ability to be produced cost-effectively on a large scale (Jain, 2005). Although several researchers are averse to ribonucleic acid (RNA) nanotechnology, due to the susceptibility of RNA to RNase degradation and serum instability, Shu and colleagues have developed a toolkit to obtain stable RNA NPs (Shu et al., 2013). In this work, 14 homogeneous RNA NPs were obtained, which targeted cancer exclusively *in vivo* without accumulation in normal organs and tissues. Functionalized NP aggregating fluorescence imaging techniques, known as quantum dots, have the potential for real-time and noninvasive visualization of biological events *in vivo*. The quantum dots are provided with a layer of lipids or polymers to prevent heavy metals from being released. The NPs can provide a solid support for sensing assays with several kinds of ligand molecules attached to each NP, simplifying assay design (Wang et al., 2013). They can also withstand a significantly larger number of cycles of excitation and light emissions than typical organic molecules, which more readily decompose, increasing the labeling ratio for higher sensitivity in complex biological systems (Fakruddin et al., 2012). Another advantage is that these miniaturized fluorescent NPs can also be easily taken up by cells through endocytosis and subsequently used for site-specific intracellular measurements and long-term tracking of biomolecules in real time (Mallick et al., 2004). This tool is being studied by several groups around the world (Kairdolf et al., 2013; Palacios et al., 2013).

2.5.2 NANOPARTICLES AS DRUG DELIVERY AGENTS

Molecular recognition is one of the central processes in drug action. The comparison and detection of binding sites are key steps in the prediction of potential interactions. Various biological macromolecules that interact with the drug molecule are enzymes, receptors, nucleic acids, membranes, carriers, proteins, etc. These are all targets for the drug action, and the structures of these targets are used to model and display the interactions with the drug structure. These interactions are crucial and must be understood before the design of a new molecule for better interactions. The intricate process of using the information contained in the three-dimensional structure of a macromolecular target and related ligand–target complexes to design novel drugs is alternately called as structure-based drug design. Several approaches directed toward this task are available.

Chemotherapy is an important part of cancer treatment. In this, anticancer drugs are (Ozin and Yang, 2001) introduced into the circulatory system to transport anticancer drugs to the tumors. However, there are negative side effects of this treatment such as nonspecificity and toxicity of the drug. The drug can attack healthy cells and organs as well as the cancerous cells. Therefore, targeted drug delivery is being developed as one alternative to chemotherapy treatment to direct the drug to the specific area where the tumor is located, thereby increasing the amount of drug delivered at the tumor site and reducing the side effects. This is accomplished by the use of magnetic NPs to deliver the drug to its specific location. Generally, the magnetic NPs are coated with a biocompatible layer, such as gold or polymers, so that the anticancer drug can either be conjugated to the surface or encapsulated in the NP, as shown in Figure 2.2. Once the drug has been administered, external magnetic field guides the complex to the specific tumor. The drug is released by enzyme activity or by changes in pH, temperature, or osmolality (Tartaj et al., 2003; Chatterjee et al., 2014).

R = OCH₃, OH

Common drug delivery materials including liposomes (Al-Jamal and Kostarelos, 2011), micelles (Ma et al., 2013), organic polymers (Yang et al., 1975), dendrimers (Patri et al., 2005), and CDs (Sun et al., 2012) have been developed. On the one hand, these materials have addressed some problems caused by small-molecule drugs to some degree. On the other hand, the accumulation of these materials within malignant lesions is generally attributed to passive targeting as a consequence of enhanced permeability and retention. An interesting tool being developed today to be utilized in tumor diagnosis is RNA NPs (Zhou et al., 2011).

Biological molecules like proteins, antibodies, nucleic acids have many attributes that make them highly suitable for nanotechnology applications. For example, proteins fold into precisely defined three-dimensional shapes, and nucleic acids assemble according to well-understood rules. Antibodies are highly specific in recognizing and binding their ligands, and biological assemblies such as molecular motors can perform transport operations. From the tripeptide glutathione (cellular antioxidant) to oligopeptides (hormones) and polypeptides (proteinaceous enzymes), the peptides of various lengths play vital metabolic roles. These peptides need to aggregate or, more specifically, self-assemble into superstructures to perform their cellular roles. The result is the self-assembly of peptides, like actin, to form the elongated fibrillar structures of the cytoskeleton. Collagen is an important component of cells and tissues both. The same peptide molecules are assembled due to molecular recognition, binding them together through noncovalent interactions, namely, electrostatic interactions, hydrogen bonding, and van der Waals interactions. Simple peptide molecules can hence self-assemble into much more complex and functional superstructures, which inspires efforts to design more of such peptides for a wide range of biomedical applications.

The self-assembly hierarchy of biological materials begins with monomer molecules (e.g., nucleotides and nucleosides, amino acids, and lipids), which form polymers (e.g., DNA, RNA, proteins, and polysaccharides), then assemblies (e.g., membranes, organelles), and finally cells, organs, organisms, and even populations (Rousseau and Jelinski, 1991). Currently, research is going on developing chips based on the base pairing of DNA to build molecular-sized, complex, three-dimensional objects. Seeman and coworkers have reported the formation of complex 2-D and 3-D periodic structures with defined topologies. DNA is ideal for building molecular nanotechnology objects, as it offers synthetic control, predictability of interactions, and well-controlled "sticky ends" that assemble in a highly specific fashion.

DNA chips are currently widely used in scientific, biomedical research. The chips comprise an inert support, which carries microarrays of hundreds to thousands of single-strand DNA molecules with different base sequences. DNA from a tissue sample that has been labeled with a radioactive or fluorescent material can be identified on the basis of the place on the chip where it binds with the chip DNA. Similar chips are being developed for the diagnosis of leukemias (Valk et al., 2004; Lowenberg et al., 2005) and mouth and throat tumors (Roepman et al., 2005). A new nanotechnological analytical method uses quantum dots. DNA in a sample is identified on the basis of its bonding to DNA molecules of a known composition embedded in micrometer-sized polymer spheres containing various mixtures of quantum dots, each of which provides a unique spectral bar code (color code) (Han et al., 2001).

A chip to detect prostate cancer has been developed by American researchers. The chip contains around 100 cantilever sensors (micrometer-sized, nanometre-thick minuscule levers), coated on one side with antibodies to prostate-specific antigen (PSA), a biomarker for that disease. The bonding of PSA with a sample placed on the chip bends the cantilevers by several nanometers, which can be detected optically. This enables clinically relevant concentrations of PSA to be measured (Wu et al., 2001; Majumdar, 2002). Antibodies placed on nanowires can be used in a similar way to detect viruses, in a blood sample, for example (Patolsky et al., 2004).

Nanoscale systems can be used to facilitate the delivery of incompatible drugs. They can also be used in theranostics, in which the particle is used as a device to diagnose and treat the disease at the same time (Shi et al., 2010). These techniques are also used for the liberation of pharmacological agents against several diseases, such as bacterial infection (Kim et al., 2004), inflammations (Camenzind et al., 1997), and principally cancer (Amna et al., 2012), among others. They are also being investigated as a tool for the delivery of drugs through the blood–brain barrier. Combining the delivery systems with contrast agents, fluorescent or radioactive substances, also makes it possible to use imaging techniques to monitor how successful the selective transport to the destination has been (Yong et al., 2007; Quintana et al., 2002). Once it has reached the target area, the active substance has to be released from the carrier at the correct rate. This can occur spontaneously by gradual diffusion, in combination with the delivery system's degradation or otherwise. It may also occur as a result of special conditions at the destination, such as a different acidity level (Hong et al., 2002), salt concentration, temperature, or the presence of certain enzymes. The accumulation of the delivery system and/or the release of the active substance at the right place can also be controlled from outside by influencing conditions in the target organs or tissues by means of magnetic fields (Alexiou et al., 2003; Saiyed et al., 2003), ultrasonic vibrations (Nelson et al., 2002), or heat (Kong et al., 2001).

The drug delivery systems require longer their residence time of the drug in the blood accumulation in the target tissues; availability of sufficient active substance and longer shelf life allow their storage and distribution. Besides acting as a delivery system, in some cases, NPs can act as an active substance. Once they have found their way through the bloodstream into a tumor, or have been injected directly into it, metal-containing NPs can be heated using near-infrared radiation (Sharma et al., 2012) and oscillating magnetic field (Hirch, 2003) so that the tumor cells die. These cells can then be killed by using near-infrared radiation to heat the tubes.

2.6   CONCLUSION

In this chapter, we presented a detailed overview of NPs, their types, synthesis, properties, and applications in the fields of supramolecular chemistry, green reactions, and biomedical measurements. NPs have size ranging from a few nanometers to 500 nm and have a large surface area, which make them a suitable candidate for various applications. Besides this, the optical properties of NPs also increase their importance in photocatalytic applications. In this chapter, the synthesis of NPs by various green methods is discussed. The applications of nanomaterials in biomedical conditions are also discussed. This information can be used to design new methodologies in nanotechnology for the benefit of the environment and humanity.

KEYWORDS

- **nanotechnology**
- **self-assembly**
- **supramolecules**
- **green chemistry**
- **drug delivery**
- **biomedical measurements**

REFERENCES

Aaron, J., Nitin, N., Travis, K., Kumar, S., Collier, T., Park, S. Y., Jose-Yacaman, M., Coghlan, L., Follen, M., Richards-Kortum, R., Sokolov, K. (2007). "Plasmon resonance coupling of metal nanoparticles for molecular imaging of carcinogenesis in vivo," *J. Biomed. Opt. 12*(3), 003407.

Abu-Youssef, M. A., Soliman, S. M., Langer, V., Gohar, Y. M., Hasanen, A. A., Makhyoun, M. A., Zaky, A. H., Ohrstrom, L. R. (2010). "Synthesis, crystal structure, quantum chemical calculations, DNA interactions, and antimicrobial activity of [Ag(2-amino-3-methylpyridine) (2)]NO(3) and [Ag(pyridine-2-carboxaldoxime)NO(3)]," *Inorg. Chem. 49*, 9788–9797.

Ahmad, A., Senapati, S., Khan, M. I., Kumar, R., Ramani, R., Srinivas, V., Sastry, M. (2003). "Intracellular synthesis of gold nanoparticles by a novel alkalotolerant actiniomycete *Rhodococcus* species," *Nanotechnology, 14*, 824–828.

Ahmed, S., Annu, S., Yudha, S. S. (2016). "Biosynthesis of goldnanoparticles: A green approach," *J. Photochem. Photobiol. B: Biol. 161*, 141–153.

Alexiou, C., Jurgons, R., Schmid, R. J. (2003). "Magnetic drug targeting-biodistribution of the magnetic carrier and the chemotherapeutic agent mitoxantrone after locoregional cancer treatment," *J. Drug Target. 11*(3),139–149.

Alexis, F., Pridgen, E., Molnar, L. K., Farokhzad, O. C. (2008). "Factors affecting the clearance and biodistribution of polymericnanoparticles," *Mol. Pharm. 5*, 505–515

Al-Jamal, W. T., Kostarelos, K. (2011). "Liposomes: From a clinically established drug delivery system to a nanoparticle platform for theranostic nanomedicine," *Acc. Chem. Res.* 44, 1094–1104.

Amna, T., Hassan, M. S., Nam, K. T., Bing, Y. Y., Barakat, N. A., Khil, M. S., Kim, H. Y. (2012). "Preparation, characterization, and cytotoxicity of CPT/Fe$_2$O$_3$-embedded PLGA ultrafine composite fibers: A synergistic approach to develop promising anticancer material," *Int. J. Nanomed. 7*, 1659–1670.

Anastas, P. T., Werner, J. C. (1998), *Green Chemistry—Theory and Practice*. Oxford University Press, New York.

Andeani, J. K., Kazemi, H., Mohsenzadeh, S., Safavi, A. (2011). "Biosynthesis of gold nanoparticles using dried flowers extract of *Achillea wilhelmsii* plant, *Dig. J. Nanomater. Biostruct. 6*, 1011–1017.

Armendariz, V., Herrera, I., Peralta-Videa, J. R., Jose-Yacaman, M., Troiani, H., Santiago, P., Gardea-Torresdey, J. L. (2004). *J. Nano. Res. 6*, 377–382.

Aslam, M., Fu, L., Su, M., Vijayamohanan, K., Dravid, V. P. (2004). "Novel one-step synthesis of amine-stabilized aqueous colloidal gold nanoparticles," *J. Mater. Chem. 14*, 1795–1797.

Atwood, J. L., Davies, J. E. D., Mac Nicol, D. D., Vogtle, F. (1996). *Comprehensive Supramolecular Chemistry*. Elsevier, Oxford.

Bankar, A., Joshi, B., Kumar, A. R., Zinjarde, S. (2010). "Banana peel extract mediated synthesis of gold nanoparticles," *Colloids Surf. B. 80*, 45–50.

Banoee, M., Ehsanfar, Z., Mokhtari, N., Khoshayand, M. R., Akhavan Sepahi, A., Jafari Fesharaki, P., Monsef-Esfahani, H. R., Shahverdi, A. R. (2010). "The green synthesis of gold nanoparticles using the ethanol extract of black tea and its tannin free fraction," *Iran. J. Mater. Sci. Eng. 7*, 48–53.

Bansal, V., Rautaray, D., Bharde, A. (2005). "Fungus mediated biosynthesis of silica and and titania particles," *J. Mater. Chem. 15*, 2583–2589.

Beer, P. D., Gale, P. A., Smith, D. K. (1999). *Supramolecular Chemistry*. Oxford University Press.

Bhattacharya, R., Patra, C. R., Earl, A., Wang, S., Katarya, A., Lu, L., Kizhakkedathu, J. N., Yaszemski, M. J., Greipp, P. R., Mukhopadhyay, D., Mukherjee, P. (2007). "Attaching folic acid on gold nanoparticles using noncovalent interaction via different polyethylene glycol backbones and targeting of cancer cells" *Nanomed. Nanotechnol. Biol. Med. 3*, 2244–238.

Buttacavoli, M., Albanese, N. N., Cara, G. D., Alduina, R., Faleri, C., Gallo, M., Pizzolanti, G., Gallo, G., Feo, S., Baldi, F., Patrizia Cancem, P. (2018). "Anticancer activity of biogenerated silver nanoparticles: An integrated proteomic investigation," *Oncotarget 9*(11), 9685–9705.

Cademartiri, L., Ozin, G. A. (2009). *Concepts of Nanochemistry*. Wiley-VCH, Weinheim.

Cai, F., Li, J., Sun, J. Ji, Y. (2011). "Biosynthesis of gold nanoparticles by biosorption using *Magnetospirillum gryphiswaldense* MSR—1," *Chem. Eng. J. 175*, 70–75.

Camenzind, E., Bakker, W. H., Reijs, A., van Geijlswijk, I. M., Foley, D., Krenning, E. P., Roelandt, J. R., Serruys, P. W. (1997). "Site-specific intravascular administration of drugs: History of a method applicable in humans," *Cathet. Cardiovasc. Diagn. 41*, 342–347.

Cavicchioli, M., Massabni, A. C., Heinrich, T. A., Costa-Neto, C. M., Abrão, E. P., Fonseca, B. A. L. (2010). "Pt(II) and Ag(I) complexes with acesulfame: Crystal structure and a study of their antitumoral, antimicrobial and antiviral activities," *J. Inorg. Biochem. 104*, 533–540.

Chatterjee, K., Sarkar, S., Jagajjanani Rao, K., and Paria, S. (2014). "Core/shell nanoparticles in biomedical applications," *Adv. Colloid Interface Sci., 209*, 8–39.

Chen, J. C., Lin Z. H., Ma, X. X. (2003). "Evidence of the production of silver nanoparticles via pretreatment of Phoma sp.3.2883 with silver nitrate," *Lett. Appl. Microbiol. 37*, 105–108.

Chen, Z. P., Zhang, Y., Zhang, S., Xia, J. G., Liu, J. W., Xu, K., Gu, N., (2008). "Preparation and characterization of water-soluble monodisperse magnetic iron oxide nanoparticles via surface double-exchange with DMSA," *Colloids Surf. A 316*, 210–216.

Collins, P. G., Bradley, K., Ishigami, M. (2000) "Extreme oxygen sensitivity of electronic properties of carbon nanotubes," *Science 287*, 1801–1804.

Crespo-Biel, O., Dordi, B., Reinhoudt, D. N., Huskens, J. (2005). "Supramolecular layer-by-layer assembly: Alternating adsorptions of guest- and host-functionalized molecules and particles using multivalent supramolecular interactions," *J. Am. Chem. Soc. 127*, 7594–7600.

De Jong, W. H., Borm, P. J. (2008). "Drug delivery and nanoparticles: Applications and hazards," *Int. J. Nanomed. 3*, 133–149.

Decher, G. (1997). "Fuzzy nanoassemblies: Toward layered polymeric" *Multicompon. Sci. 277*, 1232–1237.

Denkov, N. D., Velev, O. D., Kralchevsky, P. A., Ivanov, I. B., Yoshimura, H., Nagayama, K. (1993). "Two dimensional crystallization," *Nature 361*, 26.

Deplanche, K., Macaskie, L. E. (2008). "Biorecovery of gold by *Escherichia coli* and *Desulfovibrio desulfuricans*," *Biotechnol. Bioeng. 99*, 1055–1064

Du, Y., Ren, W., Li, Y., Zhang, Q., Zeng, L., Chi, C., Wu, A., Tian, J. (2015). "The enhanced chemotherapeutic effects of doxorubicin loaded PEG coated TiO_2 nanocarriers in an orthotopic breast tumor bearing mouse model," *J. Mater. Chem. B 3*, 1518–1528.

Ehrlich, P. (1906). *Studies on Immunity*. Wiley, New York.

Fakruddin, M., Hossain, Z., Afroz, H. (2012). "Prospects and applications of nanobiotechnology: A medical perspective," *J. Nanobiotechnol. 10*, 31.

Fischer, E. (1894). "Ber. Deutsch. Einfluss der Configuration auf die Wirkung der Enzyme," *Chem. Ges. 27*, 2985.

Franco-Molina, M. A., Mendoza-Gamboa, E., Sierra-Rivera, C. A., Gómez-Flores, R.A., Zapata-Benavides, P., Castillo-Tello, P., Alcocer-González, J. M., Miranda-Hernández, D. F., Tamez-Gurra, R. S., Rodríguez-Padilla, C. (2010). "Antitumor activity of colloidal silver on MCF-7 human breast cancer cells," *J. Exp. Clin. Cancer Res. 29*, 148.

Ganachari, S. V., Bhat, R., Deshpande, R., Venkataraman, A. (2012). *Bionano Sci. 2*, 316—321.

Gardea-Torresdey, J. L., Parsons, J. G., Gomez, E., Peralta-Videa, E., Troiani, H. E., Santiago, P., Yacaman, M. J. (2002). "Formation and growth of Au nanoparticles inside live alfalfa plants," *Nano Lett. 2*, 397–401.

Garrigue, P., Delville, M.-H., Labruge`re, C., Cloutet, E., Kulesza, P. J., Morand, J. P., Kuhn, A. (2004). "Top-down approach for the preparation of colloidal carbon nanoparticles," *Chem. Mater. 16*, 2984–2986.

Ghodake, G., Lee, D. S. (2011). "1).ake, G.,synthesis of gold nanoparticles using aqueous extract of the brown algae *Laminaria Japonica*," *J. Nanoelectron. Optoelectron. 6*, 1–4.

Ghosh, S., Patil, S., Ahire, M., Kitture, R., Kale, S., Pardesi, K., Cameotra, S. S., Bellare, J., Dhavale, D. D., Jabgunde, A., Chopade, B. A. (2012). "Synthesis of silver nanoparticles using Dioscorea bulbifera tuber extract and evaluation of its synergistic potential in combination with antimicrobial agents," *Int. J. Nanomed. 7*,483–496.

Gleiter, H. (2000). Nanostructured materials: Basic concepts and microstructures. *Acta Mater. 48*, 1–29.

Govindaraju, K., Tamilselvan, S., Kiruthiga, V., Singaravelu, G. (2010). "Biogenic silver nanoparticles by *Solanum torvum* and their promising antimicrobial activity," *J. Biopest.3*, 394–399.

Greque de Morais, M., Martins, V. G., Steffens, D., Pranke, P., da Costa1, J. A. V. (2014). "Biological applications of nanobiotechnology," *J. Nanosci. Nanotechnol. 14*, 1007–1017.

Guajardo-Pacheco, M. J., Morales-Sanchez, J. E., Gonzalez-Hernandez, J., Ruiz, F. (2010). "Synthesis of copper nanoparticles using soybeans as a chelant agent," *Mater. Lett. 64*, 1361–1364.

Guo, D., Zhu, L., Huang, Z., Zhou, H., Ge, Y., Ma, W., Wu, J., Zhang, X., Zhou, X., Zhang, Y., Zhao, Y., Gu, N. (2013). "Anti-leukemia activity of PVP-coated silver nanoparticles via generation of reactive oxygen species and release of silver ions," *Biomaterials 34*, 7884–7894.

Gurunanthan S., Park J. H., Han J. W., Kim J. H. (2015). "Compartive assessment of the apopotic potential of silver nanoparticles synthesized by *Bacillus tenquilensis* and *Calocybe indica* in MDA-MB-231 human breast cancer cells: Targeting p53 for antixancer therapy," *Int. J. Nanomed. 10*, 4203–4222.

Han, M., Gao, X., Su, J. Z., Nie, S. (2001). "Quantum-dot-tagged microbeads for multiplexed optical coding of biomolecules," *Nature Biotechnol. 19*(7), 631–635.

Harne, S., Sharma, A., Dhaygude, M., Joglekar, S., Kodam, K., Hudlikar, M. (2012). "Novel route for rapid biosynthesis of copper nanoparticles using aqueous extract of *Calotropis procera* L. latex and their cytotoxicity on tumor cells," *Colloids Surf. B 95*, 284–288.

Hasan, S. S., Singh, S., Parikh, R. Y., Dharne, M. S., Patole, M. S., Prasad, B. L. V., Shouche, Y. S. (2008). "Bacterial synthesis of copper/copper," *J. Nanosci. Nanotechnol. 8*, 3191–3196.

Hirsch, L. R., Stafford, R. J., Bankson, J. A. (2003), "Nanoshell-mediated near-infrared thermal therapy of tumors under magnetic resonance guidance," *Proc. Natl. Acad. Sci. USA 100*(23), 13549–13554.

Hof, F., Craig, S., Nuckolls, C., Rebek, J. (2002). "Molecular encapsulation," *Angew. Chem. Int. Ed. Engl. 41*, 1488–1508.

Honary, S., Barabad, H., Gharaeifathabad, E., Naghibi, F. (2012). "Green synthesis of copper oxide nanoparticles u sing penicillium aurantiogriseum, *Peicillium citrium* and *Penicillium waksmani*," *Dig. J. Nanomater. Biostruc. 7*, 999–1005.

Hong, M.-S., Lim, S.-J., Oh, Y.-K., Kim, C.-K. (2002). "pH-sensitive, serum-stable and long-circulating liposomes as a new drug delivery system," *J. Pharm. Pharmacol. 54*(1), 151–58.

Huang, K. J., Liu, Y. J., Wang, H. B., Wang, Y. Y. (2014). "A sensitive electrochemical DNA biosensor based on silver nanoparticle-polydopamine@graphene composite," *Electrochim. Acta 118*, 130–137.

Huang, X., Jain, P.K., El-Sayed, I.H., El-Sayed, M.A. (2007). "Gold nanoparticles: Interesting optical properties and recent applications in cancer diagnostics and therapy," *Nanomedicine, 2*, 681–693.

Huang, X.-C., Zhang, J.-P., Chen X.-M. (2004). "A new route to supramolecular isomer via molecular templating: Nanosized molecular polygons of copper(I) 2-methylimidazolates," *J. Am. Chem. Soc. 126*(41), 13218–13219

Husseiny, M. I., El-Aziz, M. A., Badr, Y., Mahmoud, M. A. (2007). "Biosynthesis of gold nanoparticles using *Pseudomonas aeruginosa*," *Spectrochim. Acta A: Mol. Biomol. Spectrosc. 67*, 1003–1006.

Iravani, S. (2011). "Green synthesis of metal nanoparticles using plants," *Green Chem. 13*, 2638.

Jain, D., Kumar Daima, H., Kachhwaha, S., Kothari, S. L. (2009). "Synthesis of plant-mediated silver nanoparticles using papaya fruit extract and evaluation of their anti microbial activities," *Dig. J. Nanomater. Biostruct. 4*, 557–563.

Jain, K. K. (2005). "The role of nanobiotechnology in drug discovery," *Drug Discov. Today 10*, 1435–1442.

Jain, P. K., Huang, X., El-Sayed, I. H., El-Sayed, M. A. (2008). "Noble metals on the nanoscale: optical and photothermal properties and some applications in imaging, sensing, biology, and medicine," *Acc. Chem. Res. 41*(12), 1578–1586.

Jehn, J. M. (1995). *Supramolecular Chemistry*. Weinheim, VCH.

Jha, A. K., Prasad, K., Kulkarni, A. R. (2009). "Synthesis of TiO$_2$ nanoparticles using micro-organisms," *Colloids Surf. B 71*, 226–229.

Jha, A. K., Vikash, K., Prasad, K. (2011). "Biosynthesis of metal and oxide nanoparticles using orange juice," *J. Bionanosci. 5*, 162–166.

Jobitha, G. D. G., Annadurai, G., Kannan, C. (2012). "A facile phyto-assisted synthesis of silver nanoparticles using the flower extract of Cassia auriculata and assessment of its antimicrobial activity," *IJPSR, 3*,323–330.

Kairdolf, B. A., Smith, A. M., Stokes, T. D., Wang, M.D., Young, A. N., Nie, S. (2013). "Semiconductor quantum dots for bioimaging and biodiagnostic applications," *Annu. Rev. Anal. Chem. 6*, 143–162.

Karimi, Z., Karimi, L., Shokrollahi, H. (2013). "Nanomagnetic particles used in biomedicine core and coating materials," *Mater. Sci. Eng. C 33*,2465–2475.

Kathiresan, K., Manivanan, S., Nabeel, M. A., Dhivya, B. (2009). "Studies on silveer nanoparticles synthesized by a marine fungus, *Pencillium fellutanum* isolated from a coastal mangrove sediment," *Colloids Surf. B. 71*, 133–137.

Kim, K., Luu, Y. K., Chang, C., Fang, D., Hsiao, B. S., Chu, B., Hadjiargyrou, M. (2004). "Incorporation and controlled release of a hydrophilic antibiotic using poly(lactide-co-glycolide)-based electrospun nanofibrous scaffolds," *J. Control Release 98*, 47–56.

Kompella, U. B., Amrite, A. C., Ravi, R. P., Durazo, S. A. (2013). "Nanomedicine for the back of eye drug delivery, gene delivery and imaging," *Prog. Retin. Eye Res. 36*, 172–198.

Kong, G., Braun, R. D., Dewhirst, M. W. (2001). "Characterization of the effect of hyperthermia on nanoparticle extravasation from tumor vasculature," *Cancer Res. 61*(7), 3027–3032.

Kooyman, R. P. H. (2008). *Handbook of Surface Plasmon Resonance*, Schasfoort R. B. M., Tudos A. J. (Eds.). Royal Society of Chemistry Publishing, Cambridge, vol. 15.

Krishnaraj, C., Jagan, E. G., Rajasekar, S., Selvakumar, P., Kalaichelvan, P. T., Mohan, N. (2010). "Synthesis of silver nanoparticles using *Acalypha indica* leaf extracts and its antibacterial activity against water borne pathogens," *Colloids Surf. B 76*, 50–56.

Lahav, M., Shipway, A. N., Willner, I., Nielsen, M. B., Stoddart, J. F. (2000). "An enlarged bis-bipyridinium cyclophane-Au nanoparticle superstructure for selective electrochemical sensing applications," *J. Electroanal. Chem. 482*, 217–221.

Lahav, M., Shipway, A. N., Willner, I., Nielsen, M. B., Stoddart, J. F. (2000). "An enlarged bis-pyridium cyclophane–Au nanoparticle superstructure for selective electrochemical applications," *J. Electroanal. Chem. 482*, 217–221.

Lee, H. J., Lee, G., Jang, N. R., Yun, J. H., Song, J. Y., Kim, B. S. (2011). "Biological synthesis of copper nanoparticles," *Nanotechnology 1*, 371–374.

Leszczynski, J. (2010). "Bionanoscience: Nano meets bio at the interface," *Nat. Nanotechnol. 5*, 633–634.

Li, W. R., Xie, X. B., Shi, Q. S., Zeng, H. Y., Ou-Yang, Y. S., Chen, Y. B. (2010). "Antibacterial activity and mechanism of silver nanoparticles on *Escherichia coli*," *Appl. Microbiol. Biotechnol. 8*, 1115–1122.

Liu, J., Zhao, Y., Guo, Q., Wang, Z., Wang, H., Yang, Y. (2012). "TAT mofified nanosilver for combating multidrug resistant cancer," *Huang Biomater. 33*, 6155–6161.

Lowenberg, B., Delwel, H. R., Valk, P. J. M. (2005). "The diagnosis of acute myeloid leukaemia enhanced by using DNA microarrays," *Nederlands Tijdschrift voor Geneeskunde 149*(12), 623–625.

Ma, Y., Liang, X., Tong, S., Bao, G., Ren, Q., Dai, Z. (2013). "Gold nanoshell nanomicelles for potential magnetic resonance imaging, light-triggered drug release, and photothermal therapy," *Adv. Funct. Mater. 23*, 815–822.

Majumdar, A. (2002). "Bioassays based on molecular nanomechanics," *Dis. Mark. 18*(4), 167–174.

Mallick, K., Witcomb, M. J., Scurrel, M. S. (2004). "Polymer stabilized silver nanoparticles: A photoychemical synthesis route," *J. Mater. Sci. 39*, 4459–4463.

Maury, P., Escalante, M., Reinhoudt, D. N., Huskens, J. (2005). "Directed assembly of nanoparticles onto polymer-imprinted or chemically patterned templates fabricated by nanoimprint lithography," *Adv. Mater. 17*, 2718–2723.

McQueenie, R., Stevenson, R., Benson, R., MacRitchie, N., McInnes, I., Maffia, P., Faulds, K., Graham, D., Brewer, J., Garside, P. (2012). "Detection of inflammation in vivo by surface-enhanced Raman scattering provides higher sensitivity than conventional fluorescence imaging," *Anal. Chem. 84*, 5968–5975.

Meitl, M. A., Zhu, Z. T., Kumar, V., Lee, K. J., Feng, X., Huang, Y. Y., Adesida, I., Nuzzo, R. G., Rogers, J. A. (2006). "Transfer printing by kinetic control of adhesion to an elastomeric stamp," *Nat. Mater. 5*, 33–38.

Mohanpuria, P., Rana, N. K., Yadav, S. K. (2008). "Biosynthesis of nanoparticles: Technological concepts and future applications," *J. Nanopart. Res. 10*, 507–517.

Muddineti, O. S., Ghosh, B., Biswas, S. (2015). "Current trends in using polymer coated gold nanoparticles for cancer therapy," *Int. J. Pharma. 484*, 252 – 67.

Mukherjee, P., Ahmad, A., Mandal, D., Senapati, S., Sainkar, S. R., Khan, M. I., Ramani, R., Parischa, R., Kumar, P. A. V., Alam, M., Sastry, M., Kumar, R. (2001). "Bioreduction of AuCl[4] ions by the fungus *Verticillium* sp. and surface tapping of gold nanoparticles formed," *Angew. Chem. Int. Ed. 40*, 3585–3588.

Mukherjee, P., Ahmad, A., Mandal, D., Senapati, S., Sainkar, S. S. R., Khan, M. I., Parishcha, R., Ajaykumar, P. P. V., Alam, M., Kumar, R., Sastry, M. (2012). "Fungus mediated synthesis of silver nanoparticles and their immobilization in mycelial matrix: A novel biological approach to nanoparticle synthesis," *Nano Lett. 1*, 515–519.

Nagaraj, B., Krishnamurthy, N. B., Liny, P., Divya, T. K., Dinesh, R. (2011). "Biosynthesis of gold nanoparticles of *Ixora coccinea* flower extract & their antimicrobial activities," *Int. J. Pharm. Biol. Sci. 2*, 557–565.

Nangia, Y., Wangoo, N., Sharma, S., Wu, J. S., Dravid, V., Shekhawat, G. S., Suri, C. (2009). "Facile biosynthesis of phosphate capped gold nanoparticles by a bacterial isolate Stenotrophomonas maltophilia," *Appl. Phys.Lett. 94*, 233901–233903.

Narayanan K. B., Sakthivel, B. (2008). "Coriander leaf mediated biosynthesis of gold nanoparticles," *Mater. Lett. 62*, 4588–4590.

Narayanan, K. B., Sakthivel, N. (2010). Photosynthesis of gold nanoparticles using leaf extract of *Coleus ambolinicus* Lour," *Mater. Charact.61*, 1232–1238.

Narayanan, K. B., Sakthivel, N. (2011). "Facile green synthesis of gold nanoparticles in NADPH dependent enzyme from the extract of *Sclerorium rolfsii*," *Colloids Surf. A 380*, 156–161.

Nelson, J. L., Roeder, B. L., Carmen, J. C., Roloff, F., Pitt, W. G. (2002). "Ultrasonically activated chemotherapeutic drug delivery in a rat model," *Cancer Res. 62*(24), 7280–7283.

Nesakumar, N., Sethuraman, S., Krishnan, U. M., Rayappan, J. B. B. (2013). "Fabrication of lactate biosensor based on lactate dehydrogenase immobilized on cerium oxide nanoparticles," *J. Colloid Interface Sci. 410*, 158–164.

Noruzi, M., Zare, D., Davoodi, D. (2012). "A rapid biosynthesis route for the preparation of gold nanoparticles by aqueous extract of cypress leaves at room temperature," *Spectrochim. Acta, Part A 94*, 84–88.

Noruzi, M., Zare, D., Davoodi, D., Khoshnevisan, K. (2011). "Rapid green synthesis of gold nanoparticles using *Rosa hybrida* petal extract at room temperature," *Spectrochim. Acta, Part A 79*, 1461–1465.

Ogi, T., Saitoh, N., Nomura, T., Konishi, Y. (2010). "Room-temperature synthesis of gold nanoparticles and nano-plates using *Shewanella algae* cell extract," *J. Nanopart. Res. 12*, 2531–2539.

Ozin, G. A., Yang, S. M. (2001). " The race for the photonic chip: Colloidal crystal assembly in silicon wafers," *Adv. Funct. Mater. 11*, 95–104.

Palacios, M. A., Lacy, M. M., Schubert, S. M., Manesse, M., Walt, D. R. (2013). "Assessing the stochastic intermittency of single quantum dot luminescence for robust quantification of biomolecules," *Anal. Chem. 85*, 6639–6645.

Panáček, A., Kolář, M., Večeřová, R., Prucek, R., Soukupová, J., Kryštof, V., Hamal, P., Zbořil, R., Kvítek, L. (2009). "Antifungal activity of silver nanoparticles against *Candida* spp," *Biomaterials 30*, 6333–6340.

Paraschiv, V., Zapotoczny, S., de Jong, M. R., Vancso, G. J., Huskens, J., Reinhoudt, D. N. (2002). "Functional group transfer from gold nanoparticles to flat gold surfaces for the creation of molecular anchoring points on surfaces," *Adv. Mater. 14*, 722–726.

Park, J. I., Lee, W. R., Bae, S. S., Kim, Y. J., Yoo, K. H., J. Cheon, S. Kim. (2005). "Langmuir monolayers of Co nanoparticles and their patterning by microcontact printing," *J. Phys. Chem. B 109*, 13119–13123.

Park, J., Joo, J., Kwon, S. G., Jang, Y., Hyeon, T. (2007). "Synthesis of monodisperse spherical nanocrystals," *Angew. Chem., Int. Ed. 46*, 4630–4660.

Parveen, K., Banse, V., Ledwani, L. (2016). "Green synthesis of nanoparticles: Their advantages and disadvantages," *Acta Nat. 1724*, 20048.

Patolsky, F., Zheng, G., Hayden, O., Lakadamyali, M., Zhuang, X., Lieber, C. M. (2004). "Electrical detection of single viruses," *Proc. Natl. Acad. Sci. USA 101*(39), 14017–14022.

Patri, A. K., Kukowska-Latallo, J. F., Baker Jr., J. R. (2005). "Targeted drug delivery with dendrimers: comparison of the release kinetics of covalently conjugated drug and non-covalent drug inclusion complex," *Adv. Drug Deliv. Rev. 57*, 2203–2214.

Pedersen, C. J. (1998). "The discovery of Crown ethers (Nobel Lecture)," *Angew. Chem. Int. Ed. Engl. 27*, 1021.

Prasad, K. S., Pathak, D., Patel, A., Dalwadi, P., Prasad, R., Patel, P., Selvaraj, K. (2011). "Biogenic synthesis of silver nanoparticles using *Nicotiana tobaccum* leaf extract and study of their antibacterial effect," *Afr. J. Biotechnol. 10*, 8122–8130.

Puzyn, T., Leszczynska, D., Leszczynski, J. (2009)."Toward the development of "Nano-QSARs": Advances and challenges," *Small 5*, 2494–2509.

Qi, L., Ma, J., Shen, J. (1997). "Synthesis of copper nanoparticles in nonionic water-in-oil microemulsions," *J. Colloid Interface Sci. 186*, 498–500.

Quintana, A., Raczka, E., Piehler, L. et al. (2002). "Design and function of a dendrimer-based therapeutic nanodevice targeted to tumor cells through the folate receptor," *Pharmaceut. Res. 19*(9), 1310–1316.

Ragaseema, V. M., Unnikrishnan, S., Krishnan, V. K., Krishnan, L. K. (2012). "The antithrombotic and antimicrobial properties of PEG-protected silver nanoparticle coated surfaces," *Biomaterials 33*, 3083–3092.

Rajakumar, G., Rahuman, A. A., Priyamvada, B., Khanna, V. G., Kumar, D. K., Sujin, P. J. (2012). "Eclipta prostate leave aqueous extract mediated synthesis of titanium dioxide nanoparticles," *Mater. Lett. 68*, 115–117.

Rajakumar, G., Rahuman, A. A., Roopan, S. M., Khanna, V. G., Elango, G., Kamaraj, C., Zahir, A. A. (2012). "Fungus-mediated biosynthesis and characterization of TiO₂ nanoparticles and their activity against pathogenic bacteria," *Spectrochim. Acta, Part A 91*, 23–29.

Ramanathan, R., Field, M. R., O'Mullane, A. P., Smooker, P. M., Bhargava, S. K., Bansal, V. (2013). "Aqueous phase synthesis of copper nanoparticles: A link between heavy metal resistance and nanoparticle synthesis ability in bacterial systems," *Nanoscale 5*, 2300–2306.

Ren, W., Zeng, L., Shen, Z., Xiang, L., Gong, A., Zhang, J., Mao, C., Li, A., Paunesku, T., Woloschak, G. E., Hosmane, N. S., Wu, A. (2013). "Enhanced doxorubicin transport to multidrug resistant breast cancer cells via TiO₂ nanocarriers" *RSC Adv. 3*, 20855–20861.

Renu, G., Rani, V. V., Nair, S. V., Subramanian, K. R. V., Lakshmanan, V. K. (2012). "Development of cerium oxide nanoparticles and its cytotoxicity in prostate cancer cells," *Adv. Sci. Lett. 6*, 17–25.

Roepman, P., Wessels, L. F. A., Kettelarij, N. et al. (2005). "An expression profile for diagnosis of lymph node metastases from primary head and neck squamous cell carcinomas," *Nature Genet. 37*(2), 182–186.

Roopan, S. M., Bharathi, A., Prabhakarn, A., Rahuman, A. A., Velayutham, K., Rajakumar, G., Padmaja, R. D., Lekshmi, M., Madhumitha, G. (2012). "Efficient phyto-synthesis and structural characterization of rutile TiO_2 nanoparticles using Annona squamosa peel extract," *Spectrochim. Acta, Part A. 98*, 86–90.

Rousseau, D. L., Jelinski, L.W. 1991. Biophysics. In *Encyclopedia of Applied Physics*, Vol. 2. VCH Publishers, New York.

Safaepour, M., Shahverdi, A. R., Shahverdi, H. R., Khorramizadeh, M. R., Gohari, A. R. Avicenna (2009). "green synthesis of small silver nanoparticles using geraniol and its cytotoxicity against Fibrosarcoma-Wehi," *J. Med. Biotechnol. 2*, 111–115.

Sahoo, S. K., Parveen, S., Panda, J. J. (2007). "The present and future of nanotechnology in human health care," *Nanomedicine 3*, 20–31.

Saiyed, Z. M., Telang, S. D., Ramchand, C. N. (2003). "Application of magnetic techniques in the field of drug discovery and biomedicine," *BioMagnetic Res. Technol. 1*, 1–8.

Samat, N. A., Nor, R. M. (2013). "Sol-gel synthesis of zinc oxide nanoparticles using *Citrus aurantifolia* extracts," *Ceramics Int. 39*, S545–S548.

Sangeetha, G., Rajeshwari, S., Venckatesh, R. (2011). "Green synthesis of zinc oxide nanoparticles by *Aloe barbadensis* miller leaf extract: Structure and optical properties," *Mater. Res. Bukk. 46*, 2560–2566.

Sanpui, P., Chattopadhyay, A., Ghosh, S. S (2011). "Induction of apoptosis in cancer cells at low silver nanoparticle concentrations using chitosan nanocarrier," *ACS Appl. Mater. Interfaces 3*, 218–228.

Saxena, A., Tripathi, R. M., Singh, R. P. (2010). "Biological synthesis of silver nanoparticles by using onion (*Allium cepa*) extract and their antibacterial activity," *Dig. J. Nanomater. Biostruct. 5*, 427–432.

Seemann, K. M., Luysberg, M., Révay, Z., Kudejova, P., Sanz, B., Cassinelli, N., Loidl, A., Ilcic, K., Multhoff, G., Schmid, T. E. (2015). "Magnetic heating properties and neutron activation of tungsten-oxide coated biocompatible FePt core-shell nanoparticles," *J. Control. Release 197*, 131–137.

Shakibaie, M., Forootanfar, H., Mollazadeh-Moghaddam, K., Bagherzadeh, Z., Nafissi-Varcheh, N., Shahverdi, A. R., Faramarzi, M. A. (2010). "Green synthesis of gold nanoparticles by the marine microalga *Tetraselmis suecica*," *Biotechnol. Appl. Biochem. 57*, 71–75.

Sharma, N., Pinnaka, A. K., Raje, M., Ashish, F. N. U., Bhattacharyya, M. S., Choudhury, A. R. (2012). *Microb. Cell Fact. 11*, 86–91.

Sheldon, R. A. (2012). "Fundamentals of green chemistry: Efficiency in reaction design," *Chem. Soc. Rev. 41*, 1437–1451.

Sherlock, S. P., Tabakman, S. M., Xie, L., Dai, H. (2011). "Photothermally enhanced drug delivery by ultra-small multifunctional FeCo/graphitic-shell nanocrystals," *ACS Nano 5*, 1505–1512.

Shi, J., Votruba, A. R., Farokhzad, O. C., Langer, R. (2010). "Nanotechnology in drug delivery and tissue engineering: From discovery to applications," *Nano Lett. 10*, 3223–3230.

Shi, Y., Lin, M., Jiang, X., Liang, S. (2015). "Recent advances in FePt nanoparticles for biomedicine", *Biomedicine 21*, 22.

Shiying, H., Zhirui, G., Zhanga, Y., Zhanga, S., Wanga, J., Ning, G. (2007). "Biosynthesis of gold nanoparticles using the bacteria *Rhodopseudomonas capsulata*," *Mater. Lett. 61*, 3984–3987.

Shokuhfar, A., Afghahi, S. S. S. (2014). "Size sontrolled synthesis of FeCo alloy nanoparticles and study of the particle size and distribution effects on magnetic properties," *Adv. Mater. Sci. Eng.1*, 1–10.

Shu, Y., Haque, F., Shu, D., Li, W., Zhu, Z., Kotb, M., Lyubchenko, Y., Guo, P. (2013). "Fabrication of 14 different RNA nanoparticles for specific tumor targeting without accumulation in normal organs," *RNA 19*(6), 77–167.

Sinani, V. A., Koktysh, D. S., Yun, B. G., Matts, R. L., Pappas, T. C., Motamedi, M., Thomas, S. N., Kotov, N. A. (2003)," Collagen coating promotes biocompatibility of semiconductor nanoparticles in stratified LBL films," *Nano Lett. 3*, 1177–1182.

Singaravelu, G., Arockiamary, J. S., Ganesh, K. V., Govindarajum, K. (2007). "A novel extracellular synthesis of monodisperse gold nanoparticles using marine alga, *Sargassum wightii* Greville," *Colloids Surf. B 57*, 97–101.

Singh, A. V., Patil, R., Anand, A., Milani, P., Gade, W. N. (2010). "Biosynthesis of copper oxide nanoparticles using *Escherichia coli*," *Curr. Nanosci. 6*, 365–369.

Singh, P., Raja, R. B. (2011). "Biological synthesis and characterization of silver nanoparticles using the fungus *Trichoderma harzianum*," *Asian J. Exp. Biol. Sci. 2*, 600–605.

Singh, R. P., Shukla, V. K., Yadav, R. S., Sharma, P. K., Singh, P. K., Pandey, A. C. (2011) "Biological approach of zinc oxide nanoparticles formation and its characterization," *Adv. Mater. Lett. 2*, 313–317.

Sondi, I, Salopek-Sondi, B. (2004). "Silver nanoparticles antimicrobial agent: A case study on *E. coli* as model for Gram-negative bacteria," *J. Colloid Interface Sci. 275*(1), 177–182.

Song, Y. J., Jang, H. K., Kim, S. B. (2004). "Biosynthesis of gold nanoparticles using *Magnolia kobus* and *Diopyrus kaki* leaves extract," *Process. Biochem. 44*, 1133–1138.

Steed, J. W., Atwood, J. L. (2009). *Supramolecular Chemistry*. John Wiley & Sons.

Stoddart, J. F. *Monographs in Supramolecular Chemistry*. RSC.

Suh, W. H., Suslick, K. S., Stucky, G. D., Suh, Y. H. (2009). "Nanotechnology, nanotoxicology, and neuroscience," *Prog. Neurobiol. 87*, 133–170.

Suh, W. H., Suslick, K. S., Stucky, G. D., Suh, Y.H. (2009). Nanotechnology, nanotoxicology, and neuroscience. *Prog. Neurobiol. 87*, 133–170.

Sulaiman, G. M., Mohammed, W. H., Marzoog, T. R., AlAmier A. A. A., Kadhum, A. A.H., Mohamad, A. B. (2013). "Green synthesis, antimicrobial and cytotoxic effects of silver nanoparticles using *Eucalyptus chapmaniana* leave extract," *Asian Pac. J. Trop. Biomed. 3*, 58–63.

Sun, C., Lee, J.S., Zhang, M. (2008). "Magnetic nanoparticles in MR imaging and drug delivery," *Adv. Drug Deliv. Rev. 60*, 1252.

Sun, S., Zeng, H. (2002). Size-controlled synthesis of magnetite nanoparticles. *J. Am. Chem. Soc. 124*, 8204–8205.

Sun, T., Guo, Q., Zhang, C., Hao, J., Xing, P., Su, J., Li, S., Hao, A., Liu, G., (2012). "Self-assembled vesicles prepared from amphiphilic cyclodextrins as drug carriers," *Langmuir 28*, 8625–8636.

Sundrarajan, M., Gowri, S. (2011). "Green synthesis of titanium dioxide nanoparticles by *Nyctanthes arbortristis* leaves extract," *Chalcogenide Lett. 8*, 447–451.

Tamiaki, H., Amakawa, M., Shimono, Y., Tanikaga, R., Holzwarth, A. R., Schaffner, K. (1996). "Synthetic zinc and magnesium chlorin aggregates as models for supramolecular antenna complexes in chlorosomes of green photosynthetic bacteria," *Photochem. Photobiol. 63*(1), 92–99.

Tartaj, P., Morales, M. P., Veintemillas-Verdaguer, S., Gonzalez-Carreno, T., Serna, C.J. (2003). "The preparation of magnetic nanoparticles for applications in biomedicine," *J. Phys. D: Appl. Phys. 36*, R182.

Taton, T. A. (2002). "Nanostructures as tailored biological probes," *Trends Biotechnol. 20,* 277–279. https://doi.org/10.1016/S0167-7799(02)01973-X.

Thakkar, K.N., Mhatre, S.S., Parikh, R.Y. (2010). "Biological synthesis of metallic nanoparticles," *Nanomed. Nanotechnol. Biol. Med. 6,* 257–262.

Thirumurugan, A., Jiflin, G. J., Rajagomathi, G., Neethu, A. T., Ramachandran S., Jaiganesh, R. (2010). "Synthesis of gold nanoparticles of *Azadirachta indica* leaf extract," *Int. J. Biol. Technol. 1,* 75–77.

Thomas, C. R., George, S., Horst, A. M., Ji, Z., Miller, R. J., Peralta-Videa, J. R., Xia, T., Pokhrel, S., Mädler, L., Gardea-Torresdey, J. L. (2011). "Nanomaterials in the environment: From materials to high-throughput screening to organisms," *ACS Nano 5,* 13–20.

Tolaymat, T. M., El Badawy, A. M., Genaidy, A., Scheckel, K. G., Luxton, T. P., Suidan, M. (2010). "An evidence-based environmental perspective of manufactured silver nanoparticle in syntheses and applications: A systematic review and critical appraisal of peer-reviewed scientific papers," *Sci. Total Environ. 408,* 999–1006.

Turkevich, J., Stevenson, P.C., Hillier, J. (1951). "A study of the nucleation and growth processes in the synthesis of colloidal gold," *Discuss. Faraday Soc. 11,* 55–75.

Valk, P. J. M., Verhaak, R. G. W., Beijen, M. A. et al. (2004), "Prognostically useful gene-expression profiles in acute myeloid leukemia," *N. Engl. J. Med. 350*(16), 1617–1628.

Vankar, P. S., Bajpai, D. (2010). "Preparation of gold nanoparticles from *Mirabilis jalapa* flowers," *Ind. J. Biochem. Biophys. 47,* 157–160

Velayutham, K., Rahuman, A. A., Rajakumar, G., Santhoshkumar, T., Marimuthu, S., Jayaseelan, C., Bagavan, A., Kirthi, A. V., Kamaraj, C., Zahir, A. A., Elango, G. (2012). "Evaluation of *Catharanthus roseus* leaf extract-mediated biosynthesis of titanium dioxide nanoparticles against *Hippobosca maculata* and *Bovicola ovis*," *Parasitol. Res. 111,* 2329–2337.

Vigneshwaran, A., Kathe, A. A., Varadarajan, P. V., Nachne, R. P., Balasubramanya, R. H. (2006). "Bioimmetics of silver nanoparticles by white rot fungus *Phaenerochaete chyrosporium*," *Colloids Surf. B 53,* 55–59.

Waghmare, S. S., Deshmukh, A. M., Kulkarni, S. W., Oswaldo, L. A. (2011). "Biosynthesis and characterization of manganese and zinc nanoparticles," *Universal J. Environ. Res. 62,* 102–104.

Wang, K., He, X., Yang, X., Shi, H. (2013). "Functionalized silica nanoparticles: A platform for fluorescence imaging at the cell and small animal levels," *Acc. Chem. Res. 46,* 1367–1376.

Wang Y., He X., Wang K., Zhang X., Tan W. (2009). "Barbarated skull cup herb extract mediated biosynthesis of gold nanoparticles and its applications in electrochemistry," *Colloids Surf. B 73,* 75–79.

Wang, Y., Zhao, Q., Han, N., Bai, L; Li, J., Liu, J., Che, E., Hu, L., Zhang, Q., Jiang, T., Wang, S. (2015). "Mesoporous silica nanoparticles in drug delivery and biomedical applications," *Nanomed. Nanotechnol. Biol. Med. 11,* 313–327.

Weare, W. W., Reed, S. M., Warner, M. G., Hutchison, J. E. (2000). "Improved synthesis of small (dCORE » 1.5 nm) phosphine-stabilized gold nanoparticles) 12890–12891," *J. Am. Chem. Soc. 122,* 12890.

Wen, L., Lin, Z., Gu, P., Zhou, J., Yao, B., Chen, G., Fu, J. (2009). "Extracellular biosynthesis of monodispersed gold nanoparticles by a SAM capping mode," *J. Nano Res., 11,* 279–288

Whitesides G.M. (2003). "The 'right' size in nanobiotechnology," *Nature Biotechnol. 21,* 1161–1165.

Wijnhoven, J., Vos, W. L. (1998). "Preparation of photonic crystals made of air spheres in titania," *Science 281,* 802–804.

Wu, G., Datar, R. H., Hansen, K. M., Thundat, T., Cote, R. J., Majumdar, A. (2001). "Bioassay of prostate-specific antigen (PSA) using microcantilevers," *Nature Biotechnol. 19*(9), 856–860.

Wu, K. C. W., Yamauchi, Y., Hong, C. Y., Yang, Y. H., Liang, Y. H., Funatsu, T., Tsunoda, M. (2011). "Biocompatible, surface functionalized mesoporous titania nanoparticles for intracellular imaging and anticancer drug delivery," *Chem. Commun. 47*, 5232–5234.

Wu, S.-H., Chen, D.-H. (2003). "Synthesis and characterization of nickel nanoparticles by hydrazine reduction in ethylene glycol," *J. Colloid Interface Sci. 259*, 282–286.

Yakovchuk, P., Protozanova, E., FrankKamenetskii, M. D. (2006). "Base-stacking and base-pairing contributions into thermal stability of the DNA double helix," *Nucleic Acids Res. 34*(2), 564–574.

Yang, H., Gao, P. F., Wu, W. B., Yang, X. X., Zeng, Q. L., Li, C., Huang, C. Z. (1975). "Antibacterials loaded electrospun composite nanofibers: Release profile and sustained antibacterial efficacy," *Polym. Chem.* 2014, *5*, 1965–1975.

Yin, Y., Lu, Y., Gates, B., Xia, Y. (2001). "Template-assisted self-assembly: A practical route to complex aggregates of monodispersed colloids with well-defined sizes, shapes, and structures," *J. Am. Chem. Soc. 123*, 8718–8729.

Yong, K.-T., Qian, J., Roy, I. et al. (2007). "Quantum rod bioconjugates as targeted probes for confocal and two-photon fluorescence imaging of cancer cells," *Nano Lett. 7*(3), 761–765.

Yoshizawa, M., Klosterman, J. K., Fujita, M. (2009). "Functional molecular flasks: new properties and reactions within discrete self assembledhosts," *Angew. Chem. Int. Ed.* (in English), *48*, 3418–3438.

Zhang, Y., Kohler, N., Zhang, M. (2002). "Surface modification of superparamagnetic magnetite nanoparticles and their intracellular uptake," *Biomaterials 23*(7), 1553–1561.

Zheng, Y., Tang, Y., Bao, Z., Wang, H., Ren, F., Guo, M., Quan, H. Jiang, C. (2015). "FePt nanoparticles as a potential X-ray activated chemotherapy agent for HeLa cells," *Int. J. Nanomed. 10*, 6435–6444.

Zhou, Y., Shu, P., Guo, D., Smith, D., Rossi, J. J. (2011). "Fabrication of polyvalent therapeutic RNA nanoparticles for specific delivery of siRNA, ribozyme and drugs to targeted cells for cancer therapy," *Method 54*, 284.

Zilong W., Song, N., Menz, R., Bharadwaj, P., Ying-Wei, Y., Yuebing, Z. (2015). "Nanoparticles functionalized with supramolecular host-guest systems for nanomedicine and healthcare," *Nanomedicine 10*(9), 1491–1514.

Zirbs, R., Kienberger, F., Hinterdorfer, P., Binder, W. H. (2005). "Directed assembly of Au nanoparticles onto planar surfaces via multiple hydrogen bonds," *Langmuir 21*, 8414–8421.

CHAPTER 3

Biosynthesis and Therapeutic Potential of Silver Nanoparticles

LAVNAYA TANDON and POONAM KHULLAR*

Department of Chemistry, BBK DAV College for Women, Amritsar 143001, Punjab, India

Corresponding author. E-mail: virgo16sep2005@gmail.com

ABSTRACT

Prodigious properties of metal nanoparticles (NPs) result in their potential applications in various fields. The conventional methods involved in the synthesis of metal NPs are tedious, energy-consuming, and involve the use of toxic and hazardous chemicals. Thus, alternative methods involving the use of biological molecules such as enzymes, seed extracts, fruit extracts are the need of the hour. Such methods provide many advantages and are ecofriendly.

3.1 INTRODUCTION

Nanotechnology nowadays is impacting almost every aspect of human life due to its potential applications in the areas of electronics, material sciences, biomedical engineering, and medicine (Bhattacharyya et al., 2009). Nanoparticles (NPs) due to their large surface-to-volume ratio, surface functionalization, absorption in the UV–visible range, and controlled drug release are very important in human life. Generally, there are two types of approaches that are used for the synthesis of NPs, that is, top-down and bottom-up approaches. In the top-down approach, the bulk materials are reduced in the range of nanoscale. However, the bottom-up approach involves the joining of smaller atoms or molecules to grow to larger nanostructures.

Several synthetic methods, such as physical and chemical methods, are available in the literature for the synthesis of NPs (Bao et al., 2011). However, the chemical synthetic methods involve the use of hazardous or toxic chemicals and therefore such surface-functionalized NPs cannot be directly used for biomedical applications (Mukherjee et al., 2012).

Bioenthused synthesis of NPs provides many advantages over the traditional synthetic procedures as there is no need for harsh conditions such as high temperatures, high pressures, and toxic chemicals. However, it offers an economic, commercially viable, and ecofriendly synthetic route. Silver NPs find potential applications as antibacterial and antifungal agents in textile engineering and silver-based consumer products (Kumar et al., 2011; Zhang et al., 2008; Schrand et al., 2008).

The synthesis of metal NPs using plant extracts is a challenging aspect, but this method can be used for many therapeutic applications (Mittal et al., 2013; Mittal et al., 2012). Various types of plants have been used for the synthesis of metal NPs (Mittal et al., 2013; Li et al., 2007). Capping of the metal NPs with biological molecules offers the bioapplicability of such NPs in therapeutic activities (Iravani, 2011). The plant-mediated synthesis is quite cost-effective, simple, and provides safety (Shankar et al., 2004; Narayanan and Sakthivel, 2008; Huang et al., 2007; Li et al., 2007). The major issue is that when NPs are converted into the powder form, they agglomerate and thus lose their characteristic properties. To avoid such agglomeration, various types of capping agents are used. This chapter includes the synthesis of silver NPs using fruit extracts, germinating seeds, and seed extracts and the identification of the active components responsible for the reduction of metal precursors.

3.2 SYNTHESIS OF AG NPS USING FRUIT EXTRACT

One of the common methods for the synthesis of silver nanocolloids involves the reduction of an aqueous solution of silver nitrate (Sun and Xia, 2002). The synthesis of Ag NPs was characterized by the appearance of an absorption peak at 420 nm corresponding to the formation of various Ag NPs in the size range of 20–100 nm (Vigneshwaran et al., 2007). For the synthesis of Ag NPs, various parameters such as the pH, temperature, concentration of the fruit extract, concentration of the metal precursor, and reaction time were optimized (Mittal et al., 2014). Out of the above-referred parameters, it was observed that it is the pH of the reduction medium that plays a crucial role in the synthesis of Ag NPs.

Upon varying the pH, it was summarized that it is the neutral pH, that is, 7, where the increase in the yield of NPs was observed; thereafter, that is, at pH 8, it decreases and a further increase in the pH results in the agglomeration of Ag NPs (Table 3.1). Thus, pH 8 is the optimal pH for nanoparticle synthesis (Mittal et al., 2014) and it agrees very well with other reported literature works (Krishnaraj et al., 2012; Khalil et al., 2012).

TABLE 3.1 Summary of Various Optimized Parameters for the Synthesis of Silver Nanoparticles

Optimization of Various Parameters	Corresponding Values
Concentration of fruit extract	50 µL/50 mL (from stock solution)
Concentration of silver nitrate	2 mM
pH of the medium	8
Temperature of the medium	35 °C
Reaction time	12 h

The next parameter is the temperature, and it is observed that when the incubation temperature was increased from 25 °C to 35 °C, the absorption peak increased and thereafter it decreased. Thus, 35 °C was the optimal temperature required for the synthesis of Ag NPs. The incubation time, that is, the time interval required, was observed to be 12 h. It was observed that after 24 h, due to the instability of Ag NPs, agglomeration occurs, and these results were in good agreement with the reported literature values (Khalil et al., 2012; Veerasamy et al., 2011). The stability of thus-synthesized Ag NPs was tested for 60 days with a 10-day interval using a UV–visible spectrophotometer. It was observed that the peak at 420 nm showed no alteration, thus confirming the stability of biosynthesized Ag NPs.

The size and zeta potential measurements of such particles indicated the Z-average value of 35 nm with a low polydispersity index (PDI) of 0.31 and a zeta potential value of −19.5. Thus, the negative zeta potential value clearly indicates that Ag NPs are mainly comprised of negatively charged groups and are responsible for the average stability of the NPs.

The transmission electrom microscopy (TEM) micrograph of the synthesized NPs is obtained at 5–10 nm scales (Figure 3.1) and exhibits a spherical shape. The size range is 5–20 nm with a mean diameter of 12.5 ± 2.5 nm. Energy-dispersive X-ray spectroscopy (EDS) analysis also confirmed the presence of elemental silver in the NPs. X-ray diffraction (XRD) analysis confirmed the crystalline nature of the synthesized Ag NPs. It showed peaks at 2θ values of 38.2°, 44.4°, 64.6°, and 77.5°, which are assigned to the [111], [200], [220], and [311] crystalline planes of the face-centered cubic (fcc) crystalline structure of metallic Ag NPs, respectively (Narayanan and Sakthivel, 2011).

Fourier transform infrared (FT-IR) analysis shows strong absorption bands located at 1083, 1260, 1436, 1617, and 3373 cm⁻¹. These absorption peaks were assigned to the strong stretching vibrations of C–N aromatic and aliphatic amines (1083 and 1260 cm⁻¹), amide I band of proteins due to carbonyl stretching in proteins (1617 cm⁻¹), and OH stretching in alcohols, flavonoids, and phenols (3373 cm⁻¹), respectively. FT-IR analysis indicated the presence of flavonoid coating on the surface of NPs.

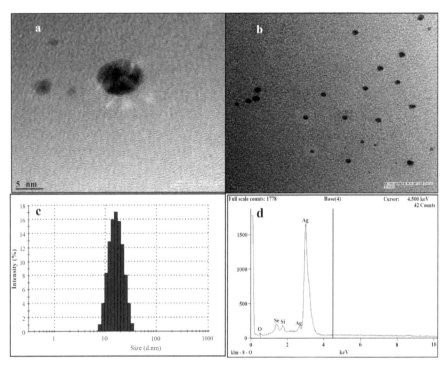

FIGURE 3.1   Enlarged TEM image (a) 5 nm scale and (b) 10 nm scale. (c) Histogram of silver NP distribution. (d) EDS spectra of silver NPs synthesized by fruit extract [Reprinted with permission from Mittal et al. (2014). © Elsevier.]

To characterize the active compounds responsible for the reduction of silver ions, the solvent extraction method was used. It was concluded that flavanoids are mainly responsible for the synthesis of NPs. The following plausible mechanism (Figure 3.2) is proposed for the reduction of Ag^+ ions to metallic silver through redox activity. The proposed mechanism agrees very well with the other proposed reports (Durán et al., 2005).

Antioxidant activity means the inhibition of the oxidation of the molecules by inhibiting the initiation step of the oxidative chain reactions that lead to the formation of the stable radicals that are quite nonreactive. Polyphenolic and flavanoid compounds present in plants possess strong antioxidant activity that protects the cells from various free radicals created as a result of oxidative stress.

The antioxidant activity of the synthesized Ag NPs was tested by various assays such as DPPH, ABTS, and MTT. The DPPH assay demonstrated an IC50 value of 22.19 µg/mL, while the MTT assay showed the antioxidant activity of 61% at the same concentration. The antioxidant property is due to the capping of Ag NPs by the flavanoids present in the plant extract.

Ag NPs were also tested for their cytotoxic effects on Dalton lymphoma (DL) cells using the MTT assay (Mittal et al., 2014). The Ag NPs showed a viability of 10% at 50 µg/mL, whereas a viability of 46% was observed at 100 µg/mL. Thus, they showed dose-dependent cytotoxicity. Such Ag NPs showed toxicity only on the tumor cells and not on the normal cells. Thus, they have the potential to be used as nanomedicine against various types of cancers. Thus, the applications of such ecofriendly NPs in bactericidal, wound healing, and electronic applications create the urge of large-scale synthesis of other metallic NPs.

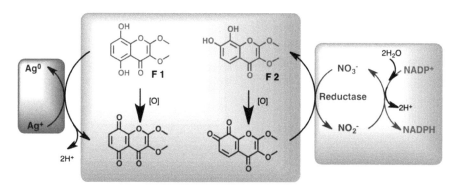

FIGURE 3.2 Prospective mechanism for silver NP biosynthesis (structures of isolated flavonoid molecules are shown) by the fruit extract of *Syzygium cumini* [Reprinted with permission from Mittal et al. (2014). © Elsevier.]

3.3   SYNTHESIS OF AG NPS USING SEED EXTRACT

Green synthesis of NPs using *Cydonia oblong* seed extracts provides an alternative cost-effective and environmentally benign technique (Zia et al., 2016). It is observed that biomolecules act as an effective capping and reducing agent. The Ag NPs prepared show temperature-dependent size variation. At high temperatures, a hypsochromic shift was observed and the wavelength shifted to a lower value as the temperature was increased from 50 °C to 90 °C through 70 °C. It clearly indicated the formation of smaller Ag NPs at high temperatures. The high temperature favored the formation of Ag NPs rapidly, that is, the reactants are consumed rapidly. Thus, 70 °C is the optimum temperature for the formation of monodispersed Ag NPs.

An optimum pH value is also important for the synthesis of metal NPs. As in the case of Ag NPs, at a low pH value of 3, practically no NPs are formed. However, as the pH was increased to 9, monodisperse Ag NPs were prepared, which show their characteristic surface plasmon resonance peak at 410 nm. The intensity of this peak increases with time, which shows that the rate of reduction is directly proportional to time.

FT-IR spectra of Ag NPs show a peak at 1728.22 cm^{-1}, which is clearly indicative of flavanones or terpenoids. This surface adsorption occurs because of the interaction of p-electrons in the carbonyl groups due to the lack of sufficient concentration of chelating agents. A few other studies (Veeraputhiran, 2013) also confirmed the capping and stabilizing property of proteins and amino acids due to the presence of carbonyl groups.

The XRD pattern showed the number of Bragg's reflection at $2q$ diffraction angles of 32.053°, 38.002°, 44.109°, 46.056°, and 64.414°, corresponding to planes [111], [200], [300], and [220], respectively, which confirmed that the crystalline spherical NPs have an fcc structure. SEM images also confirmed the formation of Ag NPs and the complete consumption of Ag$^+$ ions. Thus, NP synthesis can be successfully achieved by adopting the green protocol.

3.4   SYNTHESIS OF AG NPS USING ENZYMES

Another green approach for the synthesis of Ag NPs involves the use of enzymes such as lysozyme (Eby et al., 2009). Here, lysozyme acts as the sole reducing agent in the presence of light. Such lysozyme–Ag NPs have retained the hydrolase function of the enzyme and were found effective for inhibiting the growth of various strains of bacteria such as *Escherichia coli*, *Staphylococcus aureus*, *Bacillus* anthracis, and *Candida* albicans. These NPs were proved to be quite effective against the silver-resistant *Proteus mirabilis* strains and pMG101, which is a recombinant *E. coli* strain containing multiple antibiotic and silver-resistant plasmids.

It was observed that at lower Ag^+/lysozyme molar ratios, monodispersed, yellow, small-sized Ag NPs were obtained, which show a symmetrical SPR band at 420 nm. However, at higher Ag^+/lysozyme molar ratios, red Ag NPs were obtained, which show a broad absorption band (Figure 3.3).

Dynamic light scattering (DLS) was used to get an idea about the quantitative size distribution and to check the monodispersity of the metal NPs obtained. DLS studies confirmed the formation of Ag NPs in the range of 2–10 nm at lower Ag^+/lysozyme molar ratios, whereas the formation of larger Ag NPs in the range of 8–12 nm at higher Ag^+/lysozyme molar ratios. The dispersity in size with regard to molar ratios can be explained on the basis of the Ostwald ripening process in which smaller unstable particles dissipate and coalesce to form larger stable particles. At lower Ag^+/lysozyme molar ratios, lysozyme effectively controls the size of the resulting NPs and therefore smaller NPs are obtained. However, at higher Ag^+/lysozyme molar ratios, the concentration of Ag^+ ions get increased, which result in an increased number of Ag NPs that undergo agglomeration and therefore results in nonsymmetrical masses (Figure 3.3c).

Thus, the Ag^+/lysozyme molar ratio is the deciding factor for the morphology of the resulting NPs. Thus, lysozyme shows excellent capping behavior for the stabilization of the colloidal suspension and therefore prevents agglomeration. Lysozyme is an amphiphillic protein and cationic in nature, and being ionic, it adheres on the hydrophobic and ionic metal surfaces (Haggerty and Lenhoff, 1993; Schmidt et al., 1990; Sethuraman and Belfort, 2005; Luckarift et al., 2007). When lysozyme-coated Ag NPs were treated to dissociate or separate proteins, it was observed that NPs get agglomerate, indicating that lysozyme acts as an excellent capping agent.

The zeta potential value can be used as an indicator to predict the colloidal stability of the particles. Thus, a greater zeta potential value suggests that there exists a greater electrostatic repulsion among particles, which prevents the agglomeration and hence colloidal stability. The theoretical limit to colloidal stability is estimated to be about (Luckarift et al., 2007) mV. The zeta potential values of −20 and 40 mV have been measured for Ag–Lys in a methanolic solution.

To measure the thickness of the protein coating on metal NPs, the average particle sizes are taken independently from DLS and TEM image analyses. As we know that the DLS measurement is based on the relationship between the speed of Brownian motion and the particle size, it accurately gives the

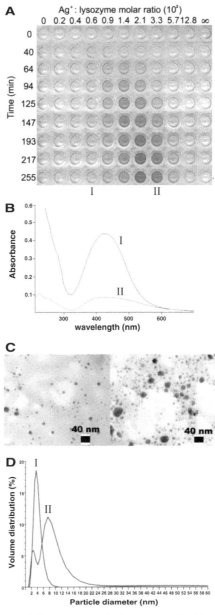

FIGURE 3.3   Lysozyme and silver acetate solutions were mixed in varying molar ratios and exposed to light over the course of 255 min. At set time intervals, reactions were added to a row of microtiter plate wells and imaged using a document scanner (A). Representative samples of yellow (I) and red (II) NPs from the synthesis assay were used in absorbance spectroscopy (B), TEM imaging (C), and dynamic light scattering to determine the nanoparticle size distribution (D) [Reprinted with permission from Eby et al. (2009). © American Chemical Society.

measurement of the size of the particle in well-dispersed colloidal solutions through fluctuations in the light scattering intensity. However, TEM can image only metal NPs and show the protein coating as nearly transparent. Therefore, the difference in particle sizes between TEM and DLS will give the thickness of the protein coating on the NP surface.

The size of the lysozyme is $3.0 \times 3.0 \times 4.5$ nm, and the protein monolayer on the surface of the metal NPs would be twice, that is, 6 nm (Radmacher et al., 1994). The size measurements clearly indicates that the protein (Lys) forms a thick coating on the metal NP surface, which inhibits Ostwald ripening and thus provides the colloidal suspension more stability.

Protein-coated NPs still show hydrolase activity of the enzyme that shows that the protein retain its native hydrolytic activity. However, in many cases, it is observed that Ag^+ ions bind nonspecifically to proteins by interacting with many different amino acid side groups (Lansdown, 2006; Schierholz et al., 1998). Sometimes, such interactions results in the loss of enzyme activity due to the denaturation of the protein.

The antimicrobial activity of Ag–Lys NPs prepared was compared with that of Ag alone as well as lysozyme alone. It was observed that lysozyme alone had no effect against the three representative bacterial strains and one fungal strain. Thus, the potency of Ag NPs was comparable to that of Ag–Lys NPs, which clearly emphasizes the important role in the antimicrobial mechanism of the NPs. These results also indicate that there was no relation between the particle size or charge and the potency.

The Ag–Lys NPs were further tested for *in vitro* tissue culture assays, and the reults concluded that the NP composite materials did not result in appreciable toxicological effects against mammalian cells at concentrations necessary to inhibit bacterial growth (Figure 3.4). For these studies, human epidermal keratinocyte cell line (HaCaT) was selected.

Thus, lysozyme in methanol attains a unique amphipathic form that causes silver reduction as well as acts as a capping agent that results in a stable colloidal solution. Proteins attains different conformations in methanol and water medium. The aqueous medium does not result in a stable colloidal suspension, which is due to the presence of strong ionic interactions among NPs. Thus, here, the hydrophobicity of methanol plays an important role in stabilizing silver colloids as a result of decreasing adsorption and agglomeration of NPs. In the present case, lysozyme proves to be an exception in which the protein retains its hydrolytic activity even after exposure to methanol and even after NP synthesis. The use of lysozyme for the synthesis of Ag NPS offer many benefits, namely, a simple, inexpensive method that results in stable silver colloids as compared to the conventional synthetic procedure.

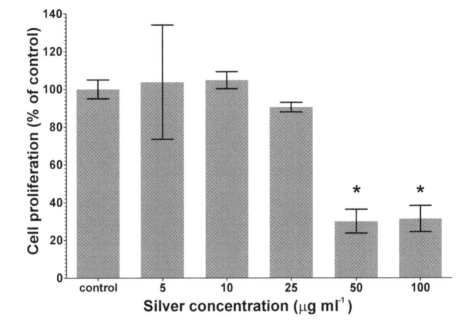

FIGURE 3.4   Viability of HaCaT cells treated with Ag⁺/Lys NPs. Cell proliferation was determined by measuring the metabolic activity of cell cultures after 24 h incubation with NPs and compared against non-NP control cultures. The asterisk (*) denotes significance compared to the control values ($p < 0.05$). [Reprinted with permission from Eby et al. (2009). © American Chemical Society.]

3.5   FUTURE PERSPECTIVES

Thus, greener synthesis of Ag NPs using natural extracts with less time leads to the future studies on Ag NP toxicity without risking contaminants from hazardous reagents and capping agents.

KEYWORDS

- metal NPs
- toxic
- enzymes
- seed extracts

REFERENCES

Bao, Q., Zhang, D., Qi, P. *J. Colloid Interface Sci.* 360 (2011) 463–470.

Bhattacharyya, D., Singh, S., Satnalika, N., Khandelwal, A., Jeon, S.H. *Nanotechnology* (2009) 229–238.

Durán, N., Marcato, P. D., Alves, O. L., De Souza, G. I. Esposito, E. *J. Nanobiotechnol.* 3 (2005) 1–7.

Eby, D. M., Schaeublin, N. M., Farrington, K. E., Hussain, S. M., Johnson, G. R. Lysozyme catalyzes the formation of antimicrobial silver nanoparticles. *ACS Nano* 3 (2009) 984–994.

Haggerty, L., Lenhoff, A. M. Analysis of ordered arrays of adsorbed lysozyme by scanning tunneling microscopy. *Biophys. J.* 64 (1993) 886–895.

Huang, J., Li, Q., Sun, D., Lu, Y., Su, Y., Yang, X., Wang, H., Wang, Y., Shao, W., He, N., Hong, J., Chen, C. *Nanotechnology* 18 (2007) 105104.

Iravani, S. *Green Chem.* 13 (2011) 2638–2650.

Khalil, M. M., Ismail, E. H., El-Magdoub, F. *Arab J. Chem.* 5 (2012) 431–437.

Krishnaraj, C., Ramachandran, R., Mohan, K., Kalaichelvan, P. T. *Spectrochim. Acta A* 93 (2012) 95–99.

Kumar, K. P., Paul, W., Sharma, C. P. *Process Biochem.* 46 (2011) 2007–2013.

Lansdown, A. Silver in health care: Antimicrobial effects and safety in use. *Curr. Probl. Dermatol.* 33 (2006) 17–34.

Li, S., Shen, Y., Xie, A., Yu, X., Qui, L., Zhang, L., Zhang, Q. *Green Chem.* 9 (2007) 852.

Li, S., Shen, Y., Xie, A., Yu, X., Zhang, X., Yang, L., Li, C. *Nanotechnology* 18 (2007) 405101.

Luckarift, H. R., Balasubramanian, S., Paliwal, S., Johnson, G. R., Simonian, A. L. Enzyme-encapsulated silica monolayers for rapid functionalization of a gold surface. *Colloids Surf. B* 58 (2007) 28–33.

Mittal, A. K., Bhaumik, J., Kumar, S., Banerjee, U. C. J. *Colloid Interface Sci.* 415 (2014) 39–47.

Mittal, A. K., Chisti, Y., Banerjee, U. C. *Biotechnol. Adv.* 31 (2013) 346–356.

Mittal, A. K., Kaler, A., Banerjee, U. C. Rhododendron dauricum. *Nano Biomed. Eng.* 4 (2012) 118–124.

Mittal, A. K., Kaler, A., Mulay, A. V., Banerjee, U. C. Synthesis of gold nanoparticles using whole cells of *Geotrichum candidum. J. Nanopart.* 2013 (2013) 150414.

Mukherjee, S., Sushma, V., Patra, S., Barui, A.K., Bhadra, M.P., Sreedhar, B., Patra, C.R. *Nanotechnology* 23 (2012) 455103.

Narayanan, K. B. Sakthivel, N. *Colloids Surf. A* 380 (2011) 156–161.

Narayanan, K. B. Sakthivel, N. *Mater. Lett.* 68 (2008) 4588.

Radmacher, M., Fritz, M., Hansma, H. G., Hansma, P. K. Direct observation of enzyme activity with the atomic force microscope. *Science* 265 (1994) 1577–1579.

Schierholz, J. M., Lucas, L. J., Rump, A., Pulverer, G. Efficacy of silver-coated medical devices. *J. Hosp. Infect.* 40 (1998) 257–262.

Schmidt, C. F., Zimmermann, R. M., Gaub, H. E. Multilayer adsorption of lysozyme on a hydrophobic substrate. *Biophys. J.* 57 (1990) 577–588.

Schrand, A. M., Braydich-Stolle, L. K., Schlager, J. J., Dai, L., Hussain, S. M. *Nanotechnology* 19 (2008) 235104.

Sethuraman, A., Belfort, G. Protein structural perturbation and aggregation on homogeneous surfaces. *Biophys. J.* 88 (2005) 1322–1333.

Shankar, S. S., Rai, A., Ahmad, A., Sastry, M. *JCIS* 275 (2004) 496.

Sun, Y., Xia, Y. *Science* 298 (2002) 2176–2179.

Veeraputhiran, V. Bio-catalytic synthesis of silver nanoparticles. *Int. J. Chem. Tech. Res.* 5 (2013) 2555–256.

Veerasamy, R., Xin, T. Z. Gunasagaran, S., Xiang, T. F. W., Yang, E. F. C., Jeyakumar, N., Dhanaraj, S. A. *J. Saudi Chem. Soc.* 15 (2011) 113–120.

Vigneshwaran, N., Ashtaputre, N. M., Varadarajan, P. V., Nachane, K. M., Paralikar, R. P., Balasubramanya, R. H. *Mater. Lett.* 61 (2007) 1413–1418.

Zhang, Y., Peng, H., Huang, W., Zhou, Y., Yan, D. *J. Colloid Interface Sci.* 325 (2008) 371–376.

Zia, F., Ghafoor, N., Iqbal, M., Mehboob, S. *Appl. Nanosci.* 6 (2016) 1023–1029.

CHAPTER 4

Pomegranate Bacterial Blight: *Abutilon indicum, Prosopis juliflora,* and *Acacia arabica* as Antibacterial Agents for *Xanthomonas axonopodis* pv. *punicae*

A. ANDHARE AISHWARYA[1], RAVINDRA S. SHINDE[1*], and AMOL J. DESHMUKH[2]

[1]*Department of Microbiology, Biotechnology and Chemistry, Dayanand Science College, Latur 413512, Maharashtra, India*

[2]*Department of Plant Pathology, College of Agriculture, Navsari Agricultural University, Waghai 394730, Gujarat, India*

*Corresponding author. E-mail: rshinde.33381@gmail.com

ABSTRACT

Xanthomonas axonopodis pv. *punicae* causes bacterial blight disease in pomegranate. A complete range of symptoms of bacterial blight caused by *X. axonopodis* pv. *punicae* appear on various pomegranate plant parts, except roots. The present investigation was initiated to find a suitable alternative to synthetic antibiotics for the management of plant diseases caused by bacteria. The study was aimed to use wild plant species, viz., *Abutilon indicum, Prosopis juliflora,* and *Acacia arabica* as antibacterial agents against *X. axonopodis* pv. *punicae*. The aqueous extracts of *A. indicum, P. juliflora,* and *A. arabica* plants has antibacterial activity against *X. axonopodis* pv. *punicae*. The antibacterial activity was tested by a well diffusion assay, minimum inhibitory concentration (MIC), and minimum bactericidal concentration (MBC). The maximum activity was recorded by *P. juliflora* (MIC = 1.03 mg mL^{-1} and MBC = 0.15 mg mL^{-1}) and *A. arabica* (MIC = 1.00372 mg mL^{-1} and MBC = 2.58 mg mL^{-1}) against

X. axonopodis pv. *punicae*, while the lowest activity was recorded by *A. indicum* (MIC = 0.619 mg mL^{-1} and MBC = 0.923 mg mL^{-1}). The largest zone of inhibition (ZOI) was shown by *P. juliflora*, while the shortest ZOI was shown by *A. indicum*. The results infer that the extracts of *P. juliflora* and *A. arabica* are highly sensitive to *X. axonopodis* pv. *punicae*. The plant extracts exhibited antibacterial activity with the potential to be used in the management of many plant diseases as an alternative to chemical antibiotics. A further phytochemical analysis is required to identify the bioactive compounds responsible for antibacterial activity.

4.1 INTRODUCTION

Pomegranate (*Punica granatum* L.) is an ancient fruit, belonging to the smallest botanical family *Punicaceae*. Pomegranate is a native of Iran, where it was first cultivated in about 2000 BC but spread to the Mediterranean countries at an early date. It is extensively cultivated in Spain, Morocco, and other countries around the Mediterranean, Egypt, Iran, Afghanistan, Arabia, and Baluchistan. Pomegranate is a good source of carbohydrates and minerals such as calcium, iron, and sulfur. It is rich in vitamin C, and citric acid is the most predominant organic acid in pomegranate (Malhotra et al., 1983). Apart from the fleshy portion of the fruit, the crop residues are also finding a place in industries. The rind of the fruit is a good source of dye, which gives yellowish-brown color to khaki shades and is being used for dying wool and silk. The flower and buds yield light red dye, which is used for dying of cloths in India. The bark of the stem and root contains a number of alkaloids belonging to the pyridine group. The bark is used as a tanning material, especially in Mediterranean countries in the East (Bose, 1985). Pomegranate is regarded as the "Fruit of Paradise." It is one of the most adaptable subtropical minor fruit crops, and its cultivation is increasing very rapidly. In India, it is regarded as a "vital cash crop," grown in an area of 116,000 ha with a production of 89,000 million tons and an average productivity of 7.3 million tons (Srivastsva and Umesh, 2008). The successful cultivation of pomegranate in recent years has met with different traumas such as pests and diseases. Among the diseases, bacterial blight caused by *X. axonopodis* pv. *Punicae* is a major threat. Since 2002, the disease has reached an alarming stage and is hampering the Indian economy and the export of quality fruits. The disease accounted for up to 70–100% during 2006 in Karnataka and

Maharashtra, resulting in the wipeout of the pomegranate. During the year 2007, the total output of pomegranate production in India was down by 60% (Raghavan, 2007).

The causal organism of blight is *X. axonopodis* pv. *punicae* (Hingorani and Singh, 1959) Vauterin, Hoste, Kersters and Swings (Hingorani and Mehta, 1952; Vauterin et al., 1995). The plant is susceptible to a blight during all stages of growth and results in huge economic loss. Bacterial blight primarily affects the above-ground plant parts, especially leaves, twigs, and fruits. While the leaves how early water-soaked lesions to late necrotic blighting, the fruits show isolated or coalesced water-soaked lesions followed by necrosis with small cracks and splitting of the entire fruit (Petersen et al., 2010). Stems show lesions around nodes or injuries, forming cankers in later stages. Suspected symptoms on floral parts have also been reported (Chand and Kishun, 1991; Rani et al., 2001). It is also presumed that the stem canker could be an outcome of the systemic spread of bacterium from the leaf (Chand and Kishun, 1993). However, it is reported by Chand and Kishun (1991) and Rani et al. (2001) that attempts to reproduce the field symptoms of blight on detached leaves, twigs, and fruits were unsuccessful in artificial inoculations. The management of bacterial blight of pomegranate is a major concern. This disease could not be effectively managed by conventional antibiotics like streptocycline in field conditions. Thus, this investigation is carried out on the management of the disease by aqueous extracts of *Abutilon indicum*, *Prosopis juliflora*, and *Acacia arabica* plants. Continual and indiscriminate use of synthetic antibiotics to control bacterial diseases of crop plants has caused health hazards in animals and humans due to their residual toxicity (Raghavendra et al., 2006). A bioactive principle isolated from plants appears to be one of an alternative for controlling plant and human pathogens developing resistance to antibiotics. Plant-originated antibacterial compounds can be one approach to plant disease management because of their ecofriendly nature (Bolkan and Reinert, 1994).

4.2 MATERIALS AND METHODS

4.2.1 PLANT MATERIAL

The commonly available weed plants were collected from the surrounding areas of Latur District, Maharashtra. The plants used for this study are *A. indicum*, *P. juliflora*, and *A. arabica*.

4.2.2   PREPARATION OF PLANT EXTRACTS

The plant leaves were washed with tap water followed by sterile distilled water and then air-dried at room temperature. Dried leaves were powdered; 50 g of powdered material was percolated with 250 mL of water for 72 h. The percolate was mixed thoroughly for every 12 h. The percolate was filtered through a double-layered muslin cloth, followed by Whatman no. 1 filter paper. The filtrate was concentrated at 35 °C and stored at 4 °C until further use (Cowan, 1999; Alade and Irobi, 1993).

4.2.3   FIELD VISIT AND COLLECTION OF BACTERIAL SAMPLES

The field visit was undertaken in major pomegranate growing regions of Latur, Maharashtra, India, that is, Murud and Harangul, from June to August. 2018. During the field survey, the randomly selected plant parts were inspected in the fields for the incidence of bacterial blight. The distribution of bacterial blight of pomegranate was recorded in these areas. Plants were diagnosed as infected based on the typical symptoms of bacterial blight, namely, yellow water-soaked lesions at early stages and corky, dark oily spots at later stages of infection. The suspected plant leaves and fruits (Figure 4.1) were collected and transferred into sterilized plastic bags and brought to the research laboratory of Microbiology Department of Dayanand Science College, Latur, for further studies.

4.2.4   STORAGE OF SAMPLES

The samples were surface-sterilized and stored in laboratory conditions.

4.2.5   ISOLATION OF THE BACTERIA

The bacteria were isolated from the infected leaves and fruits of pomegranate plants collected from regions of Latur, Maharashtra, India, that is, Murud and Harangul, from June to August. 2018. These samples were washed, air-dried, and then disinfected with 0.1% $HgCl_2$ for about 30–60 s and washed thrice with sterile water to remove traces of $HgCl_2$. They were macerated with a sterilized blade in a sterile Petri dish containing a few drops of sterile distilled water to allow the bacteria to diffuse out. A loop full of suspension was then transferred with the help of a sterilized bacteriological needle to sterilized Petri plates filled with nutrient agar (NA) medium and incubated at 28 °C for 24–72 h. After 2–3 days, the incubated plates were observed for the presence of typical pale yellow, glistening colonies (Figure 4.2), which were transferred to the NA slants and maintained in laboratory conditions for further studies.

FIGURE 4.1 Bacterial blight symptoms on pomegranate fruits and leaves: (A) bacterial blight symptoms on the pomegranate fruits, (B) bacterial blight symptoms on the fruit, (C) bacterial blight symptoms on the pomegranate fruit, and (D) bacterial blight symptoms on the fruit and leaves.

4.2.6   IDENTIFICATION AND CONFIRMATION OF ISOLATE

The identification and confirmation of *X. axonopodis* pv. *punicae* were done by using the following methods.

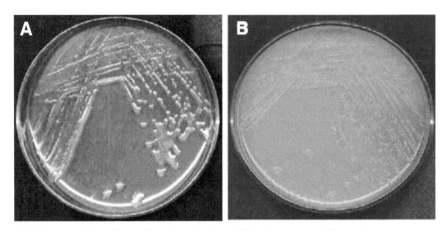

FIGURE 4.2   Pale yellow, glistening colonies of *Xanthomonas* on NA medium

4.2.6.1   MORPHOLOGICAL CHARACTERS

The morphological characters such as the shape, Gram reaction, and pigmentation characters were studied as described by the Society of American Bacteriologists, Bradbury (1970) (Table 4.1).

TABLE 4.1   Cultural and Morphological Characteristics of *X. axonopodis* pv. *punicae* Isolate on NA Media

Sr. No.	Colony Characters	X. axonopodis pv. punicae
1	Color	Yellowish
2	Size of colonies	Medium to large
3	Shape of colonies	Small circular colonies
4	Cell shape	Single rod
5	Appearance	Slightly raised, glistening
6	Elevation	Convex
7	Margin	Entire margin
8	Texture	Highly mucoid

4.2.6.2   GROWTH RATE AT 28 °C AND 37 °C

The effect of varying temperature levels on the growth of *X.s axonopodis* pv. *punicae* was studied, and the data so obtained is presented in Table 4.2.

The isolates of *X. axonopodis* pv. *punicae* were tested at temperatures 28 °C and 37 °C on NA medium. The data clearly indicated that the temperature of 28 °C was found optimum for the growth of the pathogen as a significantly maximum number of colonies was observed at this temperature. The isolates grew well at 28 °C, but no growth was observed at a temperature of 37 °C. The growth of *X. axonopodis* pv. *punicae* started 72 h after incubation. Maximum growth was observed after 120 h of incubation at 28 °C, but no growth was observed after 48 h of incubation. At 37 °C, no growth was observed up to 120 h. Similar work on temperature requirements was carried out by Hingorani and Mehta (1952). They found that the pomegranate bacterium grows well at a cordial temperature of 30 °C and can tolerate minimum and maximum temperatures of 50 °C and 40 °C, respectively. Gour et al. (2000) also got similar results while working with *Xanthomonas axonopodis* pv. *vignicola*, the causal agent of leaf blight of cowpea. The maximum growth of pathogen *Xap* at a temperature level of 30 °C, whereas Manjula (2002) recorded the highest number of colonies of *Xap* at a temperature of 27 °C.

TABLE 4.2   Growth Rate of *X. axonopodis* pv. *punicae* Isolates

Sr. No.	Isolate	Growth Rrate							
		28 °C				37 °C			
		48 h	72 h	96 h	120 h	48 h	72 h	96 h	120 h
1	*X. axonopodis* pv. *punicae*	+	++	+++	++++	−	−	−	−

Notes: Colony growth: −, no growth; +, less growth; ++, moderate growth; +++, maximum growth.

4.3   BIOCHEMICAL VARIABILITY

The methodology followed for the following experiments on biochemical variability is as per Schaad (1992).

4.3.1   *LACTOSE UTILIZATION*

The carbon source (lactose) was filter-sterilized and mixed with autoclaved, cooled Dye's medium along with 1.2% purified agar. The pH was adjusted to 7.2. Bacterial isolates were spot-inoculated with the replica plating method and incubated at 30 °C for 3, 7, and 14 days. The growth was compared with control, where the carbon source was not supplemented (Schaad, 1992).

4.3.2   STARCH HYDROLYSIS

The medium used for starch hydrolysis was sterilized by autoclaving and poured into sterilized Petri plates. These plates were inoculated and incubated at 30 °C for 7 days. The plates were flooded with Lugol's iodine and allowed to act for a few minutes. The presence of starch hydrolysis was indicated by the presence of clear zones and vice versa. The hydrolyzed zone was measured for each isolate.

4.3.3   ACID PRODUCTION FROM SUCROSE, MALTOSE, AND DEXTROSE

Acid production by the isolate of *X. axonopodis* pv. *punicae* was tested by using peptone water medium of Dye. In total, 10 mL of medium was dispensed in each test tube. This medium was sterilized in an autoclave for 15 min. To these tubes, filter-sterilized carbohydrates, viz., sucrose, maltose, and dextrose were added at 0.14% concentration. The tubes were inoculated with 0.1 mL of 24-h-old bacterial culture and incubated at room temperature for three days. A change in the color of the medium confirmed acid production (Figure 4.3).

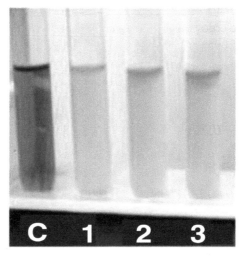

FIGURE 4.3   Acid production from sucrose, maltose, and dextrose (C, Control; 1, sucrose; 2, maltose; and 3, dextrose).

4.3.4   TESTING THE ANTIBACTERIAL ACTIVITY

After confirmation of *X. axonopodis* pv. *punicae*, the antibacterial activity of the plant extract was preliminarily screened by a well diffusion assay. Luria–Bertani (LB) agar plates were spread-plated with 20 µL of bacterial strain (1×10^8 cfu/mL). Wells of 6 mm diameter were made in the agar plates. Each plant extract was tested for antibacterial activity by adding 40 µL of extracts with different concentrations, viz., 50, 100, and 200 mg mL^{-1}. The experiment was repeated thrice. The plates were incubated at 37 °C for 24 h. Subsequently, the plates were examined for the zone of inhibition (ZOI) and the diameter was measured (in mm) after subtracting the well diameter (Ahmad et al., 1998). To determine the minimum inhibitory concentration (MIC), the required quantities of extracts were added to the LB broth of 4 mL to bring the initial concentration of 20 mg mL^{-1}. In each test tube, 0.1 mL of standardized inoculum (1×10^8 cfu/mL) was added. Two control tubes were maintained for each test batch, namely, extract control (tube containing plant extract and LB medium without inoculum) and organism control (tube containing LB medium and inoculum). The test tubes were incubated at 37 °C for 24 h. The lowest concentration (highest dilution) of plant extract that produced no visible growth (no turbidity) was recorded as MIC. The minimum bactericidal concentration (MBC) was assayed by subculturing test dilutions on a drug-free solid medium. The plates were incubated for 24 h at 37 °C. The lowest concentration of the antimicrobial at which no single colony observed after subculturing is regarded as MBC (Akinyemi et al., 2005).

4.4   RESULTS AND DISCUSSION

4.4.1   WELL DIFFUSION ASSAY

The preliminary screening of the selected three plant extracts against the *X. axonopodis* pv. *punicae* was done using a well diffusion method. The ZOI greater than 5 mm diameter is found to have significant activity against particular bacteria (Palombo and Semple, 2001). The aqueous extracts were sensitive to *X. axonopodis* pv. *punicae* tested at different concentrations (Table 4.3).

The extract of *A. arabica* was sensitive to *X. axonopodis* pv. *punicae* tested at different concentrations. The same extract was much effective at a low concentration of 50 mg mL^{-1}, that is, 11 mm ZOI. *P. juliflora* extract arrested the growth of *X. axonopodis* pv. *punicae* (24 mm) at 200 mg mL^{-1}.

The *P. juliflora* extract shows ZOIs of about 21 mm and 19 mm at 100 mg mL^{-1} and 50 mg mL^{-1}, respectively. Based on ZOIs, Satish et al. (1999) reported significant antibacterial activities *of P. juliflora* and *A. arabica* against *Xanthomonas campestris* pathovars. The extracts of *A. indicum* exhibited the antibacterial activity of 11 mm ZOI at 200 mg mL^{-1}.

TABLE 4.3 Antibacterial Activity of Plant Extracts Showing ZOI

Sr. No.	Plant Species	Conc. (mg mL^{-1})	ZOI (mm)*
1	*A. arabica*	50	11
		100	15
		200	16
2	*P. juliflora*	50	19
		100	21
		200	24
3	*A. indicum*	50	6
		100	9
		200	11

4.4.2 MINIMUM INHIBITORY AND BACTERICIDAL CONCENTRATIONS

The antibacterial activities of *A. arabica* and *P. juliflora* extracts are found to be high against all of the bacteria. The maximum activity was recorded by *P. juliflora* (MIC = 1.03 mg mL^{-1} and MBC = 0.15 mg mL^{-1}) and *A. Arabica* (MIC = 1.00372 mg mL^{-1} and MBC = 2.58 mg mL^{-1}) against *X. axonopodis* pv. *punicae*. Raghavendra et al. (2006) reported the significant antibacterial activity of *Acacia nilotica* extracts against *Xanthomonas* pathovars and human pathogenic bacteria tested, while the lowest activity was recorded by *A. indicum* (MIC = 0.619 mg mL^{-1} and MBC = 0.923 mg mL^{-1}). These findings indicate that the extracts of *P. juliflora* and *A. arabica* have the potential to be used in the management of plant diseases. Further phytochemical analysis is required to identify the active components of plant extracts showing antimicrobial activity (Table 4.4).

TABLE 4.4 MIC and MBC of Plant Extracts Against *X. axonopodis* pv. *Punicae*

Sr. No.	Plant Species	MIC (mg mL^{-1})	MBC (mg mL^{-1})
1	*A. arabica*	1.00372	2.58
2	*P. juliflora*	1.03	0.15
3	*A. indicum*	0.619	0.923

4.5 CONCLUSION

Many synthetic antibiotics are used to control several phytopathogens. The increased awareness of environmental problems with these chemical antibiotics has led to the search for nonconventional chemicals of biological origin for the management of these diseases. Bactericides of plant origin can be one approach toward disease management because of their ecofriendly nature (Bolkan and Reinert, 1994). The products of plant origin are of greater advantage to users, the public and radical environmentalists. Laboratory screening of plant extracts has given encouraging results, indicating their potential use in the management of diseases caused by *Xanthomonas* species. Plant extracts showing antibacterial activity have the potential to be used in the management of plant diseases as an alternative to chemical antibiotics. A further phytochemical analysis is required to identify the bioactive compounds responsible for antibacterial activity. Yet, the results obtained by these methods provide simple data, which makes it possible to classify extracts with respect to their antioxidant potential. As can be observed also from the present data, antibiotic activity does not necessarily correlate with high amounts of phytochemicals, and that is why both the phytochemical content and antibiotic activity information must be discussed when evaluating the phytochemical potential of extracts. On the basis of this study, the most potent Finnish plant sources for natural antibiotic agents are medicinal plants, vegetable peels, and different tree materials. Further work is underway to confirm the antibiotic effect of these promising plant extracts by using other types of bacterial strains and to characterize the active phytochemicals, their mechanisms of action, and their possible interactive effects together with the harmful effects to the environment.

KEYWORDS

- *Prosopis juliflora*
- minimum bactericidal concentration
- *Acacia arabica*
- synthetic antibiotics

REFERENCES

Akinyemi, K.O.; Oladapo, O.; Okwara, C.E.; Ibe, C.C.; Fasure, K.A. (2005) Screening of crude extracts of six medicinal plants used in southwest Nigerian unorthodox medicine for anti-methicillin resistant *Staphylococcus aureus* activity. *BMC Complement. Altern. Med.* 31, 252–268.

Bolkan, H.A.; Reinert, W.R. (1994) Developing and implementing IPM strategies to assist farmers, an industry approach. *Plant Dis.* 78, 545–550.

Bose, T.K. (1985) Pomegranate, In: *Fruit of India—Tropical and Subtropical*, Mitra, B. (Ed.). Naya Prakash Publications, Calcutta, pp. 637–640.

Bradbury, J.F. (1970) Isolation and preliminary study of bacteria from plants. *Rev. Plant Pathol.*, 49, 213–218.

Chand, R.; Kishun, R. (1991) Studies on bacterial blight (*Xanthomonas campestris* pv. *punicae*) on pomegranate. *Indian Phytopathol.* 44, 370–372.

Gour, H.N.; Ashiya, J.; Mali, B.L.; Ranjan N. (2000) Influence of temperature and pH on the growth and toxin production by *Xanthomonas axonopodis* pv. *vignicola* inciting leaf blight of cowpea. *J. Mycol. Plant Pathol.* 30, 389–392.

Hingorani, M.K.; Mehta, P.P. (1952) Bacterial leaf spot of pomegranate. *Indian Phytopathol.* 5, 55–56.

Hingorani, M.K.; Singh, N.J. (1959) *Xanthomonaspunicae*sp. Nov. on pomegranate (*Punica-granatum* L.). *Indian J. Agric. Sci.*, 29, 45–48.

Kishun, R. (1993) Bacterial diseases of fruits. In: Chadha, K.L.; Parek, O.P. (Eds.), *Advances in Horticulture*, vol. 3, Malhotra Publishing House, New Delhi, India, pp. 1389–1404.

Malhotra, N.K.; Khajuria, H.N.; Jawanda (1983) Studies on physio-chemical characters of pomegranate cultivars II. Chemical characters. *Punjab Hortic. J.* 23, 158–162.

Manjula, C.P. (2002) Studies on bacterial blight of pomegranate (*Punicagranatum* L.) caused by *Xanthomonas axonopodis* pv. *punicae.* M.Sc. (Agric.) Thesis, submitted to the University of Agricultural Sciences, Bangalore, 98.

Palombo, E.A.; Semple, S.J. (2001) Antibacterial activity of traditional Australian medicinal plants. *J. Ethanopharmacol.* 77, 151–157.

Petersen, Y.; Mansvelt, E.L.; Venter, E.; Langenhoven, W.E. (2010) Detection of *Xanthomonas axonopodis* pv. *punicae* causing bacterial blight on pomegranate in South Africa. *Australas. Plant Pathol.* 39, 544–546.

Raghavan, R. (2007) Oily spot of pomegranate in India (Maharashtra). Express India.

Raghavendra, M.P.; Satish, S.; Raveesha, K.A. (2006) *In vitro* evaluation of antibacterial spectrum and phytochemical analysis of *Acacia nilotica*. *J. Agric. Technol.* 2, 77–88.

Rani, U.; Verma, K.S.; Sharma, K.K. (2001). Pathogenic potential of *Xanthomonas axonopodis* pv. *punicae* and field response of different pomegranate cultivars. *Plant Dis. Res.* 16,198–202.

Rani, U.; Verma, K.S. (2002) Occurrence of black spot of pomegranate in Punjab. *Plant Dis. Res.* 17, 93–96.

Schaad, N.W. (1992) Laboratory Guide for the Identification of Plant Pathogenic Bacteria. 2nd ed. American Phytopathological Society, St. Paul, MN, USA, pp. 138–142.

Srivastava, U.C. (2008) Horticulture in India; a success story. *Crop Care*, 34, 23–32.

Vauterin, L.; Haste, B.; Kersters, K.; Swings, J. (1995) Reclassification of *Xanthomonas*. *Int. J. Syst. Evol. Microbiol.* 45, 475–489.

CHAPTER 5

Biocontaminants in Occupational Perspectives: Basic Concepts and Methods

DEBARSHI KAR MAHAPATRA[1*], and DEBASISH KAR MAHAPATRA[2]

[1]*Department of Pharmaceutical Chemistry, Dadasaheb Balpande College of Pharmacy, Nagpur 440037, Maharashtra, India*

[2]*Medical and Occupational Health Department, Western Coalfields Limited, Nagpur 440037, Maharashtra, India*

Corresponding author. E-mail: dkmbsp@gmail.com

ABSTRACT

Biological hazards are major occupational hazards concerned with several issues. These hazards are very dangerous and have to be taken seriously by employers and employees. The present chapter deals with the basic concepts, regulatory aspects, and methods related to the prevention of biocontaminants. The center of attraction of this chapter includes focusing on nosocomial infections, Legionnaires' disease, sick building syndrome, etc. The role of engineering, administrative, workplace hazardous material information, personal protective equipment, standard precautions, and ISO standards for avoiding risks to all types of biohazardous material are mentioned in a comprehensive manner.

5.1 INTRODUCTION

Biological hazards can enter the body in several ways. When determining the appropriate protective measures, measures are a clear understanding of how the hazards can enter the body (Stetzenbach et al., 2004). It may be through inhalation (breathing), ingestion (swallowing), injection (puncture), and absorption (through direct contact) (Górny et al., 2002). The risk is divided into four stages (Figure 5.1):

- BSL-1: Agents not known to consistently cause disease in a healthy adult human.
- BSL-2: Moderate potential threat to personnel and the environment, including bacteria and virus that cause only mild disease to the human.
- BSL-3: Microbes there can be either indigenous or exotic, and they can cause serious or potentially lethal disease through respiratory transmission, for example, HIV, H1N1 flu, rabies, SARS, ricketts, tuberculosis, etc.
- BSL-4: Dangerous and exotic, posing a high risk of aerosol-trans-mitted infections. Infections caused by these microbes are frequently fatal and without treatment or vaccines (Quinlan et al., 2010).

A product or a process may become contaminated at the various processes of manufacturing either by inanimate or animate matter. Here, we focus on any product that becomes contaminated by bacteria, viruses, or fungi. It can be the organism itself or its toxins that can be hazardous to human health if inhaled, swallowed, or absorbed into the body. The microorganisms such as bacteria, fungi, viruses, or their toxic products may contaminate through water, air, insects, soil, workers, and failure of good manufacturing process (GMP) (Frick et al., 2000). Several types of biocontaminants are found in the environment. The bacterial biocontaminant maybe Gram-positive or Gram-negative according to the laboratory method of staining by Gram stain. As different types of microbes grow in different environments, various microbial particulates can become a biocontaminant, which can become an indicator that can be monitored (Alli, 2008). Mycotoxins are toxic chemical products of a fungal metabolite that become biocontaminants as fungus affects the biological product being manufactured (Table 5.1).

5.2 SOME COMMON DISEASES CAUSED BY BIOCONTAMINANTS

5.2.1 *LEGIONNAIRES' DISEASE*

The American Legion organized its annual convention in a hotel in 1976. Three days after the convention ended and within a week, more than a hundred participants had been hospitalized due to cough, high fever, and breathlessness along with some other ailments, leading to atypical pneu-monia. About a quarter of them died. The Centers for Disease Control and

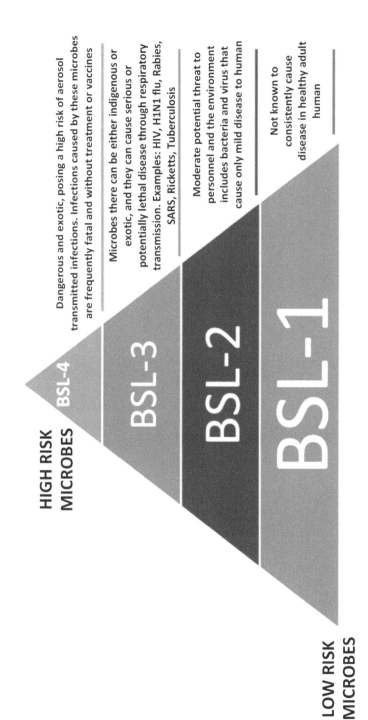

FIGURE 5.1 Levels of biocontaminant risks in nature.

TABLE 5.1 Characteristic Features and Sources of Common Biocontaminants

Living Sources	Units	Source Examples	Effect on Humans	Lifestyles	Principal Indoor Sources
Bacteria	Spore	Thermoactinomyces	Hypersensitivity pneumonitis	Saprophytes	Hot water sources
	Organism	Legionella	Pneumonia	Facultative parasites	Cooling towers
	Product	Proteases	Asthma	–	Industrial processes
		Endotoxin	Chills, fevers	–	Stagnant water reservoirs
Fungi	Spore	Alternaria	Rhinitis, asthma	Saprophytes	Outdoor air
	Organism	Sporobolomyces	Hypersensitivity pneumonitis	Saprophytes	Damp environmental surfaces
Protozoa	Antigen	Acanthamoeba	Hypersensitivity pneumonitis	–	Contaminated water reservoirs
	Organism	Naegleria	Infection	Facultative parasites	Contaminated water reservoirs
Algae	Organism	Chlorococus	Rhinitis, asthma	Autotrophic	Outdoor air
Spores	Antigen	Glycoprotein	Rhinitis, asthma	–	Outdoor air
	Volatile	Aldehyde	Mucous membrane irritants	–	Damp surfaces
	Toxin	Aflatoxin	Cancer	–	Damp surfaces
	Histoplasma	Systemic infection	Facultative parasites	Bird droppings	–
Virus	Organism	Influenza	Respiratory infection	Obligate parasites	Human host
Arthropod	Feces	Dermatophagoides		Phagotrophic	–
Plant	Pollen	Ambrosia	Rhinitis, asthma	Autotrophic	Outdoor air
Mammal	Saliva	Cat	Rhinitis, asthma	Phagotrophic	Cat
	Saliva scale	Horse	Rhinitis, asthma	Phagotrophic	Horse

Prevention investigated the outbreak from all aspects, including the hotel environment. In 1977, they identified *Legionella pneumophila*, a bacterium that was flourishing in the cooling tower of the air conditioning system of the hotel. The bacteria had spread through the duct system of the air conditioner throughout the hotel rooms. Those affected were the elderly with poor immune status who were inhaling the infected humid air. It normally does not spread through person to person contact (Rom and Markowitz, 2007). Water remained an imperative factor for the growth of microorganisms (Table 5.2).

TABLE 5.2 Water as an Imperative Source for Proliferating Biocontaminants

Outdoor contaminants	Swamp Coolers	Small Lakes	Fountains
	Poorly kept landscaping	Under house crawlspaces	Agricultural storage
	Air conditioners	Water softeners	Wooden structures
	Drainage ditches	Cooling towers	Compost piles
	Industrial HVAC exhaust	Wells or potable water storage containers	Sewer drains
Indoor contaminants	Improperly placed vapor barriers	Ventilation ducts and insulation	Water-damaged carpets
	Refrigerator pans	Humidifiers	Leaky appliances
	Leaky roofs	Subterranean rooms	Leaky plumbing
	Ceiling tiles	Air conditioners	Crawlspaces
	Bathroom shower and tub	Coldwater pipes	Ground-level cement slabs and walls
	Furniture	Wallpapers	Window coverings
	Standing water	Wet or damp materials	Hot water tanks
	Plants, animals, birds, and humans	Evaporative coolers	Pillows, bedding, and house dust

5.2.2 NOSOCOMIAL INFECTION

Nosocomial infection is also known as hospital-acquired infection (HAI). These infections develop in hospitalized patients who were neither suffering from that disease nor were they in the incubation period when admitted (Table 5.3). This infection has been acquired from a hospital or a healthcare facility. The source of such infection may be exogenous from hospital staff, other patients, or even inanimate objects like bed sheets, linen, contaminated equipment or endogenous from the patient's own microflora that may have been introduced by an operative or investigative procedure, instrumentation, or during nursing care (Gatchel and Schultz, 2012). The predisposing factors may be contaminated dressing, air, water, food, multidrug-resistant microflora, improperly screened blood and blood products, intravenous fluid, needle stick injury, droplet transmission, etc. The causative organisms may be Gram-positive or Gram-negative bacteria, viruses, fungi, or parasites. Preventive measures include proper sterilization, donor screening, isolation, proper handwashing, use of sterile gloves and gowns by staff, etc. (Schilling, 2013).

TABLE 5.3   Potential Biological Hazards

Biohazard	Disease	Spread	Control
Bacteria	Pink eye	Human-to-human contact	Do not share eye makeup, wash hands
Virus	Hepatitis A	Human-to-human contact	Do not ingest contaminated food or water, avoid direct contact with infected people
Virus	Hepatitis B	Human-to-human contact	Immunization, avoid direct contact with infected people, avoid tattooing and body piercing, dispose of sharps in disposal containers
Virus	Hepatitis C	Human-to-human contact	Avoid direct contact with infected people, avoid tattooing and body piercing, follow standard precautions
Virus	Measles	Human-to-human contact spread by cough and nasal droplets	Immunization, avoid direct contact with infected people

Biofilms are formed when a group of microbes, bacteria, fungus, and protista, adhere to each other on a living or nonliving surface that becomes protected by a layer of the self-generated matrix of extracellular polymeric substance made up of proteins, lipids, polysaccharides, and others. Bacterial biofilms are infectious and are a common cause of nosocomial infections. Biofilms may form on both living and nonliving surfaces (Levy, 2006). So, they may be seen in hospitals or industrial settings as well as in humans or animals, the commonest being the dental plaque affecting teeth and gum. Biofilms are biological systems that are developed by bacteria by organizing into functional community that symbiotically share nutrition, growth factors, and other chemicals and shelter themselves from harmful environmental factors, antibiotics, as well as the host's immune system, which are essential for their survival. Hence, because of such characteristics, biofilms become adherent to biomedical devices like catheters, prosthesis, ventilators, etc., causing HAIs, which are difficult to eradicate (Wilcock, 2006).

5.2.3 SICK BUILDING SYNDROME

The sick building syndrome describes a condition in which employees or residents experience nonspecific illnesses, like headache, nausea, throat irritation, difficulty in concentration, fatigue, allergies, etc., whose cause cannot be specifically identified (Waldron, 2013). This is often classified as an occupational disease (Table 5.4).

The factors suspected include the following:

i)   Biological contaminants like bacteria, viruses, fungi, pollen, etc. These contaminants flourish in stagnant water in humidifiers, ducts, drainpipes, insulations, upholstery, as well as insect droppings.
ii) Inadequate ventilation with little outdoor airflow.
iii) Chemicals from outside like vehicle exhaust, dust, paint, products of combustion, and volatile compounds like pesticides, cleaning agents, etc.
iv) Poor ergonomics, improper lighting, humidity, incorrect acoustics, etc.

The importance of this condition is due to the reduced work efficiency and increased absenteeism among workers, leading to a fall in productivity.

TABLE 5.4   Occupational Hazards

Occupation	Contact Source	Disease
Doctor, dentist, nurse, healthcare volunteer	Blood and other constituents of patients	Hepatitis, cold, flu
Childcare worker	Body fluid of children, diaper, cut	Cold, flu, eye infection, head lice
Food service	Undercooked food	*Salmonella, E. coli*
Waste disposal personnel	Broken glass, contaminated needle	HIV, hepatitis B, and hepatitis C

5.3   PREVENTIVE AND REGULATORY ROLES

5.3.1   ISO BIOCONTAMINATION STANDARDS

Part 1, General principles: Describes the principles and basic methodologies for a formal system to access and control biocontamination.

Part 2, Evaluation and interpretation of biocontamination data: Gives guidance on basic principles and methodological requirements for all microbiological data evaluation and the estimation of biocontamination data obtained from sampling for viable particles in zones at risk, as specified by the system selected (Slote, 1987).

5.3.2 CONTROL OF BIOLOGICAL HAZARDS

There are three approaches to control hazards. The first consideration for controlling biological hazards is to look at engineering controls. If a hazard cannot be eliminated through engineering processes, a second approach to controlling a hazard is administrative. If the exposure to a hazard cannot be prevented by either engineering or administrative controls, then prevention is necessary.

- *Engineering controls:* These controls are the first line of defense and include built-in protection in the building, work areas, equipment, or supplies. Examples include ventilation systems and construction seals to create negative pressure rooms and biosafety hoods with specific ventilation systems (Mosey, 1981) (Table 5.5).
- *Administrative controls:* These controls are steps in work procedures or work processes that minimize the risk of exposure to a hazard. This type of control does not eliminate a hazard but can significantly reduce the risk of injury. Examples include worker training, rules that require regular hand washing, etc. (Brune and Edling, 1989).
- *Personal protective equipment (PPE):* When a hazard poses a threat even after engineering and administrative controls have been implemented, PPE is necessary. Some PPE include proper masks, eye protection gears, and latex gloves for biohazards. For a PPE to be effective, it must be worn correctly and must be comfortable and fitted for each person. Workers must be trained properly so that it is worn when needed (Zenz, 1988).

TABLE 5.5 Biocontamination of Air in Pharmaceutical Areas

Grade	Maximum Number of Particle Permitted per m³		Maximum Number of Viable Microorganisms per m³
	0.5–5 µm	>5 µm	
A	3500	None	<1
B	3500	None	5
C	3,50,000	2000	100
D	3,500,000	20,000	500

5.3.3 ROLE OF THE HEALTH AND SAFETY REPRESENTATIVES

- The role includes proactive work to eliminate biological hazards from the workplace as much as educating the workers about occupational hazards.
- Elimination of hazards from the work environment is always the prime choice for disease prevention.
- When hazards cannot be eliminated completed, they must be controlled. Workers should press management to introduce effective controls in the workplace such as engineering control, administrative control, etc.
- Pay attention to new workers and visitors as they are newly exposed to a workplace and can tell you if they have health problems only when they come into the workplace (Karvonen et al., 1986).
- Development of a risk assessment system for the rapid identification and implementation of preventive measures (Figure 5.2).

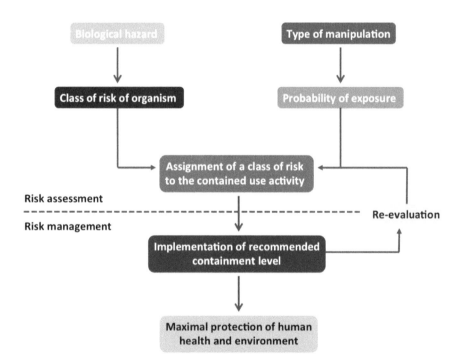

FIGURE 5.2 Overview of the biological risk assessment system.

5.4 CONCLUSION

Biological hazards cause health issues to humans and animals. These hazards are very dangerous and have to be taken seriously by employers and employees. There are several ways to reduce contaminants. Engineering, administrative, workplace hazardous material information, personal protective equipment, standard precautions, and ISO standards are necessary for avoiding risk to all types of biohazardous materials.

KEYWORDS

- biocontamination
- biological hazards
- pharmaceuticals
- control engineering controls
- administrative controls
- personal protective

REFERENCES

Alli, B. O. (2008). *Fundamental principles of occupational health and safety*, 2nd edition. International Labour Office, Geneva.

Brune, D. K., & Edling, C. (1989). *Occupational hazards in the health professions*. CRC Press, FL, USA.

Frick, K., Jensen, P. L., Quinlan, M., & Wilthagen, T. (2000). *Systematic occupational health and safety management: Perspectives on an international development*. Pergamon Press, Oxford.

Gatchel, R. J., & Schultz, I. Z. (Eds.). (2012). *Handbook of occupational health and wellness*. Springer Science & Business Media, Berlin.

Górny, R. L., Reponen, T., Willeke, K., Schmechel, D., Robine, E., Boissier, M., & Grinshpun, S. A. (2002). Fungal fragments as indoor air biocontaminants. *Appl. Environ. Microbiol.*, *68*(7), 3522–3531.

Karvonen, M., Mikheev, M. I., & World Health Organization. (1986). *Epidemiology of occupational health*. World Health Organization, Regional Office for Europe.

Levy, B. S. (Ed.). (2006). *Occupational and environmental health: recognizing and preventing disease and injury*. Lippincott Williams & Wilkins, Philadelphia.

Mosey, A. C. (1981). *Occupational therapy: Configuration of a profession* (vol. 63, pp. 67–69). Raven Press, New York.

Quinlan, M., Bohle, P., & Lamm, F. (2010). *Managing occupational health and safety*. Palgrave Macmillan, New York.

Rom, W. N., & Markowitz, S. B. (Eds.). (2007). *Environmental and occupational medicine*. Lippincott Williams & Wilkins, Philadelphia.

Schilling, R. S. F. (Ed.). (2013). *Occupational health practice*. Butterworth-Heinemann, Massachusetts.

Slote, L. (1987). *Handbook of occupational safety and health*, Wiley Interscience, New Jersey.

Stetzenbach, L. D., Buttner, M. P., & Cruz, P. (2004). Detection and enumeration of airborne biocontaminants. *Curr. Opin. Biotechnol. 15*(3), 170–174.

Waldron, H. A. (2013). *Occupational health practice*. Butterworth-Heinemann, Massachusetts.

Wilcock, A. A. (2006). *An occupational perspective of health*. Slack Incorporated, New Jersey.

Zenz, C. (1988). *Occupational medicine: principles and practical applications*. Year Book Medical Publishers, Missouri.

CHAPTER 6

General and Chemical Perspectives and Studies on Tannins as Natural Phenolic Compounds for Some Ecoefficient Applications

HUSSEIN ALI SHNAWA[1*], MOAYAD NAEEM KHALAF[2],
ABED ALAMER HUSSEIN TAOBI[2], BINDU PANAMPILLY[3], and
SABU THOMAS[*3]

[1]*Polymer Research Center, University of Basrah, Basrah, Iraq*

[2]*Chemistry Department, College of Science, University of Basrah, Basrah, Iraq*

[3]*School of Chemical Sciences, Mahatma Gandhi University, Priyadarshini Hills P.O., Kottayam 686560, Kerala, India*

[*]*Corresponding author. E-mail: hussanqi@yahoo.com*

ABSTRACT

Tannins belong to the phenolic classes of secondary plant metabolites and are widely distributed in nature. Their chemical structures may vary greatly consisting of simple, oligomeric, and polymeric compounds that have multiple structure units with multiple free phenolic hydroxyl groups. From the chemistry point of view, this review discusses and presents general definitions, classifications, chemical structures, significance, and some ecoefficient applications of tannins as well as the building units of tannins, which are mainly gallic and ellagic acids of hydrolyzable tannins, the first type of tannins. Also, in contrast, procyanidins and profisetinidins are converted into the second one, which is named as condensed tannins. Tannins as well as other natural phenolic products have found ascending interest in many scientific areas, and recently, more investigations and projects are carried out that are especially related to the pharmaceutical effects, environment, dyes, food industries, green adhesives, natural antioxidants, anti-inflammatory and antibacterial agents, cosmetics, biobased polymers additives, etc.

6.1 INTRODUCTION

6.1.1 NATURAL PHENOLIC COMPOUNDS

Naturally occurring phenolic compounds are a large and heterogeneous group of plant secondary metabolites that are distributed throughout the plant kingdom. They have been implicated in various bioactivities such as protective roles in plants and recently in human diet. The three most important plant phenolics are flavonoids such as anthocyanins (as plant pigments), phenolic acids such as coumaric and salicylic acids, and stilbenes such as resveratrol in red grapes. All of these phenolic products can be found and used as natural polyphenolic or oligophenolic molecules (Maestri et al., 2006).

Essentially, many polyphenols were found to have multiple functions mainly for the protection of plants from environmental hazards such as pollution, photo-oxidation, disease resistance, and pathogenic attacks and they also provide pigments and flavors. Furthermore, natural phenolic compounds have been found interesting in many scientific areas, and more investigations and scientific projects are carried out in these directions, especially that are related to pharmaceutical effects, environment, dyes, food industries, adhesives, antioxidants, antiinflammatory and antibacterial agents, cosmetics, etc. (Mamta et al., 2013; Jin and Russell, 2010).

6.1.2 TANNINS: DEFINITIONS, STRUCTURES, AND CLASSIFICATIONS

Tannins, also known as plant polyphenols, are a group of natural polyphenols that are found in numerous different plants, often in high concentrations in the outer bark of hardwood. Basically, they have been known and utilized by man for centuries even without full knowledge of their chemical structures and physical and chemical properties. Their name is derived from the French word "tan or tannin," meaning the oak bark and other trees used in leather tanning; in other words, this name comes from the historical practice of extracting compounds from oak bark to tan leathers. This denomination is used for a range of natural phenolic compounds with a variety of molecular weights and complexities (Mohamed and Alessandro, 2008; Horvath, 1981).

The most persuasive general descript of the tannins was the one proposed by Horvath (1981): the tannins are defined as "any phenolic compound of sufficiently high molecular weight containing sufficient phenolic hydroxyls and other suitable chemical groups (i.e., carboxyl) to form effectively strong complexes with proteins or other macromolecules under the particular environmental conditions being studied." Comparable to other natural phenolic products, tannins are extremely found in the plant tissues and pertain to the phenolic molecules of secondary plant metabolites. With multiunits or polymeric structures, tannins having multiple building blocks with plentiful hydroxyl groups. According to the Bate-Smith definition, additing to the previous definition, tannins can be explained as "water-soluble phenolic compounds having molecular weight between 500 and 3000 Da, and besides giving the usual phenolic reactions, they have special properties such as the ability to precipitate alkaloids, gelatins and other proteins" (Versari et al., 2013; Bate-Smith and Swein, 1962).

The first classification of tannins depends on their chemical structure, which was made in the 1920s by Freudenberg, who split it into two types: hydrolyzable tannins (HTs) and condensed tannins (CTs); the first type received their name from the fact that they can be hydrolyzed in acidic or basic medium to sugar units, simple phenols, and their carboxylic acids as other components, whereas the second class of tannins has been referred to as proanthocyanidins in many scientific journals related to healthy aspects. They obtained their name from the fact that in an acidic medium they break up to form anthocyanidins, which appear as red compounds or pigments (Nonhlanhla, 2011).

Some works such as Takuo and Hideyuki (2011) and Mamta et al. (2013) divided tannins according to their final chemical structures and characteristics into four major groups: gallotannins, ellagitannins, complex tannins, and CTs, as mentioned in Figure 6.1.

1. *Gallotannins* belong to HTs and include all those tannins in which galloyl units or their *meta*-depsidic derivatives are bound to diverse polyol-, catechin-, or tri-terpenoid units. Generally, they are made from polyester of gallic acid with sugar (Mamta et al., 2013).

2. *Ellagitannins* are one class of HTs in which at least two galloyl units are C–C coupled to each other and do not contain glycosidically linked catechin units.

3. *Complex tannins*, such as camellia tannins, have a catechin unit bound glycosidically to agallotannin or ellagitannin units. In other words, the structures of the complex tannins are made of a gallotannin unit or an ellagitannin unit and a catechin unit, designated as number 3 in Figure 6.1.

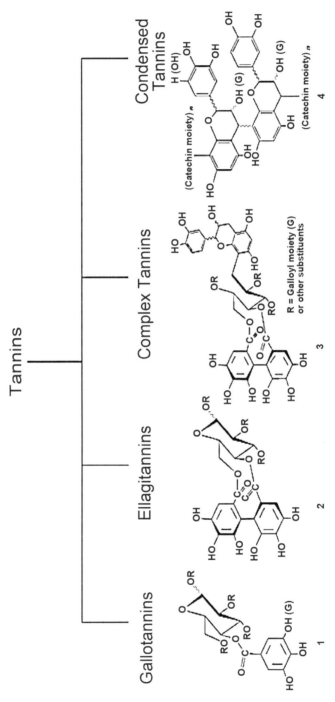

FIGURE 6.1 Chemical structures and main classes of tannins.

4. *CTs* are referred to as polyflavanols or proanthocynidines; they are all oligomeric and polymeric proanthocyanidins formed by linking catechin (or epicatechin) with the other monomeric catechins (or epicatechins) units.

"The chemical complexity and heterogeneity of plant tannins means that they do not lend themselves to ready quantitative assessment, and this has produced a confused picture of their real significance—both evolutionary and ecological…" (Haslam, 1988).

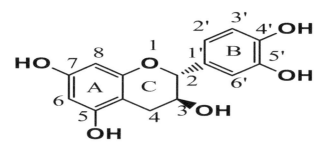

FIGURE 6.2   Basic building block of flavonoids and flavan-3-ol unit.

6.2   HYDROLYZABLE TANNINS

Based on the chemical structure and susceptibility to acid hydrolysis, HTs, one of the main groups of tannins, are a mixture of simple phenols, such as pyrogallol and ellagic acid, and esters of sugars with gallic acid and digallic acid. The key feature of HTs is a formation with a carbohydrate core, in most cases, glucose, whose hydroxyl groups are partially or totally esterified with phenolic groups of gallic acid or ellagic acid (hexahydroxydiphenic acid). The common examples of HTs include theaflavins (from tea) and daidzein.

The HTs can be considered as polyesters derived from glucose, which can be categorized into (1) gallotannins, which release gallic acid and its derivatives when submitted to acid hydrolysis, and (2) ellagitannins, which upon hydrolysis release ellagic acid and valonic acid (Bossu, 2006; Karamali and Teunis, 2011). HTs are readily hydrolyzed by acids and bases (or enzymes) into sugar or related polyhydric alcohol and phenolic carboxylic acid. Through the process of hydrolysis, the HTs release phenolic acid; if the acid component is gallic acid, then the tannins are called gallotannins (commercially called tannic acid), and if the acid is ellagic acid, then the tannins are called ellagitannins (Figure 6.3) (Mark, 2010).

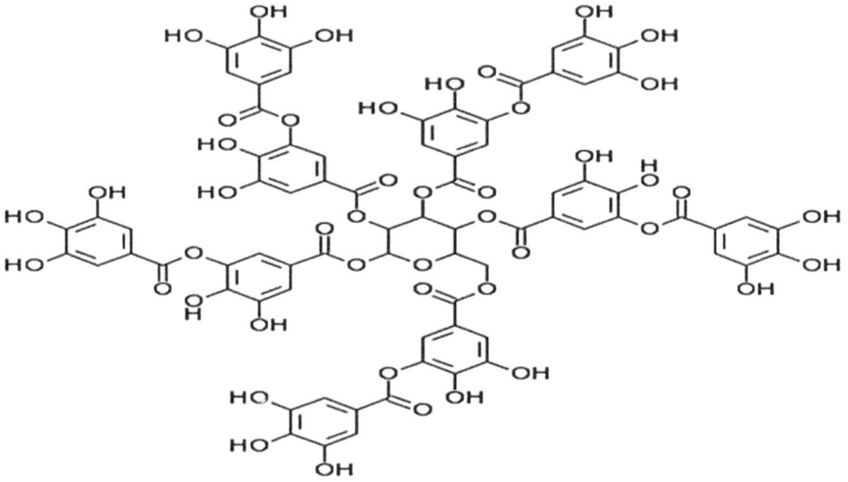

FIGURE 6.3 Hydrolyzable tannins.

Informally, HTs are divided into two types, the first one being those composed of mixtures of oligomeric or simple phenols, such as gallic acid and ellagic acid, and the second consists of esters of sugars (mainly glucose) with gallic acid and digallic acid; the more complex structures of the latter may contain ellagic acid, and the two types are known as gallotannins and ellagitannins, respectively. In other words, HTs are further subdivided into two subclasses: gallotannins and ellagitannins (Ascacio-Valdés, 2011; Ramakrishnan and Krishnan, 1994).

6.2.1 GALLOTANNINS

Esters of gallic acid and sugars, gallotannins, also commercially called tannic acid, are considered to be the simplest hydrolyzable type since they yield gallic acid on hydrolysis. The typical sample for gallotannins is pentagallolyl glucose or gallic acid, which has five identical gallolyl groups esterified with a glucose core. It has been reported that the general physical and chemical properties of HTs and gallotannins, as an example, are bonding with proteins and gelatin, bitter flavor, blue–black complex formation when treated with ferric salts, and forming leather from animal skins. They are soluble in water, alcohol, and ethyl acetate but insoluble in other organic solvents such as ethers and carbon disulfide. They can be converted into acetyl derivative, a crystalline derivative that has a melting point of 137 °C when reacted with acetic anhydride (Ramakrishnan and Krishnan, 1994; Stephane and Ken, 1996).

6.2.2 ELLAGITANNINS

As previously demonstrated, ellagitannins (esters of hexahydroxy diphenic (ellagic) acid and sugars) are characterized by the monomers building ellagitannins, which are hexahydroxyl diphenyl (from hexa hydroxyl diphenic acid HHDP) ester groups with sugars, besides polygalloyl esters. The hydrolytic release of HHDP ester groups leads to their facile and unavoidable conversion by transesterification into ellagic acid, based on which these natural products are named. Ellagitannins yield ellagic acid on hydrolysis, in addition to other phenolic compounds, namely, chebulic acid, chloroellagic acid, etc., which have the phenolic chemical structures derived from ellagic acid.

The oligomerization of ellagitannin molecules is presumed to arise primarily from oxidative C–O coupling between galloyl and hexahydroxy-diphenoyl moieties of appropriate monomeric precursors. Over about 150 or recently more structurally characterized dimeric to tetrameric ellagitannins have been isolated and classified according to the type of monomeric fragments involved and the regiochemistry of the attachment process (Stephane and Ken, 1996).

6.3 CONDENSED TANNINS

In the same manner, CTs are complex structures consisting of oligomers and polymers of flavonoid units (flavan-3-ol and flavan-3,4-diol); these flavonoid monomer units are repeated 2–11 or 1–30 times or more (i.e., 2–50 or more; Andersen and Markham, 2006) to create the structure of CTs such as mimosa and quebracho tannins. Comparatively, they usually have molecular weight higher than HTs and they do not contain a sugar moiety (Jaganathi and Alan, 2010). The monomer units that make up CTs are distinguished by the number of –OH groups on the B-ring, the stereochemistry of C2, C3, and C4 of the central ring, and the distinct location and stereochemistry of the interflavanoid linkages; for example, procyanidins possess a di-hydroxy B ring, while prodelphinidins possess a tri-hydroxy B ring.

6.4   PHYSICAL AND CHEMICAL PROPERTIES OF TANNINS

Generally, tannins from different plants species have different properties and reactivities. However, CTs and HTs have some native physical and chemical properties, such as they have a three-dimensional structure and noncrystallized amorphous and nonnitrogenous plants products. But, in particular, tannins have astringent and bitter taste properties and some of them can be dissolved in water, especially the HTs. Their aqueous solutions show an acidic reaction and a sharp astringent taste. Most of their compounds cause precipitation of proteins, alkaloids, glycosides, gelatin, and heavy metal salts (Frutos et al., 2004; Takuo and Hideyuki, 2011).

Multiple phenolic hydroxyl tannins can create many hydrogen bonding with carbonyl groups of some macromolecules such as proteins. Figure 6.4 shows the sum of the general chemical reactivity of CTs. Nucleophilic substitution reactions occur at the hydroxyl groups introducing either acyl or alkyl groups. Acylation is commonly achieved by reaction with acid chlorides or anhydrides, while the alkylation reaction is commonly achieved through reaction with alkyl halides or epoxides (Warren et al., 2013; Helene et al., 2011).

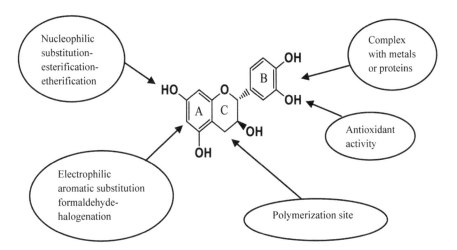

Tannins chemistry for some eco-efficient applications

FIGURE 6.4   Chemistry of tannins.

The chemical reactivity of tannins within the electrophilic reaction is high because of high hydroxyl group content. This functional group is a relativity strong electron donor in basic media, whereas phenoxide salts are formed during this reaction. Two examples of electrophilic reactions are halogenations and the reaction with aldehydes; the site of reaction with aldehydes is the A ring and only the B ring in very alkaline conditions. The nucleophilic center on the A rings of flavonoid units tends to be more reactive than that found on the B rings. This reaction is well studied due to its applications in adhesives systems (Pizzi, 2003).

6.5 SOURCES AND SIGNIFICANCE OF TANNINS

Essentially, the sources of tannins are numerous, as there are a large number of trees and shrubs that contain tannins. There are many species rich in tannins, which contain both HTs and CTs. Notable for either their present or past economic and industrial importance, tannins resources are black-wattle or black mimosa bark, quebracho wood, oak bark, chestnut wood, mangrove, eucalyptus bark, tara, barks of several pines, and many more. Moreover, tannins can range between 2% and 40% of the dry bark of many trees species; such values may render the product economically viable for industrial utilization. Tannins are most commonly obtained from the wood, bark, and leaves of different plants; the bark of *Picea abies* can contain up to 15% tannins, some mangrove species can possess up to 35% tannins, and the tannins content varied between 30% and 35% in the case of the quebracho tree (Michel et al., 2011).

Tannins from the bark of the black-wattle and quebracho trees are considered to be the most important sources for CTs when produced at industrial scales. In Brazil, works involved with tannins, including wood gluing, were initiated in the late 1970s and the early 1980s. The authors in the previous reference (i.e., Michel et al., 2011) reported that, according to *Sociedade Brasileira de Silvicultura*, SBS, approximately 3000 companies in Brazil use pine wood in their production processes, concentrating on sawn wood, cellulose, papers, and panels (plywood, MDF, and OSB); pine and other trees barks are a residue for the majority of the wood industries, presenting an environmental and economic challenges upon disposal.

The international markets for tannin extracts of vegetable origin produce about 160,000 tons per annum (TPA), including approximately 100,000 TPA originating from acacia plantation and the remainder is sourced from other species, such as castanheira (Italy), quebracho (Argentina), and tara (Peruvian) (Michel et al., 2011; Antwi-Boasiako and Animapauh, 2012). Practically, tannins are extractable from wood and bark by soaking or percolating in water or in some aqueous solutions or by organic solvent extraction. Researchers have extracted tannins from the bark of *Pinus radiate* and *Acacia*, leaves of sumac, seeds of pigeon pea, flowers and leaves of *Crataegus*, membrane of pecan nut, bark of mimosa, and wood of quebracho.

In Malaysia, tannins have been extracted from *Phyllanthus niruri* or known as "dukung anak" and from mangrove bark of *Rhizophora opiculata*. The bark of mangrove is a waste product from the charcoal industry in Malaysia. By using the solid–liquid extraction process with about 70% aqueous acetone solution, 34.68 wt% of tannins can be isolated from the mangrove bark; therefore, the barks of these trees can be considered a potentially affordable and abundantly available source of tannins (Raquez et al., 2010; Chuan Wei, 2008).

Tannins are extracted using different solvents such as methanol, ethanol, water, ethers, NaOH, sodium carbonate or bisulfite solutions, and other alkali. Many techniques such as water extraction with the addition of chemicals, solvent extraction, and fractionation extraction with ultrafiltration have also been developed in an attempt to increase the yield of bark tannins and improve their quality for wood adhesives. However, only the water extraction processes, with or without the addition of chemicals, is relatively economically feasible (Antwi-Boasiako and Animapauh, 2012).

The objectives of study, which is carried out by Michel et al. (2011), were to evaluate the tannin extractions from the bark of *Pinus occarpa* by using water and water solutions containing sodium carbonate and sodium bisulfate; the results of that study have revealed that it is possible to obtain high yield (~39%) of tannins from the bark by using extraction when coupled with the addition of sodium salts. Generally, the solid–liquid extraction is still the most common procedure to extract and analyze the phenolic fraction of plant sources due to its simplicity, high efficiency, and easy tunability. Eucalyptus bark is separated as a waste product from eucalyptus wood when used as a raw material, mainly to produce cellulose pulp and secondary to produce panels and boards, and is also used as a fuel. So, this bark represents an important and industrial resource for tannin production at industrial scales (Onifade, 2001).

6.6   TANNIN ACTIVITY AND APPLICATIONS

We have seen that the extracted tannins have several commercial applications, including leather tanning, adhesive additives, dispersants, asphalt additives, and pharmaceuticals. Leather tanning is the traditional and oldest application field of tannins; tannins interact with and precipitate proteins to preserve the hide, hence forming leather. During the past decades, the explosive growth of biotechnology and green chemistry has been extended to chemical industries and organic polymers chemistry, which leads, in nature, to rapidly increase the awareness of the need for green chemistry in chemical industries and to create new polymeric materials with improved performance.

For example, phenolic resins retain industrial and commercial interest a century after their introduction.

Because of innovative research, new phenolic products and applications continue to emerge, demonstrating their versatility and potential to cope with the ever-changing requirements and challenges of advanced technology (Raquez et al., 2010). Among different renewable resources, tannins, due to their similar structure to that of phenols, have been considered as a natural and renewable raw material in the design of phenolic resins and are also considered as an excellent renewable resource for replacing petroleum-derived phenolic compounds. However, the major applications of tannins and their derivatives have been manly as adhesives, mostly wood adhesives (Lan et al., 2012; Frederique et al., 2012). The reactivity of tannins is due to their A-ring phloroglucinole or resorcinol nuclei, determining the reactivity toward formaldehyde and consequently their potential to cross-link, as shown in Figure 6.5. Simple phenols and tannins can react with formaldehyde to produce prepolymers or resins that have the ability to polymerize through methylene bridge linkages at reactive positions of the flavonoid molecules mainly at A rings. The free C6 or C8 sites on the A ring react with formaldehyde, due to their strong nucleophilicity, to form adhesives.

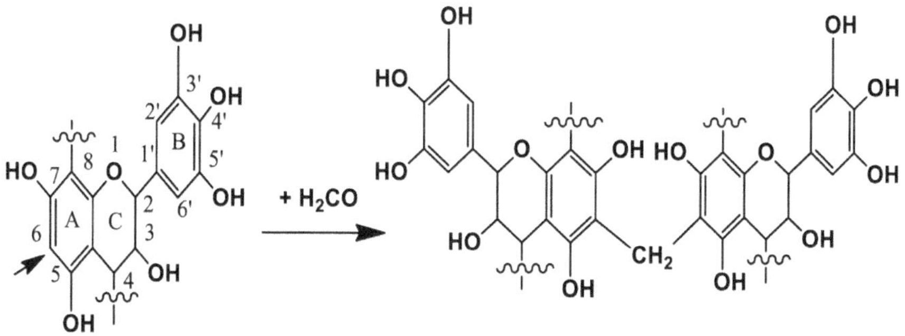

FIGURE 6.5 Reaction of a tannin with formaldehyde.

Tannins also can react with formaldehyde via phenolic nuclei. Their A-ring phloroglucinolic or resorcinolic nuclei may be main reason for this reactivity. However, tannins can react with formaldehyde about 10 to 50 times faster than that with typical phenols. Because of this characteristic, low formaldehyde emission and more environmentally friendly benefits were obtained from tannin-based adhesives in comparison with other wood adhesives, such as phenol-formaldehyde (PF), urea-formaldehyde (UF), etc. As a result, for example, tannins- based adhesives production rate will enhanced in comparison to those of the PF (Pizzi and Tekely, 1995; Li et al., 2016).

The modification of PF with a high tannin content enables the recovery of resins with almost the same technical specifications as those of petroleum-based Novolac resins. Wood-based composites and solid wood are often bonded with synthetic adhesives that contain formaldehyde or other chemicals that are harmful to the human health and environment; formaldehyde-based resins including UF, PF, and melamine-urea-formaldehydee (MUF) are the three most commonly used binders in wood composite industries. In the past decades, several attempts have been made to develop environmentally friendly adhesives from natural materials or to reduce or replace formaldehyde content in adhesive formulations (Sumin and Hyun-Joong, 2003; Terry and George, 2004).

Although the behavior of HTs is analogous to that of simple phenols and can easily substitute phenol in PF resins as an extender, it has numerous disadvantages such as high viscosity, low strength, and poor water resistance. The need for modification and the limited worldwide production decrease their chemical and economic interest in this field due to their low reactivity toward formaldehyde or other electrophilic compounds needed for the network formation. For these reasons, there is not much information in the literature about HTs and their applications as adhesives (Terry and George, 2004; Pizzi, 2006). HTs have many applications in the fields of medicines and coating, but limited information is available in the literature that deals with their uses in adhesive industeries. Hills (1979) point out in his study that the HTs can be used as additives in phenolic resin technologies.

A key feature in tannins is that they are phenolic in nature and undergo the same well-known reactions of phenols with formaldehyde that are either base- or acid-catalyzed, mainly weak base-catalyzed for industrial applications. However, being 30–50 times faster than phenol, formaldehyde reacts with tannins to produce methylene bridge linkages mainly with the A rings of flavonoids, having reactivity compared to or slightly lower than that of resorcinol. Furfuryl aldehyde is also found to be a very good cross-linker and plasticizer when coupled with formaldehyde (Pizzi and Scharfetter, 2003).

Tannins can also react with glyoxal and benzaldehyde since the tannin molecules are generally large; the rate of molecular growth is high so that the tannin adhesives tend to have short pot lives. Glutaraldehyde and furan have been shown to react with tannins to produce a slow-forming precipitate, whereas the precipitates with formaldehyde form much faster. Generally, most tannin-based adhesives are fortified with a synthetic polymer system, such as commercial UF, commercial PF, resorcinol-formaldehyde, and isocyanate (Pizzi and Scharfetter, 2003; Electra and Zoi, 2011). Copolymerization of urea, formaldehyde, and tannins is reported at pH 5.5–6.5, where the probability of copolymerization and self-polymerization could occur in a ratio of 50:50. However, the composition applied to industries is in fact mixtures of UF resin with tannin solution rather than copolymers (Electra & Zoi, 2011).

Vazquez et al. (2005) studied the curing of polymers and adhesives prepared from the copolymerization of 5–17% pine bark tannins with phenol-urea-formaldehyde (PUF) prepolymers. Differential scanning calorimetry (DSC) and dynamic mechanical analysis (DMA) techniques have been used to study the curing processes and overall. The author reports that the DSC curves of the PUF prepolymers show an exothermic peak between 146 and 149 °C and a second peak at higher temperatures (186–196) °C. Both peaks shift to lower temperatures in the PUFT adhesives. The enthalpy of curing also decreases when increasing the tannin content in the PUFT adhesives. Also, the gel time decreased significantly when the adhesive tannin content was increased from 10% to 17%. However, the maximum rigidity, corresponding to the finishing point of the resin curing reaction, decreased. The data of this study show that these resins allow for shorter hot press time and therefore increase productivity. This is a great advantage from the industrial point of view, as per the author's statement.

Herbert (1989) reported that about 10 KT of tannins (CTs) from acacia trees barks extracts were utilized as adhesives in comparison to 300–400 KT of phenol consumed for the same purpose. Herbert further stated that 30–50% tannin replacement of the resin solids in amino resins (UF and melamine-formaldehyde) and phenolic resins (PF and resorcinol-formaldehyde) has been formulated into wood adhesives. The aims of Amine et al. (2013, 2010) in their studies about tannins adhesives were to replace partially or totally the petrochemical-based adhesives such as UF, PF, and melamine-formadehyde-UF, which can generate under certain conditions and have human health and environmental hazards, with environmentally friendly adhesives derived from corn starch and tannins; their investigation on the use of these materials to prepare natural wood adhesives shown that particle board panels bonded with these adhesives showed comparable mechanical properties to the panels made with the commercial UF resins.

Since tannins are natural polymers with more than two hydroxyl groups per molecule, they can be used as polyol for rigid PU preparation, which opens a new and another important application for tannin activity. Recently, broad studies show that tannins can be used for the preparation of various PU and other polymer resins. Essentially, a major development in the use of tannins as macromolecular materials has taken place and increased in the last few years. CTs were found to cross-link in polycondensation reactions with furfuryl alcohol and small amount of formaldehyde, thus making these compositions almost totally based on renewable resources, when these reactions were conducted in the presence of volatile additives, and as a result foaming took place giving rice to rigid cellular materials. These foams displayed remarkable properties in terms of insulation, fire resistance, chemical inertness, and metal ion sequestration (Tondi et al., 2014; Basso et al., 2011).

Further works on these remarkable materials involved synthesis of a new kind of polyurethane foam from wattle tannin, by the work of Jinjie et al. (2003), which has good prospects of applications for its biodegradation and antimicrobial properties, and it was found by this study that wattle tannins could contribute not only to bacteriostatic activity but also to biodegradability during soil microorganism treatment for polyurethane foams. Moreover, bioresourced materials are much more acceptable from the society point of view than synthetic ones, especially for in-house applications.

Basso et al. (2011) clearly showed the possibility of the prepared green and formaldehyde free rigid foams for thermal insulation based on renewable chemicals: tannins and furfuryl alcohol are very easy to produce and are totally stable, have low density, low thermal conductivity, low hydrophilicity, and have an expected interest for the thermal insulation of buildings. Polyphenols, and more specifically CTs, can be an alternative to bisphenol-A to produce epoxy prepolymers. Helene et al. (2011) synthesized bio-based epoxy prepolymers from catechin, a repeating unit in tannins.

Natural phenolic compounds can provide important complex building blocks; gallic acid is one of the main components of HTs and can be used as renewable monomers for thermosetting applications. Aouf et al. (2013) and Benyahya et al. (2014) described the production of bio-based epoxy resin by using novel glycidyl derivatives of gallic acid, pyrogallol and green tea tannins; the chemical structures of these epoxy products were determined by electrospray ionization-mass spectrometry analyses, which demonstrated the feasibility of formulating epoxy thermosets based completely on natural phenolic products. In their conclusions, Benyahya et al. (2014) show that the functionalization of the natural green tea tannin extract with epichloro-hydrin led to the corresponding epoxy prepolymers with an epoxy equiva-lent weight of 7.2 mmol/g. From DSC and thermogravimetric analysis, it clearly appears that this material exhibits a high cross-linking density with isophorone diamine curing agent in a 1:1 molar ratio of epoxy group. It have totally insoluble and high thermal stability properties. In the last decades, the research studies have been focused on low environment impact, recyclable and eco-sustainable products, natural substances obtained from wastes, or by products of agro-food industry, which have attracted particular attention.

Grigsby et al. (2013) evaluated the potential of esterified tannins by using anhydrides as a route to synthesize tannin esters, possessing varying ester chain lengths and degrees of substitution, and then used them as plastic additives in poly(lactic acid) (PLA), and the researchers reported that the esterification of tannins by acid anhydrides is a preferred synthetic route to enhancing tannin hydrophobicity and that tannin derivatives exhibited melt behaviors suitable for proceeding in plastic and also the tannin ester deriva-tives may act to reduce PLA stiffness as well as confer a protection role to it. On the whole, the results of this study suggest the scope for the use of tannin esters as plastic additives. In another subsequent paper, Grigsby et al. (2014) synthesized tannin hexanoate and tannin hexanoate acetate. Esterified and native CTs have been evaluated in polypropylene and aliphatic polyester (Bionolle). While the presence of tannin additives contributes color to the plastics, their presence can provide beneficial functionality to plastics, such as antioxidant activity and ultraviolet (UV) stabilization.

A series of tannin derivatives were assessed as plastic additives in polypropylene and aliphatic polyester Bionolle and evaluated for their influence on mechanical, thermal, and aging characteristics of plastics. The results show that the condensed tannin ester additives containing hexanoate groups can be compounded into either polypropylene or aliphatic polyester up to 10% w/w content without impacting plastic processing. Both tannin esters (tannin hexanoate and tannin hexanoate acetate) act to stabilize the plastics during artificial aging by inhibiting UV degradation. Oxidative induction testing also suggests that tannin esters can act to prevent the oxidative degradation of polypropylene. Overall, this study has shown that bio-derived CTs resins can be a functional equivalent of synthetic polymer additives and have the potential to improve plastic sustainability credentials (Grigsby et al. 2014).

Heavy metal ion removal from aqueous solutions can be achieved by adsorption onto vegetable waste; absorption technology has recently become a real alternative to traditional waste water treatment due to its relative simplicity and high efficiency. Moreover, a larege number of low-cost adsorbents for wastewater treatment can be found; one class of such materials is tannins. Tannins exhibit specific affinity to metals ions and have the potential to be used as an effective and efficient alternative absorbent for the recovery of metal ions. The ability of tannins to complex with metal ions is due to orthohydroxyls present in the B rings, and it was found that ion exchange was the main adsorption mechanism as metal cations displace the adjacent phenolic protons forming chelate structures (Mehmet and Şükrü, 2006). Mahmut et al. (2006) carried out the polymerization reactions of formaldehyde with mimosa and valonia tannins to synthesize two resins and then used them as adsorbents for the removal of various metal ions from aqueous solutions. The conclusion of that study suggests that the two resins are highly stable and may be used for the removal of metal ions at room temperature.

The aims of the work carried out by Graciela et al. (2003) were to evaluate pine bark tannins, chemically modified with an acidified formaldehyde solution, as an absorbent to remove metal ions from aqueous solutions and acidic mine copper waste water, and the results show that the modified tannins absorb smaller amounts of metal ions than the modified bark. Clearly, tannins demonstrate specific affinity to metal ions and have the potential to be used as an effective and efficient alternative adsorbent for metal ion absorption and water purification from heavy metal ions. Since tannins are water soluble, this property restricts their practical applications in this field. Therefore, many approaches have been reported to immobilize tannins onto various matrixes to solve the water-soluble problem of tannins.

Tannin gel including sodium ions was made by Mitsubishi Nuclear Fuel Co. Ltd. and investigated by Yoshitake et al. (2004) as an adsorbent for heavy metal ions, such as lead, cadmium, and copper ions, and it was found that the gel shows high adsorption capacity for these ions. The results indicate that the gel can capture a few heavy metal ions selectivity from their mixture solutions, and the conclusion of their investigation points that the adsorption mechanism was based on ion-exchange reactions and the use of tannins gel as an adsorbent for heavy metal ions is beneficial.

In the project of Laurent et al. (2013), phenolic synthons (stable phenolic monomers) were produced by the depolymerixation process of CTs, which were isolated from industrial grape seed extracts at the multigrams scale by using ethanol solution consisting of 0.2 M HCl and other chemical reagents as extraction media. In the conclusions of their study, the authors show that the depolymerization of industrial tannins was obtained with an overall yield of 52% w/w of released monomers, and the residual part consists of unknown cleavage resistance structures; this method provide cheap access to bio-based phenolic monomers. The experimental results of the research work carried out by Ndazi et al. (2006) on the production of composite boards from a mixture of rice husks and wattle tannin-based resins has shown that rice husk particles when blended with alkali-catalyzed tannin resins resulted in improvements in the interfacial bond strength and stiffness of the composite panels from 0.041 to 0.200 MPa and from 1039 to 1527 MPa, respectively.

Barbosa et al. (2010) have used tannins in the preparation of tannin-phenol polymers and then used them in the preparation of bio-based composites reinforced with coir fibers. They have observed that the obtained mixture showed extreme adhesion between the fibers and matrix. The experimental data of Izod impact strength, scanning electron microscopy images, and the values of the diffusion coefficient of water that were carried out in this study showed that the bio-based composites obtained have the potential in nonstructural applications such as in the internal parts of automotive vehicles.

Elaine and Elisabete (2012) prepared tannin-phenolic resin by using condensed-type tannins. In their research, the tannin-phenolic resin was used to prepare polymer–matrix composites; according to the obtained results, tannins can be used as a macromonomer in the synthesis of the tannin-phenolic resin used for the production of composites.

In a previous work, tannins were used as a partial substitute of phenol in the synthesis of phenolic *Acacia* bark-based composites. Low water absorption and high storage modulus were the main advantageous properties of the composite (Frollini et al., 2008). Blends of HTs, cashew nut hell liquid, and UF resins have been synthesized and tested in the work carried out by Bisanda et al. (2003) to prepare particle boards from coffee husks, and they determined their mechanical and physical properties for particle board applications. The blending of HTs with UF resins has been found to reduce the formaldehyde emission levels significantly. A blend of HTs and cashew nut shell liquid has been found to possess better dimensional stability. Tannin-blend resins cure faster, which means they have shorter pot life and result in composites with better water and moisture resistance when compared with UF resins.

6.7 TANNINS AS NATURAL ANTIOXIDANTS

Plant phenols have received increasing interest due to their free radical-scavenging ability and their effects on the prevention of various oxidative reaction-associated diseases and stress. Recently, the isolation, use, and evaluation of natural phenolic compounds as pure or crude extracts from different plant resources have become a major field for health- and medical-related researches (Jin and Russell, 2010). In their work, Hagerman et al. (1998) evaluated tannins as biological antioxidants, and the results suggest that the tannins, which are found in many plants, are potentially very important biological antioxidants as they have quenching ability for peroxyl radicals about 15–30 times more than that of simple phenolics. Riedl and Hagerman (2001) determined the antioxidant activity of tannin–protein complexes, which are founded to act as scavengers and sinks for free radicals.

As an alternative antioxidant property, some phenolic compounds with dihydroxy groups can conjugate transition metals, preventing metal-induced free radical formation. Phenolic compounds with catecholate and gallate groups can inhibit metal-induced oxygen radical formation either by coordination with Fe^{2+} (or Cu^{2+}) or enhancing the autoxidation of this ion (Jin and Russell, 2010). Phenolic compounds in the bark of *Acacia mangium* were extracted, and their antioxidant activities were investigated using 1,1-diphenyl-2-picrylhydrazyl free radical-scavenging and ferric-reducing antioxidant power assays in the work of Liangliang et al. (2010). A significant linear relationship among antioxidant potency, antiradical activity, and the content of phenolic compounds of bark extracts was observed.

Generally, the action of polymer additives against a thermo-oxidative process is a main tool to preserve the desired polymer properties for a long period and at different conditions. A study on the efficiency of natural antioxidants, bio-based tannins, and other phenol byproducts, as stabilizers for polypropylene, is reported by Ambrogi et al. (2011). Their stabilizing activity was compared with that of a commercial phenolic antioxidant. The experimental results provided evidence for the effectiveness of natural stabilizers. Finally, increasing interest in the lowering or replacement of synthetic antioxidants has led to focus the research studies on natural antioxidants, especially on plant-origin materials.

6.8   CONCLUSIONS

Polyphenolic materials and specifically tannins, as this review focused, isolated or extracted from a broad variety of plant materials (from forestry or viticulture byproducts) are available in high quantities. They are historically used in the leather tanning and adhesive industries. Increasingly, CTs are polyphenolic oligomers based on a flavanyl repeat units with the hydroxylation pattern providing potent antioxidant activity. These properties, their availability, and the relative ease to chemical modification make tannins attractive for many projects and investigations such as polymer additives or modifiers. They could be modified by acetylation, hydrolysis, condensation, polymerization, etc. They may also be copolymerized with formaldehyde and aminoplast or phenolic resins to elaborate thermosetting binders for particle panels as well. They could also be formulated as adhesives by direct hydrolysis followed by polymerization. Several new emerging applications are in adhesive industries for partially replacing hazardous phenol by phenolic resins. On the other hand, the antioxidant properties of tannins are not a surprise. The antioxidant activity of polyphenols and other properties make these compounds a very interesting potential raw material for the development of green materials.

KEYWORDS

- **natural polyphenols**
- **tannins**
- **chemistry of tannins**
- **tannin applications**
- **hydrolyzable tannins**
- **condensed tannins**

REFERENCES

Andersen OM, Markham KR. 2006. Flavonoids: chemistry, biochemistry and applications. Chaps 3 and 6. Boca Raton, London: Taylor and Francis.

Alan C, Indu B Jaganath, Michael N Clifford. 2006. Phenols, polyphenols and tannins: an overview. Chap. 1. In: Alan C, Clifford MN, Hiroshi A (Editors). Plant secondary metabolites: occurrence, structure and role in the human diet. Oxford, UK: Blackwell Publishing Ltd.

Ambrogi V, Cerruti P, Carfagna C, Malinconico M, Marturano V, Perrotti M, Persico P. 2011. Natural antioxidants for polypropylene stabilization. Polymer Degradation and Stability. 96:2152–2158.

Amine M, Ahmed A, Antonio P, Fatima C, Bertrand C. 2010. Preparation and mechanical characterization of particle board made from maritime pine and glued with bio-adhesives based on cornstarch and tannins. Maderas Ciencia y Tecnologia. 12:189–197.

Amine M, Antonio P, Ahmed A, Fatima C, Bertrand C. 2013. Polenta corn starch-mimosa tannin-based urea formaldehyde adhesives for interior grade particleboard. Journal of Materials and Environment Science. 4:496–501.

Antwi-Boasiako C, Animapauh SO. 2012. Effect of solvent type on tannin extractability from three *Tetrapleura tetraptera* (Schum. and Thonn.) taubert positions for wood composite adhesive formulation. Journal of Emerging Trends in Engineering and Applied Sciences. 3:517–525.

Antwi-Boasiako C, Animapauh SO. 2012. Tannin extraction from the barks of three tropical hardwoods for the production of adhesives. Journal of Applied Sciences Research. 8: 2959–2965.

Ascacio-Valdés JA, Buenrostro-Figueroa JJ, Carbo AA, Barragán AP, Herrera RR, Aguilar CN. 2011. Ellagitannins: biosynthesis, biodegradation and biological properties. Journal of Medicinal Plants Research. 5:4696–4703.

Aouf C, Nouailhas H, Fache M, Caillol S, Boutevin B, and Fulcrand H. 2013. Multifunctionalization of gallic acid. synthesis of a novel bio-based epoxy resin. European Polymers Journal. 49:1185–1195.

Barbosa Jr V, Elaine Cristina R, Ilce Aiko TR, Elisabete F. 2010. Biobased composites from tannin-phenolic polymers reinforced with coir fibers. Industrial Crops and Products. 32:305–312.

Basso MC, Xinjun L, Vanessa F, Antonio P, Samuele G, Alain C. 2011. Green formaldehyde-free foams for thermal insulation. Advanced Material Letters. 2:378–382.

Benyahya S, Aouf C, Caillol S, Boutevin B, Pascault JP, Fulcrand H. 2014. Functionalized green tea tannins as phenolic prepolymers for bio-based epoxy resins. Industrial Crops and Products 53:296–307.

Bisanda ETN, Ogola WO, Tesha JV. 2003. Characterisation of tannin resin blends for particle board applications. Cement and Concrete Composites 25:593–598.

Bossu CM, Ferreira EC, Chaves FS, Menezes EA, Nogueira ARA. 2006. Flow injection system for hydrolysable tannin determination. Microchemical Journal. 84:88–92.

Chuan Wei OO. 2008. Preparation and evaluation of mangrove tannins-based adsorbent for the removal of heavy metal ions from aqueous solution. Ph.D. Thesis, School of Chemical Sciences, Universiti Sains Malaysia, Penang, Malaysia.

Dixon RA, De-Yu X, Shashi BS. 2005. Proanthocyanidins-a final frontier in flavonoid research. New Phytologist. 165:9–28.

Elaine CR, Elisabete F. 2012. Tannin–phenolic resins: synthesis, characterization, and application as matrix in bio based composites reinforced with sisal fibers. Composites: Part B. 43:2851–2860.

Electra P, Zoi N. 2011. D 3.3: resins that can be produced by the European bio industry. In: Non-food crops-to-industry schemes in EU27. WP1. Non-food crops, Grant agreement no. 227299, avalaible at: www.crops2industry.eu

Graciela P, Juanita F, Jaime B. 2003. Removal of metal ions by modified pinus radiata bark and tannins from water solutions. Water Research. 37:4974–4980.

Grigsby WJ, Bridson JH, Cole L, Jaime-Anne E. 2013. Esterification of condensed tannins and their impact on the properties of poly (lactic acid). Polymers. 5:344–360.

Grigsby WJ, Bridson JH, Cole L, Helena F. 2014. Evaluating modified tannin esters as functional additives in polypropylene and biodegradable aliphatic polyester. Macromolecular Materials and Engineering. 299:1251–1258.

Hagerman EA, Ken M R, Alexander Jones G, Kara NS, Nicole TR, Pual WH, Thomas R. 1998. High molecular weight plant polyphenolics (tannins) as biological antioxidant. Journal of Agricultural and Food Chemistry. 46:1887–1892.

Haslam E. 1988. Plant polyphenols (Syn. vegetable tannins) and chemical defence-a reappraisal. Journal of Chemical Ecology. 14:1789–1805.

Hergert HL. 1989. Condensed tannins in adhesives: introduction and historical perspective. Chap. 12 In: Hemingway RW, Conner AH, Branham SJ (Editors). Adhesives from renewable resources ACS symposium, Series 385. Washington, DC: American Chemical Society.

Hillis WE. 1979. Natural polyphenols (tannins) as a basis for adhesives. In: Proc of phenolic resins: chemistry and application. June 6–8. Weyerhaeuser Co., Tacoma, WA.

Horvath PJ 1981. The nutritional and ecological significance of acer-tannins and related polyphenols. M.Sc. Thesis, Cornell University, Ithaca, NY, USA.

Frederique B, Sandra TL, Antonio P, Paola N, Michel PC. 2012. Development of green adhesives for fiberboard manufacturing using tannins and lignin from pulp mill residues. Cellulose Chemistry and Technology. 46:449–455.

Frollini E, Oliveira FB, Ramires EC, Barbosa Jr V. 2008. Composites based on tannins: production, process and uses (Portuguese). Brazil Patent, BRPI0801091.

Frutos P, Hervas G, Giraldez FJ, Mantecon AR. 2004. Tannins and ruminant nutrition. Spanish Journal of Agricultural Research. 2:191–202.

Jaganathi IB, Alan C. 2010. Dietary flavonoids and phenolic compounds. Chap. 1 In: Fraga CG (Editor). Plant phenolics and human health: biochemistry, nutrition and pharmacology. John Wiley and Sons, Inc.

Jin D, Russell JM. 2010. Plant phenolics: extraction, analysis and their antioxidant and anticancer properties. Molecules. 15:7313–7352.

Jinjie G, Xinghai S, Meiqin C, Rui W, Min W. 2003. A novel biodegradable antimicrobial PU foam from wattle tannin. Journal of Applied Polymer Science. 90:2756–2763.

Karamali K, Teunis VR. 2011. Tannins: classification and definition. Natural Product Reports. 18:641–649.

Lan P, Francois G, Anttnio P, Zhou Ding G, Nicolas B. 2012. Wood adhesives from agricultural by-products: lignins and tannins for the elaboration of particle boards. Cellulose Chemistry and Technology. 46:457–462.

Laurent R, Chahinez A, Eric D, Hélène F. 2013. Depolymerisation of condensed tannins in ethanol as a gateway to bio sourced phenolic synthons. Green Chemistry. 15:3268–3275.

Liangliang Z, Jiahong C, Yongmei W, Dongmei W, Man Xu. 2010. Phenolic extracts from *acacia mangium* bark and their antioxidant activities. Molecules. 15:3567–3577.

Li J, Li C, Wang W, Zhang W, Li J. 2016. Reactivity of larch and valonia tannins in synthesis of tannin-formaldehyde resins. Bioresources. 11:2256–2268.

Mahmut OZ, Cengiz S, Ayhan SI. 2006. Studies on synthesis, characterization, and metal adsorption of mimosa and valonia tannin resins. Journal of Applied Polymer Science. 102: 786–797.

Maestri DM, Nepote V, Lamarque AL, Zygadlo JA. 2006. Natural products as antioxidants. Chap. 5 In: Filippo I (Editor). Phytochemistry: advances in Research. Kerala, India: Research Signpost.

Mamta S, Jyoti S, Rajeev N, Dharmendra S, Abhishek G. 2013. Phytochemistry of medicinal plants. Journal of Pharmacognosy and Phytochemistry 1:168–182.

Mark D. 2010. Tannin management in the vineyard. Fact Sheet, available at: www.gwrdc.com.au

Mehmet EA, Şükrü D. 2006. Removal of heavy metal ions using chemically modified adsorbents. Journal of International Environmental Application and Science. 1:27–40.

Michel CV, Roberto CC, Bruno C da Silva, Gisely de L Oliveira. 2011. Tannin extraction from the bark of *Pinus oocarpa* var. Oocarpa with sodium carbonate and sodium bisulfite. Floresta e Ambiente. 18:1–8.

Mohamed NB, Alessandro G. 2008. Monomers, polymers, and composites from renewable resources. Chap. 8, Oxford, UK: Elsevier.

Ndazi B, Tesha JV, Karlsson S, Bisanda ET. 2006. Production of rice husks composites with acacia mimosa tannin-based resin. Journal of Material Science. 41:6978–6983.

Nonhlanhla MR. 2011. Multidimensional fractionation of wood based tannins. Master Thesis, Department of Chemistry and Polymer Science, University of Stellenbosch, South Africa.

Onifade KR. 2001. Production of tannin from the bark of eucalyptus camadulensis. A.U. Journal of Technology (Assumption University of Thailand). 5:66–72.

Pizzi A. 2006. Recent developments in eco-efficient bio-based adhesives for wood bonding: opportunities and issues. Journal of Adhesion Science and Technology. 20:829–846.

Pizzi A, Scharfetter HO. 2003. The chemistry and development of tannin-based adhesives for exterior plywood. Journal of Applied Polymer Science. 22:1745–1761.

Pizzi A, Tekely P. 1995. Mechanism of polyphenolic tannin resin hardening by hexamethylenetetramine: CP–MAS ^{13}C-NMR. Journal of Applied Polymer Science. 56:1645–1650.

Pizzi A. 2003. Natural phenolic adhesives I: tannin. Chap. 27. In: Pizzi A, Mittal KL (Editors). Handbook of adhesive technology, 2nd ed. revised and expanded. NY: Marcel and Dekker.

Raquez JM, Deléglise M, Lacrampe MF, Krawczak P. 2010. Thermosetting (bio) materials derived from renewable resources: a critical review. Progress in Polymer Science. 35: 487–509.

Ramakrishnan K, Krishnan MRV. 1994. Tannin—classification, analysis and applications. Ancient Science of Life Journal. 13:232-238.

Riedl KM, Hagerman AN. 2001. Tannin-protein complexes as radical scavengers and radical sinks. Journal of Agricultural and Food Chemistry. 49:4917–4923.

Stephane Q, Ken SF. 1996. Ellagitannin chemistry. Chemistry Review. 96:475–503.

Sumin K, Hyun-Joong K. 2003. Curing behavior and viscoelastic properties of pine and wattle tannin-based adhesives studied by dynamic mechanical thermal analysis and FT-IR-ATR spectroscopy. Journal of Adhesion Science and Technology. 17:1369–1383.

Takuo O, Hideyuki I. 2011. Tannins of constant structure in medicinal and food plants-hydrolysable tannins and polyphenols related to tannins. Molecules. 16:2191–2217.

Terry Sellers Jr, George D Miller Jr. 2004. Laboratory manufacture of high moisture southern pine strand board bonded with three tannin adhesive types. Forest Products Journal. 54: 296–301.

Tondi G, Pizzi A, Olives R. 2008. Natural tannin-based rigid foams as insulation for doors and wall panels. Maderas Ciencia y Tecnología. 10:219–227.

Yoshitake S, Kentaro S, Kazuyuki C. 2004. Adsorption characteristics of tannin for heavy metal ions. Prepared for presentation at AIChE 2003 annual meeting, November 7–12, 2004, Austin Convention Center, TX. Avalaible at: www.nt.ntnu.no/users/skoge/prost/.../aiche-2004/pdffiles/.../265j.pdf.

Vazquez G, Lopez-Suevos F, Gonzalez-Alvarez J, Antorrena G. 2005. Curing process of phenol-urea-frmaldehyde-tannin (PUFT) adhesives: kinetic studies by DSC and DMA. Journal of Thermal Analysis and Calorimetry. 82:143–149.

Versari A, Du Toit W, Parpineloo GP. 2013. Oenological tannins. Australian Journal of Grape and Wine Research. 19:1–10.

Warren JG, James HB, Cole L, Jaime-Anne E. 2013. Esterification of condensed tannins and their impact on the properties of poly(lactic acid). Polymers. 5:344-360.

Yoshitake S, Kentaro S, Kazuyuki C. 2004. Adsorption characteristics of tannin for heavy metal ions. Prepared for presentation at AIChE 2003 annual meeting, November 7–12, 2004, Austin Convention Center, TX. Avalaible at: www.nt.ntnu.no/users/skoge/prost/.../ aiche-2004/pdffiles/.../265j.pdf.

CHAPTER 7

Bryophytes: Natural Biomonitors

RAJEEV SINGH[1], HEMA JOSHI[2], and ANAMIKA SINGH[3*]

[1]*Department of Environment Studies, Satayawati College, University of Delhi, Delhi, India*

[2]*Hema Joshi, Department of Botany, Hindu College, Sonipat, Hryana*

[*3]*Department of Botany, Maitreyi College, University of Delhi, Delhi, India*

Corresponding author. E-mail: arjumika@gmail.com

ABSTRACT

Bryophytes are the most successful nonflowering thallus land plants. Bryophyte life structure is the simplest, having recurring arrangements of photosynthetic tissues, which provide maximum primary production and minimum water loss. Different life forms are found, which is a specialty of bryophytes. Bryophytes are mainly known as amphibians of plant kingdom, as they can successfully survive on both land and water. Bryophytes are very much economically important. Bryophyte are proved as potential bioindicators of air pollution as they shows a wide range of habitat diversity, structural simplicity, totipotency, and fast multiplication rate. One most important feature of bryophytes is that they show high metal accumulation capacity. Bryometers are used for the measurement of phytotoxic air pollution. Bryophyta alone or along with lichens gives valuable information about air pollutant by an index of atmospheric purity (IAP), which is based on the number, frequency-coverage, and resistance factor of species. In this chapter, we will discuss about the role of bryophytes as bioindicators and the methods by which we can measure pollutants in the thallus of bryophytes.

INTRODUCTION

Bryophyta is a division of kingdom Plantae and contains a small group of lower plants, placed between algae and vascular plants. The word Bryophyta is derived from Greek words, "Bryon" meaning mass and "phyton" meaning plant. This division comprises of mosses, hornworts, and liverworts. They are groups of green plants that occupy a position between thallophytes (algae) and vascular cryptogams (pteridophytes). Bryophyta lacks a proper vascular tissues system and are not able to form seeds after fertilization; that is why they are considered as the most primitive type of embryophyta. Regarding the origin of bryophyte, two important scientific communities are there: one group believes that they are originated from algae, while the other believes that they are originated from pteridophytes. Bryophyata has alteration of generations, i.e., their life cycle contains two phases: gametophytic and sporophytic. These two generations have unique structures. Gametophytic-phase nutrients are directly absorbed by diffusion through the cells. These plants are very peculiar and sensitive and act as potent bioindicators for a few pollutants, and nowadays bioindicators are a very important field of concern. Lichens and bryophytes are two important bioindicators.

Le Blanc and Rao (1975) exploited an instrument called bryometer for the measurement of phytotoxic air pollution. Bryophyta alone or along with lichens gives valuable information about air pollutant by an index of atmospheric purity (IAP), which is based on the number, frequency coverage, and resistance factor of species.

Bryophytes can be of two types on the basis of indicators of pollution:

1. Species that are very sensitive and show visible symptoms or injuries at very low concentration of pollutants. These are a very good indicator.
2. Species that are able to absorb and retain pollutants at a very high level as compared to other plants growing in the same habitat. These plants trap and prevent the recycling of such pollutants in the ecosystem for different periods of time. Analysis of such plants gives a fair idea about the degree of metal pollution.

7.1 GASEOUS POLLUTANTS AND BIOMONITORS

Air pollution is mainly due to gases like carbon monoxide (CO), hydrogen sulfide (H_2S), fluorides, hydrocarbons (HCs), nitrogen oxides (NO_X), ozone (O_3), sulfur dioxide (SO_2), aldehydes, and lead. NO_X and NH_3 are primary gaseous pollutants that are strongly phototoxic (Greven, 1992). SO_2 and NO_2 are converted to strong acids by oxidation and react with atmospheric water to form acid rain. Nitrogen oxides, SO_2, and smoke are produced by the combustion of fuels in automobiles, which releases various metals into the atmosphere.

Dust, particles of metal oxides, coal, soot and fly ash, cement, liquid particles, heavy metals, and radioactive materials are particulate pollutants. Ozone (O_3) is a secondary pollutant formed by the action of sunlight on nitrogen dioxide and certain hydrocarbons and is more phytotoxic. Pesticides and fertilizers are also pollutants, adding various forms of aquatic pollution. These pollutants are very harmful for the flora and fauna of the earth, while there are some plants that acts as biomonitors or bioindicators of pollution. Biomonitors are actually biological indicators that can detect changes in the physical and chemical properties of the abiotic environment. They act as direct indicators of the physical and chemical properties of the environmental pollution. Biological indicator organisms or plant species shows differences in their occurrence, vitality, and responses under changed environmental conditions.

7.2 METAL POLLUTANTS AND BIOMONITORS

A few highly dangerous heavy metals like lead (Pb), zinc (Zn), mercury (Hg), and cadmium (Cd) are important metal pollutants. The list of metal pollutants also contain iron (Fe), aluminium (Al), copper (Cu), and manganese (Mn). These metals are called heavy metals as they have a specific gravity of 5 or higher, although this definition is not up to the mark, and in year 1980, Nieboer and Richardson classified the metals in Class A group of metals, which are oxygen seeking. Class B group contains nitrogen- or sulfur-seeking metals. The third and last group contains intermediate metal ions called as borderline and are generally those who are classified on the basis of atomic properties (Table 7.1).

TABLE 7.1   Classification of Metals by Nieboer and Richardson (1980)

Sr. No.	Class	Property	Metals
1.	Class A	Oxygen seeking	Cs^+, K^+, Na^+, Li^+, Ba^{2+}, Sr^{2+}, Ca^{2+}, Mg^{2+}, La^{3+}, Gd^{3+}, Y^{3+}, Lu^{3+}, Sc^{3+}, Be^{2+}, Al^{3+}
2.	Class B	Nitrogen/sulfur seeking	Cs^+, K^+, Na^+, Li^+, Ba^{2+}, Sr^{2+}, Ca^{2+}, Mg^{2+}, La^{3+}, Gd^{3+}, Y^{3+}, Lu^{3+}, Sc^{3+}, Be^{2+}, Al^{3+}
3.	Borderline	Atomic property	Pb^{2+}, Sn^{2+}, Cd^{2+}, Cu^{2+}, Co^{2+}, Fe^{2+}, Ni^{2+}, Cr^{2+}, Ti^{2+}, Zn^{2+}, Mn^{2+}, V^{2+}, Ln^{3+}, Sb(III), Fe^{3+}, Ga^{3+}, As(III), Sn(IV)

These metals are absorbed by plants and get accumulated in plant tissues. Further, these plant tissues can be analyzed to assess the level of metal accumulation and to give clear information about metal pollutants present in the environment (air, water, and soil). *Sphagnum acutifolium* accumulates Pd, Cd, and Zn. *Hylocomium splendens*, *Hypnum cupressiforme*, and *Pleurozium schreberi* are the good bioindicators of Hg, Be, and Ag, as they accumulate these metal pollutant in their cells. *Fontinalis antipyretica* and *Eurhynchium rioariodes* and the aquatic mosses are indicators of heavy metal pollution in water.

A perfect biomonitor must have a few important characteristics listed as follows:

1.   It can accumulate high levels of pollutants and these pollutants should not causes death of the plant.
2.   It should be distributed widely and abundant for repeated sampling and comparison.
3.   Life must contains various ages and stages and its life must be long for survey.
4.   It must have specified target tissues or cells for further research at the microcosmic level.
5.   It can be sampled and cultured easily in the lab.

Biomonitoring can be used by two methods:

1.   *Direct monitoring:* It is based on the amount of pollutant in a suitable organism (bioindicator spp.) or basically it is measuring the quantity of pollutants in suitable organisms.
2.   *Indirect monitoring:* It represents biological signals, which are due to changes in the environment that have an impact on plants, like changes in morphology, physiology, and cytology. This indirect method is a sampling- and data-based method and is included in reports. Although data provided by physical and chemical monitoring are indispensable for evaluating the changes in the environment, the application of an (ideal) biomonitor has many advantages compared to direct monitoring techniques. In general, the concentration of pollutants in the monitoring organism is often higher than in the system to be monitored, so accurate sampling and analysis lead to accurate results. Biological monitors mainly reflects external conditions averaged over a certain time, like a specific substance in that organism may increase or decrease with age. This is important when the monitoring levels change rapidly in time. The concentrations of pollutants in organisms give an idea about the bioavailability of that pollutant. Biological monitors are natural species and already present in the environment and monitoring continuously naturally.

In direct biomonitoring of atmospheric pollutants, most of the relevant information is missing as they have to meet the specific requirements like abundant occurrence in the area of interest, independent of local conditions, availability of species in all seasons, tolerant to pollutants at relevant levels, etc. Other important factors like sample preparation, sampling, and measurement of the amount of element accumulated at that concentration can be done easily. Many animal and plant species are used for air pollution monitoring, as they can meet most of the above requirements. Animal species usually reflect complex changes in the environment, that is, air, water, and soil pollution and element intake in their diet, so it is tough to analyze and interpret air pollution. However, a few plant species appear to be especially suitable to indicate elemental air pollution and can be used as possible biomonitors for air pollution studies.

7.3 BIOMONITORING BY MOSSES

Mosses have been used for biomonitoring in a number of different ways, and they are mainly

1. indigenous and
2. transplants.

Indigenous mosses are naturally growing mosses, and biomonitoring is mainly done by them only. Transplants are "moss bags" and mosses species that are collected from a clean environment, washed with dilute acids before use, and left at a sampling site for a desired period. Within this specific period of exposure to the new place, they show different changes in their morphology and physiology. Transplant mosses are used for local studies, not for accurate studies and high pollution zones. Epigeic mosses grow on the ground and are widely used for the regional surveys in northern and central Europe. Mosses growing on the top of a substrate mainly consisting of organic soil and decaying plant debris are clearly preferable over using *epiphytic* mosses growing on trees. Many mosses are found on the trunks or branches of trees, and it may be possible that the incoming flux of trace elements from the atmosphere, deposited in the form of wet or dry forms (Chakrabortty et al. 2006), will be substantially modified by the canopy before reaching the moss, and it is also conceivable that trace elements supplied to the tree from the soil through the root system and eventually leached, for example, from leaf surfaces, will form part of the exposure of the moss. In warmer climates, epigeic mosses cannot be used as biomonitors as they grow directly on the mineral soil and pick up soil particles to a great extent. In such situations, epiphytic species growing on stumps are the best option to detect pollution. Two types of epigeic mosses are used for sampling and survey: feather mosses such as *P. schreberi* and *H. splendens* and peat mosses such as *Sphagnum fuscum.* Bryophytes are also humidity sensitive, and many species are restricted to specific microhabitats. A very striking feature of bryophytes is that they dry out very quickly and at the same time can also absorb minute quantities of available moisture from different sources, like fog, mist, and dew. This environmental water is available in a very small quantity, and naturally, other plants are not able to absorb it, while bryophytes can. However, during droughts, all physiological processes are quickly reduced to a minimum. A few bryophytes grow successfully on calcareous substrates like *Tortella lortuosa. Racomilrium lanuginosum* grows on acidic ground, and its presence on the soil surface indicates the acidic nature of soil.

7.4   SITUATIONS WHERE BIOINDICATORS/BIOMONITORS ARE USED

1. Where the indicated environmental factor cannot be measured, like situations where environmental factors in the past are different than present and palaeo-biomonitoring can be done.
2. Where the present factor is difficult to measure the amount of pesticides and their residues or complex toxic that are highly interacting.
3. Where the environmental factor is easy to measure but difficult to interpret, e.g., whether the observed changes have ecological significance.

7.5   TYPES OF BIOMONITORS

There are different plant and animal species that act differently as biomonitors.

7.5.1   CLASSIFICATION ON THE BASIS OF THE AIMS OF BIOINDICATORS

1. *Compliance indicators:* They are used to measure the specific attributes of any population, community, or ecosystem level, for example, fish population in pond. They are used to verify he maintenance or restoration goals.

2. *Diagnostic indicators:* They are used to measure individual or subspecies biomarkars. They are used to investigate the observed environmental disturbances.

3. *Early-warning indicators:* They are very sensitive to a minute change in the environment and are able to give visible response. They can reveal the first signs of disturbance in the environment before any other species get affected.

4. *Accumulation indicators:* They store pollutants without affecting their own metabolism, and at different times, they show different levels of accumulation at different levels of organization, e.g., lichens and mosses.

7.5.2 CLASSIFICATION ACCORDING TO THE DIFFERENT APPLICATIONS OF BIOMONITORS/BIOINDICATORS

1. *Pollution indicators:* These species are able to detect the pollutant level in the environment. These bryophytes are sensitive to a few pollutants and show remarkable effect on them.

2. *Environmental indicators*: They are usually species that are able to detect changes in the environmental state. In this case, the organisms are used as early-warning devices, and they delimit the effects of a disturbance, while they may also have the ability to accumulate toxins, which may further be used as a bioassay for the pollutants (Figure 7.1).

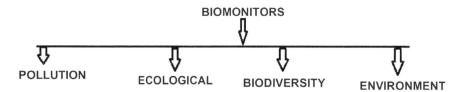

FIGURE 7.1 Biomonitor classification.

3. *Biodiversity indicators:* Biodiversity indicators are environmental or biotic attributes. A biodiversity indicator is a group of genus or tribe whose diversity reflects some measures of the diversity of other taxa in the same habitat. These measures of diversity may include characters like richness, level of endemism, genetic diversity, etc.

4. *Ecological indicators:* They include species that are sensitive to pollution, habitat fragmentation, or other stresses. These species represent the nature and sensitivity of the community. Due to the effect of pollutants, the species show responses like decline in population size, change in spatial distribution, or any number of life history changes.

7.6   MEASUREMENT TECHNIQUES USED TO MEASURE POLLUTANTS BY USING BRYOPHYTA

7.6.1   PHYTOSOCIOLOGICAL METHOD

The IAP measurement is one of the best ways to detect the changes in the quality of air in the environment. The measurement technique was suggested by Govindapyari et al. (2010). The air quality change in the environment directly reflects the change in phytophysiology. In this method, bryophytes growing naturally on the tree bark are examined along with a pollution gradient. The frequency of changes and abundance of species through a number of transects are recorded with time. The number of transects used are radiating in all directions and with increasing distance from the source of pollution. IAP is determined by applying the following formula:

$$IAP = \Sigma(Q \times f/10)n$$

where

n = total number of species present at a site,
f = frequency coverage of species at each site, and
Q = resistant factor.

7.6.2   ECOPHYSIOLOGICAL METHOD

This method includes direct exposure of bryophytes to pollutants with known concentrations. For this, the fumigation of a pollutant with different concentrations is the best way and the cultured plants were exposed to this fumigated environment. The growth and survival rate of plants, injury due to pollutant exposure, chloroplast degradation, and unusual growth of the plants are the main observations. On the basis of these observations, the level of toxicity, its effects, and the level of tolerance of plants can be determined in different species.

7.6.3 SURVEY METHOD

This method is a sampling- and survey-based method, and the abundance of the individuals or species exposed to the pollutants are used for sampling. The level of pollutants is not uniform everywhere in an environment, so the exposure and effect of pollutants may vary from place to place. Within a specified environment, different sites are selected. Among these sites, different bryophyte species are compared and readings are recorded in a survey report. This indicates the quantity of pollutants or stress on the organisms of the site. Periodic surveys are made on the native bryophytes in different sites. The number, frequency, and abundance of native species and the dominance of the growth form can be compared with the past records, reports, and periodic herbarium collection. Disappearances of already existed and reported species are marked as sensitive species, while a few tolerant species may appear in the new existed environment.

7.6.4 TRANSPLANTATION METHOD

In this method, bryophytes are taken along with their substrates from an unpolluted area and further grown in a polluted area or any industrial area with remarkable pollution. This method clearly shows visible changes in bryophytes and can be easily compared from the previous unpolluted environment to the new polluted environment. Polluted sites shows some visible responses in bryophytes, like injuries, difference in growth forms, plasmolysis and chlorophyll degradation in the leaf cells, etc. Transplanted species show a modified pattern of growth of shoots and branching and deposition of wax on the plant surface. Transplantation can be done by the following three ways.

7.6.4.1 TRANSPLANTATION IN THE SOIL

This method is based on bryophyte growth comparison in polluted and nonpolluted areas. First, polluted sites were selected and bryophytes grown in small plots were prepared in the ground. Similarly, some plots of nonpolluted sites are established as a control. Branching of plant parts, production of basal regenerative shoots, and the rate of survival are recorded periodically, which give a fair picture of pollution stress.

7.6.4.2   MOSS BAG METHODS

In this method, epiphytic mosses are widely used and muslin cloth bags (20×20 cm^2 size) are prepared and filled with epiphytic mosses. These bags are hanged in different locations of the cities; these locations may be polluted or unpolluted. A periodic analysis of the rate and ability of regeneration of these samples provides data on the trend of pollution. Similar to this, aquatic mosses filled in the bags can be kept in the water bodies and their analysis indicates the extent of pollution in the water bodies.

7.6.4.3   BRYOMETER

It is a box where bryophytes are grown in a moist chamber with their original substrata. The box has transparent sides made of thin glass with openings so that air and light can pass to the plants inside the box. The boxes are kept in different locations of the polluted area. The frequent observations of the growth and survival rate of the bryophytes indicate the trend of pollution in the sites. Spore germination pattern and protonema growth are easy to observe in this moist chamber.

7.7   FEATURES OF BRYOPHYTA AS BIOMONITORS

Biomonitors provide both qualitative and quantitative information about the environment. Liverworts and mosses are good bioindicators as they have a simple thalloid structure having one-cell thickness. They lack cuticle or epidermis, which causes higher absorption and accumulation of nutrients and pollutants. Some bryophytes grow in soils having specific pH and are indicators of the particular pH of the soil. Bryophytes are widely used as they are able to accumulate the pollutants and available throughout the year. The accumulated pollutants are easily measured by quantitative methods, and this provides the information about the level of pollutant deposition. Mosses and lichens are considered as the most appropriate as they are able to deposit atmospheric heavy metals in them. Electronic detectors and other methods are used to detect the level of pollutants, but it may also possible that if the level of pollutants is below a certain concentration level, then its estimation is very tough. Chemical analysis is strongly dependent on the time and place of sampling, while bryophytes have the ability to facilitate the detection of the elements present in very low concentrations. Mosses as natural biomonitors thrive in a humid climate. Ectohydric mosses are widely used as biomonitors.

The most important feature is that mosses can be stored for several years without any deterioration, and old specimens can easily be analyzed chemically. Bryophytes have wide distribution, ability to grow on a variety of habitats, large surface area, lack of cuticle and stomata, evergreen nature, which make them unique. Bryophytes are considered as Environmental Specimen Banks as they can be grown on a variety of habitats. Bryophytes obtain nutrients from the substances dissolved in moisture. The substances are directly absorbed from the substrata by diffusion through the cells. Pollutants reached to the cells and get deposited in the form of particles or gases. During analysis, these cells show the presence of elements and their concentration gradient in the respective substrata clearly indicated the level of pollution in the environment.

7.7.1 BRYOPHYTES AS POLLUTION CONTROLLERS

H. splendens and *P. schreberi* are used for heavy metal detection in air as they have the ability to accumulate many metals in high concentrations. *Marchantia polymorpha*, *Solenostoma crenulata*, *Ceratodon purpureus*, and *Funaria hygrometrica* (Coombes and Lepp, 1974) are metal-tolerant bryophytes. Bryophytes growing on rocks are more tolerant to pollutants as compared to bryophytes growing on tree trunks. However, epiphytic species growing on a tree base can cope with the pollution conditions better than those on tree trunks. *Brachythecium rutabulum*, *Grimmia pulvinata*, *Orthotrichum diaphanum*, *Bryum capillare*, *Bryum argenteum*, *Tortula muralis*, *Rhynchostegium confertum*, and *H. cupressiforme* show better survival growth in a polluted environment. Different life forms of bryophytes are responding differently to the level of pollutants. Similarly, rough mats, tall turfs, wefts, large cushions, and leafy liverworts are less tolerant to pollutants than smooth mats and small cushions, while short turf and thalloid forms are highly resistant to pollutants as compared to others.

The leafy gametophores are more sensitive than the mature gametophytes (Gilbert, 1969). The reproductive potential of a species determines its degree of success in a polluted environment. The survival of *B. argenteum*, *C. purpureus*, *Dicranella heteromalla*, *F. hygrometrica*, *Leptobryum pyriforme*, *Lunularia cruciata*, *M. polymorpha*, and *P. proligera* have high reproductive capacity and fast growth. These species produce spores or gemmae on a very large scale. A few terricolous species grows best on soils at a pH of 3.4

and in SO_2-polluted areas, for example, *D. heteromalla, Pohlia nutans*, and *C. purpureus*. Sulphur dioxide (SO_2)-resistant bryophytes show fast growth rate; for example, *Hypnum yokohamae* and *Glyphomitrium humillimum* were able to tolerate SO_2 concentrations of 0.04–0.05 ppm (Gilbert, 1970).

Fluoride-resistant btryophytes are *P. nutans* and *Aulacomnium androgynum, Polytrichum commune, Leucobryum glaucum, Rhytidiadelphus squarrosus. Sphagnum* and *Bryum* like terrestrial mosses are good indicators of SO_2, NO_2, fluorides, and HCl in soil. *Bryum dyffrynense* is a poikilohydrous moss, and a few grasses are sensitive pollution indicators. A few aquatic mosses are found in water sources when there is good content of calcium and nutrients in water. A few mosses are specific as they grow only in copper-rich soil and are indicators of the presence of copper in the soil, for example, *Merceya, Mielichhoferia elongata*, and *M. mielichhoferiana*.

As bryophytes lack a protective layer like epidermis or cuticle, this makes them highly sensitive. Bryophytes are widely used for the measurements of heavy metal toxicity like chromium, copper, cadmium, nickel, and vanadium particularly in the areas near power stations. Cesium is a radioactive metal, and its presence in nature can be detected by bryophytes. They dry very quickly and also absorb a very small quantity of moisture present in the atmosphere in any form like mist, fog, dew, etc. *Tortella lortuosa* is associated with calcareous substrates, while *R. lanuginosum* grows only on acidic surfaces.

7.7.2   EFFECT OF METALS ON BRYOPHYTES

Bryophytes absorb atmospheric chemicals either soluble chemicals in wet deposition or particles from dry deposition. Bryophytes absorb heavy metals without any disturbance of the normal biological pathway. This special ability makes them unique, and that is why they are successfully used as biomonitors for environmental pollutants. The efficiency of a moss to uptake the metal varies from species to species. Mostly, bryophytes take dissolved elements in ionic forms. The absorbed ions attached to the surface of mosses by physical and chemical forces. The cells of bryophytes have different retention capacities for different ions. A very large retention capacity indicates that both simple cation exchange on negative surface charges and complex formation with ligands on the surface of moss are involved. Mosses cells absorb strongly Cu and Pd as compared to Zn and Cd. *H. splendens* and *P. schreberi* are the two important species used in environmental element studies. *M. polymorpha* and *Calymperes delessertii* are good monitors of aerial lead and copper. *Pottia truncata, Polytrichum ohioense, D. heteromalla*, and *B. argenteum* are very tolerant of high tissue levels of cadmium (610 ppm), copper (2700 ppm), and zinc (55,000 ppm). *H. cupressiforme* accumulates zinc, copper, and cadmium threefold more as compared to lichens and other plants. Metal uptake by bryophytes depends upon the level of metals and the affinity of cells of bryophytes with these metals. For example, copper and lead are absorbed more as compared to nickel. The absorption of nickel is more as compared to that of cobalt, zinc, and manganese.

7.7.3 SO_2 AND ACID RAIN MONITORING BY BRYOPHYTES

Bryometers were not accepted by everyone so new technologies must be needed. In 1967 and 1968, Gilbert found that SO_2 is an important factor and decides the distribution of mosses and development of reproductive structures and capsules. He published his research article based on *G. pulvinata* as an SO_2 indicator in 1969 (Gilbert, 1969). Monitoring studies based on bryophytes developed a list of tolerant and intolerant bryophytes, which can be used as successful bioindicators. Taoda (1972), in Japan, started using epiphytic species to assess the pollution level in Tokyo city. He divided the city into five different zones, based on pollution intensity. His classification included both mosses and liverworts. This zonation is on the basis of sensitivity of bryophytes to SO_2. (1) *G. humillium* and *H. yoko-hamae*; (2) *Entodon compressus, Hypnum plumaeforme, Sematophyllum subhumile,* and *Lejeunea punctiformis; (*3) *Aulacopilum japonicum, B. argenteum, Fabronia matsumurae,* and *Venturiella sinensis*; (4) *Haplohy-menium sieboldii, Herpetineuron tocceae, Trocholejeunea sandvicensis,* and *Frullania muscicola.*

SO_2-exposed mosses are reduced in coverage. The damage may be due to direct exposure of SO_2 or by the formation of sulfuric acid. SO_2 reacts with water and forms sulfuric acid. Further, sulfuric acid breaks into hydrogen ions and makes water to become acidic. In plant cells, these free hydrogen ions replace magnesium of chlorophylls and further lead to the destruction of chlorophyll. Despite this simple mechanism, there are mosses that can protect their chlorophyll molecule from such destruction and such mosses tolerate the acidic environment. *Dicranoweisia* change SO_3^{-2} into a harmless sulfate (SO_4^{-2}) salt. It is also observed that the concentration of chlorophyll is very high in these tolerant bryophytes. Due to acid rain, the acidification of plant barks takes place and acid-tolerant mosses grow on the surface of barks. *P. schreberi* grew faster in an acidic environment having pH 4.5. At pH 3.5, its growth and chlorophyll content reduced and the production of capsule decreased.

7.8   CONCLUSION

The most striking feature of bryophytes is the accumulation and retention of pollutants, which makes them so special for the interpretation of heavy metal emission pattern. There is a great need to extend the observations on mineral location and effect to a much wider range of species. More species should be recognized to inhibit the chemical environment. Other important aspect is visible responses shown by plants in a polluted environment that are different from those in an unpolluted environment. These aspects of bryophytes also attract research on their unique mechanisms at the cellular level by which they can survive in extreme acidic and polluted environments.

KEYWORDS

- **bioindicator**
- **pollutants**
- **nonflowering plants**
- **air pollution**
- **Bryophyta**

REFERENCES

Coombes, A. J. and Lepp, N. W. (1974). The effect of Cu and Zn on the growth of *Marchantia polymorpha* and *Funaria hygrometrica*. Bryologist 77: 447–452.

Govindapyari H., Leleeka M., Nivedita M. and Uniyal P. L. (2010). Bryophytes: indicators and monitoring agents of pollution. NeBIO 1(1): 35–41.

Farmer A. M. (1992). Ecological effects of acid rain on bryophytes and lichens. In: Bates J.W. and Farmet A.M. (Eds.), Bryophytes and Lichens in a Changing Environment. Oxford: Clarendon Press, pp. 284–313.

Gilbert, O. L. (1969). The effects of SO_2 on lichens and bryophytes around Newcastle upon Tyne. In: Proceedings of the 1st European Congress Influence of Air Pollution on Plants and Animals, Wageningen, pp. 223–235.

Gilbert, O. L. (1970). Further studies on the effect of sulphur dioxide on lichens and bryophytes. New Phytologist 69: 605–627.

Greven, H. C. (1992). Changes in the Dutch Bryophyte Flora and Air Pollution. Berlin: J. Cramer.

Govindapyari H., Leleeka M., Nivedita M. and Uniyal P. L. (2010). Bryophytes: indicators and monitoring agents of pollution. NeBIO 1(1): 35.

Chakrabortty S, Tryambakro G., Paratkar. (2006). Biomonitoring of trace element air pollution using mosses. Aerosol and Air Quality Research 6(3): 247–258.

Garty J. (2001). Biomonitoring atmospheric heavy metals with lichens: theory and application. Critical Reviews in Plant Sciences 20(4): 309–371.

Nieboer E. and Richardson D. H. S. (1980). The replacement of the nondescript term "heavy metals" by a biologically and chemically significant classification of metal ions. Environmental Pollution (Series B) 1: 3–26.

CHAPTER 8

Applied Techniques for Extraction, Purification, and Characterization of Medicinal Plants Active Compounds

CRYSTEL ALEYVICK SIERRA RIVERA[1], LUIS ENRIQUE COBOS PUC[1], MARIA DEL CARMEN RODRÍGUEZ SALAZAR[1], ANNA ILINÁ[2], ELDA PATRICIA SEGURA CENICEROS[2], LAURA MARÍA SOLÍS SALAS[1], and SONIA YESENIA SILVA BELMARES[1*]

[1]Research groups of Chemist-Pharmacist-Biologist and Nano-Bioscience

[2]School of Chemistry, Autonomous University of Coahuila, Blvd. Venustiano Carranza, Col. República Oriente, Saltillo 25280, Coahuila, Mexico

*Corresponding author. E-mail: yesenia_silva@uadec.edu.mx

ABSTRACT

The consumption of medicinal plants has increased in recent years, as plants synthesize compounds that prevent diseases. However, the efficiency and innocuousness of many of them are unknown, and some countries lack regulatory mechanisms for their trade. For this reason, researchers have focused their work on developing both biological and chemical studies to identify active ingredients. Therefore, they have established processes of extraction, purification, and characterization of some active compounds of plants. Regularly, these techniques are established based on the chemical structure of the interest group. To isolate an active compound, first, the extraction of the plant material is carried out. Then, the extract is subjected to a qualitative phytochemical analysis to identify the groups of compounds. Then, a separation technique is established, which includes thin-layer chromatography, gravitational column chromatography, liquid vacuum chromatography, or liquid flash chromatography. Finally, some techniques are used to purify and identify the active compounds, such as high-performance liquid chromatography–mass spectrometry, gas chromatography coupled to mass spectrometry, nuclear magnetic resonance, infrared spectroscopy, and ultavoilet–visible spectroscopy.

8.1   INTRODUCTION

Since ancient times, humanity has searched plants to cure their diseases; consequently, it has deepened their study. Currently, there is an increase in the consumption of medicinal plants to prevent diseases since medicinal plants produce secondary metabolites that are an important source of medicines. Therefore, to increase in the production of plants that produce compounds of pharmaceutical importance is highly desirable to maintaining sustainable (Pandey et al., 2018). Additionally, rural people residing in developing countries rely on the use of medicinal plants because they have limited access to allopathic medicines (Aziz et al., 2018). For some years, the World Health Organization reported that around 80% of the world population depends on traditional medicine. For instance, in the United States, 90 medicines prescribed are obtained from plants (Al-Nahain et al., 2014). On the other hand, a large number of allopathic medicines are synthesized from plants such as codeine and morphine (*Papaver somniferum*) (Higashi et al., 2010; Weid et al., 2004). Around the world, there is a great diversity of flora; as result, different cultural groups use plants to treat their diseases according to medicinal properties. However, to understand the function of the active principles of each plant, it is necessary to carry out biological and phytochemical studies that guarantee their proper use. Therefore, it is essential to establish extraction, purification, and characterization processes when a new compound is discovered or to formulate an allopathic drug.

Regularly, these techniques are established based on the chemical structure of the group of interests. Consequently, preliminary tests are carried out to determine the solubility of the components of the plant extract. Then a qualitative phytochemical analysis is performed to identify the groups of compounds as well as a chromatographic profile using thin layer chromatography (TLC). The chromatographic profile by TLC allows selecting the eluents to be used in the work process to separate the extract by gravitational column chromatography or liquid vacuum chromatography. Some processes of selection of the active principles include bioguided assays and chemical traces (Hernández-Ocura et al., 2017). Finally, techniques are used to purify and identify the compounds that include high-performance liquid chromatography–mass spectrometry

(HPLC–MS), gas chromatography coupled to mass spectrometry (GC–MS), nuclear magnetic resonance (NMR), infrared (IR) spectroscopy, and ultavoilet–visible (UV–vis) spectroscopy.

8.2 MEDICINAL PLANTS AND PROPERTIES

Human beings have used plants since prehistoric times (Abe and Ohtani, 2013; Khan, 2014). In ancient civilizations, the plants were ingested for curative and psychotherapeutic purposes as tinctures, teas, poultices, powders, and other herbal formulations since almost 5000 years ago in India, China, and Egypt and at least 2500 years in Greece and Central Asia (Halberstein, 2005; Balunas and Kinghorn, 2005). At least, 422,000 flowering plants have been identified around the world and closely 12% are used to treat some illness (Abe and Ohtani, 2013). Many medicinal plant species have spread globally both via intentional and carefully planned transfers and as the unintentional outcome of people's movements (Aguilar-Stoen and Moe, 2007).

Mexico has an estimated 30,000 species of plants, where approximately 3000–7000 are of medicinal use (Camou-Guerrero et al., 2008). Indeed, from times before the arrival of the Spaniards, ancient Mexicans had already developed a knowledge deep on the qualities and medicinal uses of the plants as stated in colonial documents such as *Códice de la Cruz Badiano*, *Ensayo a la materia Médica vegetal de México,* and *Historia de las cosas de la Nueva España* written by Martín de la Cruz and Juan Badiano, Vicente Cervantes, and Bernardino de Sahagún, respectively (Alonso-Castro et al., 2012; Schifter Aceves, 2014). Importantly, most of the Mexican medicinal plants are gathered from the wild, and only 15% are cultivated. On the other hand, in our country, the plants represent direct inputs to satisfy the necessity of food, material for construction, fuel, or fodder, and some plant species may also contribute to the familial economy through commercialization of plant products (Camou-Guerrero et al., 2008). Furthermore, despite the vast botanical wealth of Mexico, the pharmacological, phytochemical, and toxicological effects of only a few groups of medicinal plants are known. However, almost half of the population has no access to a public health institution, whereby this population needs to use medicinal plants (Alonso-Castro et al., 2017).

In most cases, the pharmacological properties of medicinal plants are well defined by indigenous people of each place worldwide (Mahdi, 2010). Nowadays, the prevalence of use of herbal medicinal by persons around the world is high and continuously increasing, especially because some medicins are cheaper than conventional medicines (Posadzki et al., 2013). However, some plants could be inefficient in the treatment of some diseases, the use of certain medicinal plants might be toxic, and the pharmacologically active ingredients of some medicinal plants may interact with synthetic drugs, which, in turn, could endanger the health of patients (Alonso-Castro et al., 2017; Posadzki et al., 2013; Bourgaud et al., 2001). In any case, investigations are required to develop medicines based on plants.

Clinical, pharmaceutical, and chemical studies on medicinal plants were the basis of many early drugs such as aspirin (from willow bark), digoxin (from foxglove), morphine (from opium poppy), quinine (from cinchona skin), and pilocarpine (from *Maranham jaborandi*) (Halberstein, 2005; Balunas and Kinghorn, 2005; Aguilar-Stoen and Moe, 2007; Petrovska, 2012). Also, if that is not enough, the secondary metabolites of medicinal plants can be useful as cosmetics, fine chemicals, or more recently nutraceutical compounds. Indeed, 25% of the molecules used in the pharmaceutical industry are obtained from plants (Bourgaud et al., 2001). Some plant natural products are easily chemically synthesized, but many others with multiple chiral centers are difficult to synthesize and are often produced through the exploitation of native biological pathways using the natural harvest (Paek et al., 2005). Therefore, efforts toward sustainable conservation and rational utilization of biodiversity at the present and future should be implemented (Vanisree et al., 2004; Wilson and Roberts, 2014). In this sense, several biotechnological strategies are in development to enhance the production of secondary metabolites from medicinal plants in vitro; some of these include plant cell culture techniques, hairy root cultures, biotransformation, heterologous production, and others (Narayani and Srivastava, 2017).

Currently, in the world, a large number of plants are commercialized for their medicinal properties, among which are *Rosmarinus officinalis* (rosemary), *Zingiber officinale* (ginger), *Panax ginseng* (ginseng), *Garcinia cambogia* (tamarind malabar), *Curcuma longa* (curcuma), *Amphipterygium adstringens* (cuachalalate), *Aloe vera* (aloe), *Allium sativum* (garlic), *Allium nigrum* (black garlic), *Camelia sinensis* (green tea), *Valeriana officinalis* (valeriana), *Gnaphalium chartaceum* (mullein), *Ternstroemia lineata* (tila), *Eucalyptus globulus* (eucalyptus), *Matricaria recutita* (chamomile), *Lavandula angustifolia* (lavender), *Moringa oleifera* (moringa), and *Salvia mexicana* (salvia) (Al-Nahain et al., 2014; de Oliveira et al., 2019; Li et al., 2012; Liu et al., 2019; Jamila et al., 2019; Karamalakova et al., 2019; Rodríguez-García et al., 2015; López et al., 2019; Gao et al., 2019; Hassanain et al., 2010; Caudal et al., 2017; González-Burgos et al., 2018; Hajaji et al., 2017; Cardia et al., 2018; Zeng et al., 2019). The medicinal effects of plants vary; some are used in the region of origin and others in the whole world. Table 8.1 describes the properties of the main plants used in traditional medicine worldwide.

TABLE 8.1   Properties of Some Plants Used in Traditional Medicine

Plant	Properties	Reference
Rosmarinus officinalis	Anti-inflammatory, antioxidant, antimicrobial, antiproliferative, antitumor and protective, inhibitory, and attenuating activities	(de Oliveira et al., 2019)
Zingiber officinale	Anticancer, anticlotting, anti-inflammatory, analgesic, antiasthmatic, antidiabetic, antitrhombotic, and laxative	(Al-Nahain et al., 2014; Li et al., 2012)
Panax ginseng	Cardiovascular disorders and aging-related disorders	(Liu et al., 2019)
Garcinia cambogia	Antiobesogenic, antitumor, antiulceric, antihemorrhoidal, antidiarrheal, antidisenteric, antifebrifuge, antiplatelet aggregation, and antidiabetic	(Jamila et al., 2019)
Curcuma longa	Antioxidant, anti-inflammatory, and anticancer	(Karamalakova et al., 2019)
Amphipterygium adstringens	Anti-inflammatory and antiulceric	(Rodríguez-García et al., 2015)
Aloe vera	Apoptotic, hepatoprotective, antioxidant, antibacterial, antidiabetic, antihyperglycemic, and anti-inflammatory	(López et al., 2019)
Allium sativum	Antioxidant, antimicrobial, anti-inflammatory, and anticancer	(Gao et al., 2019)
Camelia sinensis	Antioxidant, anti-inflammatory, and antimutagenic	(Hassanain et al., 2010)
Valeriana officinalis	Anxiolytic, sedative, myorelaxant, gastrointestinal, and menstrual spasmolytic	(Caudal et al., 2017)
Eucalyptus globulus	Antiasthmatic, antibronchitic, antimicrobial, antihelmintic, and antidiabetic	(González-Burgos et al., 2018)
Matricaria recutita	Antimicrobial, larvicidal, anticarcinogenic, and anti-inflammatory	(Hajaji et al., 2017)
Lavandula angustifolia	Anticonvulsant, anxiolytic, antioxidant, anti-inflammatory, and antimicrobial	(Cardia et al., 2018)
Moringa oleifera	Antitumor, antibacterial, antiarthritic, antioxidant, antiasthmatic, hepatoprotective, and antipoisoning by metals	(Zeng et al., 2019)

8.3   ACTIVE COMPOUNDS ISOLATED FROM PLANTS USED IN TRADITIONAL MEDICINE AND HOMOEOPATHY

For years, hundreds of plants, due to their active ingredients, are used in homeopathic medicine (Bartomeu Costa-Amic and Aldape Barrera, 1998). The plants synthesize various compounds, among which are flavonoids and related compounds, sesquiterpene lactones, coumarins, lignans, terpenoids, steroidal compounds, carotenoids, limonoids, meliacins, simaroubalidans, cardiotonic glycosides, alkaloids, and terpenes. Some of them are specific to the treatment of particular diseases (Solís-Salas et al., 2019; Domínguez, 1979). However, the active compounds mostly used in traditional medicine belong to the phenylpropanoid, flavonoids and related compounds, terpenoids, and alkaloids. Table 8.2 includes some plants used as homeopathic medicines in the world. Additionally, Table 8.2 describes the properties of each plant and related active compounds.

On the other hand, each society has systems such as alternative, traditional, and allopathic medicine or professional health systems that are influenced by differences between cultures. At a global level, there is greater interest and demand for the use of traditional and allopathic health systems, so a professional collaboration between both health systems is desirable (van Rooyen et al., 2015).

Some countries lack the regulation of plant products (James et al., 2018), but in others, they have governmental institutions commissioned to regulate them. In this sense, from 1938, all homoeopathic medicines included in the Homeopathic Pharmacopoeia of the United States are regulated under the Food, Drug and Cosmetic Act. Additionally, since 1962, the Food and Drug Administration has requested that clinical trial data must be provided for all new drugs that support the safety and efficacy before the medicine can enter the market. In Canada, these products are regulated by the Food and Drug Act of Canada since before January 2004 (Johnson and Boon, 2007). Therefore, the pharmaceutical industry develops phytochemical and biological processes that guarantee reliability in the consumption of its products. These processes include separation and characterization extraction methods.

8.4   TECHNIQUES FOR THE EXTRACTION, ISOLATION, AND CHARACTERIZATION OF ACTIVE COMPOUNDS

Medicinal plants synthesize a wide variety of compounds with medicinal properties. Among the main groups of compounds are phenylpropanoids, flavonoids

TABLE 8.2 Active Compounds Related to the Properties of Medicinal Plants

Plant	Properties	Active Compounds	Reference
Aesculus hippocastanum	Antioxidant, anti inflammatory, venoconstrictor, immunomodulator, antitumor, antiviral, and antifungal	Escin (beta-escin) (triterpenic saponin)	(Vašková et al., 2015; Michelini et al., 2018; Patlolla and Rao, 2015; Pittler and Ernst, 2012; Felixsson et al. 2010; Sirtori; 2001)
Allium cepa	Antioxidant, antimicrobial, antimelanogenesis, antispasmodic, and antiproliferative	Phenolics, sulfur compounds, and essential oils	(Benkeblia, 2005; Benkeblia, 2004; Prakash et al., 2007; Nuutila et al., 2003; Marrelli et al., 2019)
Arnica montana	Anti-inflammatory, Antioxidant, and cytoprotective	Helenalin (sesquiterpene lactone), flavonoids, and phenolic acids	(Olioso et al., 2016; Craciunescu et al., 2012; Lyß et al., 1998)
Atropa belladonna	Sedative, local analgesic, hypertensive antiarrhythmic, antiulcer, and antiviral	Atropine (alkaloid), scopolamine (alkaloid), and hyoscyamine (alkaloid)	(Bousta et al., 2001; Kwakye et al., 2018)
Silybum marianum L. Gaertn	Antioxidant, antihepatotoxic, anti-inflammatory, antitumor, antibacterial,	Silymarin extract (polyphenolic flavonolignans): major components are silybin, taxifolin, and quercetin (flavonoids)	(Rakelly de Oliveira et al., 2015; Ebrahimpour koujan et al., 2015; Wellington and Jarvis, 2001)
Chenopodium anthelminticum Linn	Cytotoxic against tumor cell lines	Ascaridol isolated from essential oils	(Yadav et al., 2007; Efferth et al., 2002)
Cinnamomum zeylanicum	Antidiabetic, antioxidant, antimutagenic, antitumor, and antimicrobial	Essential oils, such as eugenol and cinnamaldehyde, and cynammic acetate	(Cardoso-Ugarte et al., 2016; Husain et al. 2018; Verspohl et al., 2005)
Colchicum autumnale	Antirheumatic and anti-inflammatory	Colchicine (alkaloid)	(Akram, 2012; Kurek, 2018)
Echinacea angustifolia	Anti-inflammatory, immunostimulator, antioxidant and antimicrobial	Flavonoids, lipophilic compounds, such as alkamides and ketoalkenes, and hydrophilic phenolic compounds, such as caffeic acid derivatives	(Shaffique et al., 2018; Bonomo et al., 2017; Kindscher, 2016)

TABLE 8.2 *(Continued)*

Plant	Properties	Active Compounds	Reference
Hamamelis virginiana	Astringent, anti-inflammatory, antioxidant, antiviral, and inhibits cancer cell proliferation	Tannins, such as proanthocyanidins, hamamelitannin pentagalloyl-glucose, gallotannins, and gallates	(Sánchez-Tena et al., 2012; Touriño et al., 2008; Theisen et al., 2014)
Juglans regia	Cytotoxic against various human cancer cell lines: lung (NCI-H322 and A549), breast (T47D), skin (A-431), prostate colon (Colo-205 and HCT-116), and prostate (PC-3 and DU-145)	Juglone	(Panth et al., 2016)
Lippia mexicana	Anti-inflammatory	Lippiolide and lippiolic acid	(Maldonado et al., 2010)
Lophophora williamsi	Acts as a neurotransmitter	Mescaline	(Hassan et al., 2017)
Matricaria chamomilla	Antimicrobial Fungistatic Antiseptic	α-bisabolol Umbelliferone Chamazulene and α-bisabolol	(Singh et al., 2011)
Persea americana	Anti-inflammatory Antiparasitic (against epimastigotes and trypomastigotes) Virustatic (inhibiting HIV syncytium formation and viral p24 antigen formation)	1,2,4-Trihydroxyheptadec-16-ene 1,2,4-Trihidroxiheptadec-16-eno Quercetin	(Yasir et al., 2010)
Piper nigrum	Anticancer and antimetastatic	Piperine	(Paarakh et al., 2015)
Plantago major	Antibacterial, antiallergic, anti-inflammatory, antioxidant, and enzyme inhibitory activities	Plantamajoside	(Amakura et al., 2012)
Salvia officinalis	Antifungal	Carnosol and 12-methoxy-trans-carnosic acid	(Kerkoub et al., 2018)
Taxus baccata	Anticancer	Taxol	(Kooti et al., 2017)
Ruta graveolens	Anti-inflammatory	Skimmianine	(Mancuso et al., 2015)

The list of plants was compiled from the Pharmacopoeia of the Mexican United States (Bartomeu Costa-Amic and Aldape Barrera, 1998).

and related compounds, terpenoids, and alkaloids. Phenylpropanoids represent a group of compounds synthesized by secondary plant metabolism through the shikimic acid pathway and include lignans and coumarins. Flavonoids, quinones, tetracyclines, and acetogenins are a group of compounds synthesized by the malonyl coenzyme-A pathway. Terpenoids (monoterpenes, diterpenes, sesquiterpenes, steroids, and limonoids) are synthesized through the mevalonic acid pathway and alkaloids by the three metabolic pathways (Solís-Salas et al., 2019; Domínguez, 1979; Tetali, 2019; Nabavi et al., 2018). According to its chemical structure, each group of compounds has characteristics of affinity for solvents or extraction, separation, and identification conditions (Domínguez, 1979). Therefore, some extraction, isolation, and characterization techniques of the main groups of bioactive compounds identified in medicinal plants are described below.

8.5 TECHNIQUES FOR THE EXTRACTION OF ACTIVE COMPOUNDS

The techniques for extraction, isolation, and identification of medicinal plants vary according to the group of compounds to be isolated. The extraction techniques most frequently used are extractions with water, ethanol, or their mixtures (He et al., 2019). However, also are used extractions with solvents such as methanol, ethanol, acetone, chloroform, ethyl acetate, dichloromethane, and petroleum ether (Cam et al., 2019; Ohikhena et al., 2018). The techniques most frequently used for this objective are decoction, infusion, maceration with solvents, continuous extraction soxhlet, reflux extraction, percolation, and distillation (Zhong et al., 2019; Malik et al., 2019). In recent years, variants of these methods have been developed using alternative forces such as ultrasound, microwaves, and electric pulses to increase extraction yields (Matias et al., 2019).

8.5.1 DECOCTION AND INFUSION

The decoction is an extraction technique that uses water as a solvent. In this process, 10 g of plant material is placed in 1 L of water and boiled (Rasheed et al., 2018; Pukdeekumjorn et al., 2016). The roots, leaves, flowers, and foliated stems are boiled in water for about 15 min, while the branches, bark, or trunk for up to 1 h, with replenishment of evaporated water. The infusion is an extraction technique that uses water as a solvent. In this process, 10 g of plant material is placed in 1 L of hot water until cooled to room temperature (Mirosławski and Paukszto, 2018).

8.5.2   MACERATION

The maceration technique is carried out below 50 °C regularly from 15 to 20 °C (Loh et al., 2017) and uses water or alcohol as solvents (Temrangsee et al., 2011), but white or red wines can be used during the process. On the other hand, the preparation of homeopathic tinctures uses formal ethanol (95% v/v) for maceration (Bartomeu Costa-Amic and Aldape Barrera, 1998). The amount of plant extracted is usually one part of plant per 20 g of solvent. Mucilaginous plants such as marshmallow and flax are marinated for approximately 30 min, while bitter and aromatic plants for 2–12 h.

8.5.3   SOXHLET EXTRACTION

Extraction in Soxhlet equipment is another solvent extraction technique, in which a known quantity of solvent placed in a flask. Then, the flask is boiled in the lower part of the equipment. Therefore, the solvent vapor rises through the outer tube. At that point, the solvent condenses in the receptacle containing a thimble of filter paper with the plant. When the solvent reaches the upper part, it is returned to the flask located in the lower part through a "U"-shaped siphon and is continuously concentrated (Harvey, 2004). On the other hand, the extraction of active principles from medicinal plants with solvents is based on the solubility of the compounds of interest, while in other cases, adjusting the pH is essential.

8.5.4   PERCOLATION

It is similar to cold maceration (15–20 °C) but using a circulating liquid. It is the method used to extract the substances from medicinal plants through a chromatographic support (Misra et al., 2018; Gholamhoseinian et al., 2009). The recovery of anti-inflammatory flavonoids and genins from *Agave americana* is an example since they are recovered by fractionation with 70% w/w ethanol (Misra et al., 2018).

8.5.5   DISTILLATION

Distillation is an extraction technique that is often used to recover essential oils from plants. Hydrodistillation is the most commonly used technique (Cicció,

2004), although some compounds are recovered by distillation by pyrolysis using microwaves since this technique has advantages such as reduced cost, extraction time, energy consumption, and CO_2 emissions (Cardoso-Ugarte et al., 2013). The extraction of essential oils from *Tagetes lucida* and *Abies sibirica* L. are examples as they are recovered with hydrodistillation and microwave heating methods on an extractor STARTE Microwave Extraction System, respectively (Cicció, 2004; Noreikaitė et al., 2017).

8.5.6 REFLUX EXTRACTION

The reflux extraction is carried out in a round flask containing the plant material and the solvent. Subsequently, the flask is connected to a refrigerant and heated below 50 °C for at least 30 min. The extraction time varies according to the vegetal material. Finally, the reflux system is disassembled and the extract is recovered (Tian et al., 2019).

8.5.7 SUPERCRITICAL FLUIDS EXTRACTION

The extraction of active compounds from medicinal plants using fluids in the supercritical state is carried out in two stages: (1) *Extraction stage,* in which the solvent flows through a static bed formed by solid particles and dissolves the extractable components of the solid. The extract is concentrated in the flow direction of the supercritical solvent. (2) *Separation stage*, in which the solute is precipitated and deposited at the bottom of the separation cell. Then, the solvent returns to the extraction process. This technique is useful for the green, efficient, and simultaneous extraction of both volatile and nonvolatile bioactive phytochemicals (Kuś et al., 2018). However, when this method is performed, the previous treatments (control of particle size, shape, and distribution of the particles), the shape of the bed, and the surface of the pores should be taken into account.

8.6 CONDITIONS FOR THE EXTRACTION OF ACTIVE COMPOUNDS

The extraction of active compounds from medicinal plants is carried out by different techniques that depend on the chemical structure of each compound. In any technique, the plant dries below 50 °C and is pulverized to fine grains to break the cells and vacuoles to release the compounds of interest.

8.6.1 COUMARINS

The coumarins are in free or glycosylated form in plants. Free coumarins easily extracted with water or ethanol, while the coumarin glycosides with sequential extractions, initiating the extraction with methanol or ethanol and continuing with petroleum ether (Domínguez, 1979). The coumarin aglycone recovered by acid or enzymatic hydrolysis. The isolation of coumarin from *Citrus grandis* by an ethanolic extraction with reflux is an example (Tian et al., 2019).

8.6.2 LIGNANS

Lignans are oxygenated dimers of simple phenylpropane with a β–β′ bond. These compounds have a parallel, antiparallel, and pseudo-parallel association into their chemical structure. The glycosylated lignans readily dispersed in water or ethanol. However, some are recovered with solvents such as acetone. Additionally, lignans are isolated with methanol followed by partitions with solvents of different polarities.

Lignans such as schisandrol A and schisandrol B, extracted in mixtures of ethanol–water (95%, 70%, 50%, and 30%) from *Schisandra chinensis* (Kim et al., 2011), as well as the extraction of lignans from *Herpetospermum pedunculosum*, are an example (Ma et al., 2019). Additionally, lignans such as schisandrol A, schisandrol B, gomisin G, schisantherin A, schisantherin D, schisanhenol, (+)-anwulignan, deoxysquisandrine, schisandrin B, schisandrin C, 6-O-benzoyligisin O, and interiotherin A of *Schisandra sphenanthera* fruits were extracted with ultrasound (Liu et al., 2012). Dimethylether of pinoresinol magnolinc, cpi-magnoline Λ, fargesin, and demethoxyschantin recovered in the extract of dried flower buds from *Magnolia biondii Pamp* are some examples of ethanolic extracts (Zhao et al., 2007).

8.6.3 FLAVONOIDS

Flavonoids are synthesized in medicinal plants such as aglycones or glycosides. Therefore, they are easily extracted in water and ethanol. The glycosylated flavonoids are sparingly soluble in other organic solvents since the sugar residue influences the solubility of the compound. Flavonoids contain at least one aromatic ring and a phenolic group in their chemical structure (Ávalos García and Pérez-Urria Carril, 2009). Among them are the traditionally used techniques such as extraction with hot water under pressure, solid–liquid extraction, extraction with fluid under pressure, and alternative procedures, such as extraction with supercritical fluids, ultrasound-assisted extraction, microwave-assisted extraction, and pulsed electric field extraction (Baiano, 2014). Therefore, it is necessary to establish different parameters such as flow rate, solvent, temperature, pressure, and time to obtain high performance of active compounds.

On the other hand, to extract flavonoids, microorganisms and enzymes can be applied to plant materials. This procedure decreases the use of solvents used in the traditional extraction methods, decreases the extraction time, increases the yield, and improves the quality of the obtained flavonoids (Wang et al., 2010). The fermentation of microorganisms allows modifying the plant matrix to facilitate the release of flavonoids. The microorganisms most commonly used to facilitate the extraction of flavonoids are *Bacillus pumilus*, *Bacillus subtilis*, *Lactobacillus acidophilus*, *Lactobacillus johnsonii*, *Lactobacillus reukusus*, *Aspergillus oryzae*, *Monascus purpureus*, *Aspergillus niger*, *Lentinus edodes*, *Rhizopus oryzae*, and *Rhizopus oligosporus* (Il´ina, 2013).

Enzymatic extraction of flavonoids is promising compared to solvent extraction. The enzymatic extraction of flavonoids has benefits such as the high specificity and regioselectivity of the enzymatic activity as well as the capacity of the enzymes to catalyze reactions in aqueous media under mild conditions. Additionally, the technique is a procedure that respects the environment (Il´ina, 2013). Enzymes (mainly hydrolases) of different sources such as fungi, bacteria, plant extracts, and animal organs hydrolyze the biopolymer components of cell walls to increase their permeability and the performance of the extraction of flavonoids and other physiologically active substances (Wang et al., 2010; Castro-Vazquez et al., 2016). Among them are hydrolases, glucanases, hemicellulases, cellulases, and pectinases that are used alone or in mixtures during extraction.

During the enzymatic extraction, the optimization of the parameters is fundamental. Therefore, the enzyme–substrate relationship, the treatment time, pH, and temperature must be established (Chávez-Santoscoy et al., 2016; Tomaza et al., 2016). Additionally, 90% of the glycosides of some flavonoids are transformed into aglycones (Mandalari et al., 2006), which facilitates its characterization.

However, enzymatic and microbial methods have some disadvantages. Among them are the lack of complete hydrolysis of the cell wall of the plant, the difficulty to expand enzymatic and microbial extractions on an industrial scale, and the relatively high cost of biocatalysts for large volumes of raw material treatment (Wang et al., 2010).

8.6.4   QUINONES

Quinones are unsaturated diacetones, easily reduced and converted into polyphenols that regenerate by oxidation. They are divided in benzoquinones, naphthaquinones, anthraquinones, and phenatraquinones. Quinones are synthesized in plants as aglycones or glycosides. The aglycones are extracted with nonpolar solvents such as petroleum ether and hexane, while the glycosides are extracted with water, ethanol, or mixtures of both. Some are extracted by reflux or with soxhlet equipment, while others with steam entrainment such as nonhydroxylated *p*-benzoquinones and some naphthoquinones (Domínguez, 1979).

The cytoprotective eleutherins A and B extracted in methanol from the bulbs of *Eleutherine americana* L. Merr. (Chen et al., 2019), as well as the diterpenic quinones such as tebesinone B and aegyptinone A with isolated methanolic extract from *Salvia tebesana* Bunge (Eghbaliferiz et al., 2018), are an example.

8.6.5   ACETOGENINS

Acetogenins are compounds that are synthesized exclusively in Annonaceae plants and belong to the polyketides. Their chemical structure is often characterized by an unbranched C32 or C34 fatty acid ending in a γ-lactone. Some are antiparasitic and other cytotoxic that cause the apoptosis of cancer cells. (Li et al., 2008) According to the chemical structure, they are extracted with methanol and partitioned with chloroform. The acetogenins isolated of the methanol extract from *Annona montana* are an example (Liaw et al., 2005).

8.6.6   TERPENOIDS

Terpenes are a wide and varied class of hydrocarbons, produced mainly by a wide variety of plants, particularly conifers. Terpenes and terpenoids are the constituents of the essential oils used as additives in pharmaceutical products such as ointments. According to their size, they classified as hemiterpene (C_5H_8), monoterpene ($C_{10}H_{16}$), sesquiterpene ($C_{15}H_{24}$), diterpene ($C_{20}H_{32}$), sesterterpene ($C_{25}H_{40}$), triterpene ($C_{30}H_{48}$), and tetraterpene ($C_{40}H_{56}$). The main terpenoids of pharmaceutical importance are monoterpenes, diterpenes, sesquiterpenes, steroids, and limonoids.

Monoterpenes are extracted by hydrodistillation by steam trawl. Limonene, linalool, citronellal, and citronellol extracted from *Citrus aurantifolia* are examples (Lemes et al., 2018). The diterpenes can be extracted with ethanol. Metaglyptin A (abietane diterpenoid) extracted from *Metasequoia glyptostroboidesis* is an example (Tu et al., 2019).

Los sesquiterpenoids as sesquiterpene lactones are polar compounds, so they are insoluble in petroleum ether. However, some are insoluble in water. These compounds can be dissolved in ethanol, methanol, chloroform, ethyl ether, and dichloromethane easily.

Steroidal compounds such as dioscin (steroid saponin isolated from *Dioscorea nipponica* Makino with antiarthritic effect) (Cao et al., 2019) are very polar and synthesized in plants as aglycones or glucosides. Consequently, they are extracted in water, ethanol, and mixtures of both. Sitosterol-3-O-β-D-glucoside isolated in the ethanol extract from *Moringa oleifera* seeds is an example of a steroidal compound with antipsoriasis effect (Ma et al., 2018). Calotroposide A isolated in the ethanol extract from *Calotropis gigantea* is another example of a steroidal compound with anticancer activity against WiDr colon cancer cells (Mutiah et al., 2018).

Additionally, in ethanol extracts from *Lyonia ovalifolia* var. hebecarpa was isolated triterpene glycosides of lanostane with a cytotoxic effect on cancer cell lines (SMMC-7721, HL-60, SW480, MCF-7, and A-549) (Teng et al., 2018).

8.6.7 LIMONOIDS

The limonoids can be extracted with solvents of intermediate polarity such as chloroform, ethyl acetate, and ethanol with previous degreasing. Krishnagranatins A–I limonoids isolated in the ethanol extract from *Xylocarpus granatum* are examples (Liu et al., 2018).

Additionally, these compounds are extracted by alkaline hydrolysis and deesterification. The antimethanogenic limonoids of the hexane extract from *Azadirachta indica* seeds are an example. Six of salannin-type (ohchinin, ohchinin acetate, 1-benzoyl-1-decinnamoylohchinin, 3-deacetylsalannin, salannin, and 3-deacetyl-28-oxosalannin), two of nimbin-type (6-deacetyl-nimbin and nimbin), six isosalannins (isoohchinin, isoohchinin acetate, 1-benzoyl-1-decinnamoylisoohchinin, 3-deacetylisosalannin, isosalannin, and 3-deacetyl-28-oxoisosalnnin), and two isonimbins (6-deacetylisonimbin and isonimbin) were isolated from this plant (Akihisa et al., 2016).

8.6.8   ALKALOIDS

Most alkaloids are found in plants as salts of acids or organic bases so they can be extracted depending on the pH. However, some alkaloids glycosylated with units of rhamnose, galactose, and glucose and recovered with water or polar solvents (Domínguez, 1979).

The alkaloid extraction of the acidified methanolic extract from *Chelidonium majus* using ultrasound is an example of extraction with solvents as a function of pH (Wójciak-Kosior et al., 2019). From this plant as well as *Nandina domestica*, an isoquinoline alkaloid was isolated called protopine that inhibits the proliferation of colon cancer cells HCT116 (Son et al., 2019). Additionally, bisindole alkaloids from *Tabernaemontana corymbosa* were isolated of an ethanolic extract acidified with 3% tartaric acid (Sim et al., 2019).

Table 8.3 presents some techniques for extracting active compounds isolated from medicinal plants.

8.7   TECHNIQUES FOR THE PURIFICATION AND CHARACTERIZATION OF ACTIVE COMPOUNDS

8.7.1   PURIFICATION TECNIQUES

The bioactive compounds of medicinal plants are isolated by separation and characterization techniques according to the chemical structures of the compounds of interest.

The compounds regularly purified by different separation techniques, including liquid–liquid partitioning, gravitational liquid chromatography, vacuum liquid chromatography, liquid flash chromatography, TLC, GC, ion-exchange chromatography, medium-pressure liquid chromatography (MPLC), HPLC, and ultrahigh-performance liquid chromatography (UPLC). The selection and implementation of the technique are based on the chemical structure of each compound as well as the affinity it has for both the mobile phase and the stationary phase (Cela et al., 2002).

In some types of chromatographic techniques, it is necessary to establish adequately the critical conditions for the separation process to be efficient and reproducible. Therefore, in the separation of compounds by gravitational liquid chromatography, vacuum liquid chromatography, and liquid flash chromatography, it is necessary to define the diameter of the separation column according to the amount of extract (Hernández-Ocura et al., 2017;

TABLE 8.3 Extraction Techniques Used to Extract Compounds From Medicinal Plants

Plant/Compound Group	Compound	Property	Extraction Technique/ Solvent/Conditions	Reference
Citrus grandis/coumarin	Meranzin hydrate I	Hepatoprotective activity	Ethanol extraction 79%/ reflux twice/1.5 h each time	(Tian et al., 2019)
Herpetospermum pedunculosum/lignans	Herpetosiol A, Herpetosiol C	Human cancer cell lines	Cold maceration/room temperature/acetone 70%	(Ma et al., 2019)
Agave americana/ flavonoids	Flavonoids and genins	Anti-inflammatory activity	Percolation/ethanol 70%/ room temperature	(Misra et al., 2018)
Selaginella helvetica/ flavonoids	Biflavonoids	Antioxidant	Percolation/ethanol 95%/24 h/flow rate of 3 mL/min	(Jiang et al., 2018)
Eleutherine americana L. Merr./quinones	Eleutherins A y B	Cytoprotective	Cold maceration/room temperature/methanol	(Chen et al., 2019)
Salvia tebesana Bunge./ diterpenic quinones	Tebesinone B and aegyptinone A	Cytotoxic activity on MCF-7, B16F10, PC-3, and C26 human cancer cell lines	Cold maceration/room temperature/methanol	(Eghbaliferiz et al., 2018)
Annona montana/ acetogenin	Montalicin G, montalicin H, monlicin A, monlicin B, (+)-monhexocin, ()-monhexocin, murisolin, 4-deoxyannomontacin, and muricatacin	Cytotoxicity against human hepatoma cells, Hep G2	Cold maceration/room temperature/methanol	(Liaw et al., 2005)
Citrus aurantifolia/ essential oils	Limonene, linalool, citronellal and citronellol	Antibacterial activity	Hydrodistillation/Clevenger-type apparatus/3 h	(Lemes et al., 2018)
Metasequoia glyptostroboides/abietane diterpenoid	Metaglyptin A	Cytotoxicity against MDAMB-231 cell line	Cold maceration/room temperature/ethanol 95%/24 h	(Tu et al., 2019)
Elaeagia utilis/ Sesquiterpenlactones	Loliolide	Antimicrobial activity	Sequential cold maceration for 72 h/petroleum ether and ethanol	(Aldana et al., 2013)

TABLE 8.3 *(Continued)*

Plant/Compound Group	Compound	Property	Extraction Technique/Solvent/Conditions	Reference
Moringa oleifera/steroidal compound	Sitosterol-3-*O*-β-D-glucoside	Antipsoriasis-like lesions in vivo	Cold maceration/room temperature/EtOH-H$_2$O (95:5)	(Ma et al., 2018)
Lyonia ovalifolia var. hebecarpa/glucósidos triterpénicos de lanostano	Hebecarposides C and H	Potent antiproliferative activities against MCF-7 cell line	Maceration/45 °C/ethanol 95%/24 h	(Teng et al., 2018)
Chelidonium majus/alkaloids	Protopine	Anticancer	Ultrasonic 35 kHz/methanol	(Wójciak-Kosior et al., 2019; Son et al., 2019)
Crinum latifolium/alkaloids	9-Methoxy-cripowellin B and 4-methoxy-8-hydroxy-cripowellin B	Antimicrobial, antioxidant and antiinflamatory activities	Cold maceration for 24 h at room temperature/ethanol 95%	(Chen et al., 2018)
Kopsia fruticosa/alkaloids	Kopsiafrutine C, kopsiafrutine D, and kopsiafrutine E	Cytotoxicity against all of seven tested tumor cell lines (HS-1, HS-4, SCL-1, A431, BGC-823, MCF-7, and W480) and antimicrobial and antifungal activities	Cold maceration three times for 15 h under reflux/ethanol 80%	(Long et al., 2018)
Azadirachta indica/limonoids	3-Deacetyl-28-oxosa-lannin, nimbidinol, 17-epi-17-hydroxynimboci-nol, 17-epi-17-hydroxy-15-methoxynimbocinol, and 7-deacetyl-17-epinimolicinol, in addition to one natural limonoid, ohchinin	Antimethanogenic activity	Alkaline hydrolysis and deesterification	(Akihisa et al., 2016)

Harris, 2001). In these chromatographic techniques, the separation columns with a porous glass filter with a diameter according to the amount of the extract were used. Extracts of less than 100 mg use diameters of 0.5–1 cm, extracts of 0.5–1 g use 2.5–4 cm, and extracts of 1–10 g use diameters greater than 5 cm.

The conventional height of the column used for vacuum liquid chromatography is 4–5 cm, while the height of the column to carry out gravitational liquid chromatography and liquid flash chromatography goes from 10–20 cm. If it is necessary to scale the separation process, the diameter should be modified but not the height of the column (Harris, 2001).

On the other hand, one of the critical factors to achieve a better separation in TLC is the selection of the eluent. Therefore, in most cases, separation profiles are made to choose the mixture of solvents that separate the extract. The eluent that separates the components of the extract with ratio of fronts (RF) of 0.1–0.2 will be the optimum for the separation. The RF is the ratio between the distances travelled by the extract and by the eluent. This value is measured from the area of application of the sample. RF = distance travelled by the extract/distance traveled by the eluent. Another factor to take into account is that the thickness and the measurements of the chromatographic support used in the separation are always the same (Hernández-Ocura et al., 2017; Wagner and Bladt, 2001). In ion-exchange chromatography and HPLC–UPLC, one of the most important factors is the selection of the chromatographic support. Therefore, the characteristics of the chromatographic support used in the separation should be consulted. The selection of the chromatographic support is based mainly on the particle size and in some cases on the ionogenic groups interlaced to the chromatographic support that is regularly Sephadex (Cela et al., 2002).

Currently, there are prefabricated columns filled with the chromatographic support. However, it is sometimes necessary to prepare the column manually. Therefore, it is necessary to condition the Sephadex particles in MeOH for 24 h at rest. Because of this, it calculates the amount of support according to the area of the column and the degree of expansion of the chromatographic support. This data is regularly in the labeling of the material. Once the column is packed, the extract completely dissolved is eluted (Cela et al., 2002; Harris, 2001).

Purification techniques that employ pressures, such as vacuum liquid chromatography, liquid flash chromatography, ion-exchange chromatography, HPLC, and UPLC, establish parameters that allow reproducible elution, such as the amount of sample, flow, injection, detector, among others.

A liquid–liquid partition is usually carried out, followed by a separation process in column chromatography to obtain partially purified fractions; then, the automated methods such as HPLC are used to achieve the purification of some compounds.

Currently, the development of analytical separation and purification techniques, such as HPLC, TLC, and GC, and identification techniques, such as MS, NMR, GC–MS, and liquid chromatography coupled to mass spectrometry (LC–MS), has allowed the identification of a large number of secondary metabolites: flavonoids, terpenes, phenolic compounds and their esters, sugars, hydrocarbons, and mineral elements present in various plants (Huang et al., 2014).

The HPLC has displaced the GC–MS technique for the quantification of compounds since it has the disadvantage of derivatization of components. The most used detection systems have been UV detectors, diode arrays, and mass spectrometers (Gómez Caravaca et al., 2006). Colorimetric methods have also been used (Chang, 2002); spectroscopic techniques (Popovaa et al., 2007), capillary electrophoresis in the zone capillary electrophoresis (CZE) zone, micellar electrokinetic chromatography (MECK) micellar, and IR spectroscopy (Zhang et al., 2013), having a large number of analytical tools, allow characterizing various secondary metabolites.

8.7.2 HIGH-PERFORMANCE LIQUID CHROMATOGRAPHY

It is a technique that allows the distribution of the analyte of interest between two phases, a mobile and a stationary one (Gilbert, 1987). This analytical technique is the most used for the characterization of compounds; the chromatographic conditions of HPLC methods include a reverse-phase C_{18} column, a UV–vis or diode array detector, and a binary solvent system containing acidified water and a polar organic solvent. HPLC–MS MS is the most used technique for the separation and determination of metabolites; however, it has some disadvantages in the limit of detection and sensitivity, especially for plant extracts of complex matrixes and environmental samples. Therefore, the initial preconcentration and purification of the compounds are important before instrumental analysis (Pellati et al., 2011; Falcão et al., 2013). The most common detectors in HPLC are absorbance, fluorescence, electrochemical, refractive index, conductivity, Fourier transform infrared (FT-IR) spectrometer, and light scattering and MS detectors (Gilbert, 1987).

8.7.3 GAS CHROMATOGRAPHY

The determination of components by GC allows a qualitative as well as a quantitative analysis to be carried out. However, sample preparation is complex due to the need for removing impurities from the extract, which may affect the reproducibility of the results. GC–MS is the most used method since MS detection allows the acquisition of molecular mass data and structural information that facilitates the identification of compounds (Sawaya et al., 1983). It is a technique that has a great separation capacity and offers high sensitivity and selectivity when combined with MS (Pellati et al., 2011). The common detectors in these systems are a flame ionization detector, nitrogen phosphorus detector, flame photometric detector, thermal conductivity detector, and electronic capture detector (Gilbert, 1987).

8.7.4 FOURIER TRANSFORM INFRARED SPECTROSCOPY

IR spectroscopy is an analytical technique that measures the wavelength and absorption intensity of IR light in a sample; it is possible only because molecular absorption generates changes in the state of vibrational and/or rotational energy. The resulting spectrum contains information specific to the aspects of molecular vibration of the sample, its physical properties, and its unique interaction with the instrument. The absorption data obtained in the spectrum give structural information that represents the molecular details of the sample that is measured. Each functional group of a molecule has a unique vibrational frequency that can be used to determine which functional groups are present in a sample. When the effects of all the different functional groups are together, the result is a unique "molecular fingerprint" that can be used to characterize the identity of a sample. Depending on the type of link, there are several types of molecular vibrations that absorb at unique wavelengths or near-IR (NIR) energy frequencies. Some NIR spectrometers are capable of measuring transmittance (direct and diffuse), reflectance (specular and diffuse), etc. The IR region is not especially useful, by itself, as a quantitative measurement technique. However, it is a quick, easy to perform, and nondestructive method that can record spectra of solid and liquid samples without any previous treatment. One of its main advantages is that it implies a low-cost operation and provides spectrum quickly, being able to predict physical and chemical parameters from a single spectrum.

The analytical information contained in the typically large and widely overlapping bands of NIR spectra is not selective and is influenced by a series of physical, chemical, and structural features. In addition, the differences between the samples can cause very slight spectral differences that are difficult to distinguish with a simple view (Blanco and Villaroya, 2002).

8.7.5   UV–VIS SPECTROPHOTOMETRY

At present, sufficient information can be obtained for rapid or "screening" evaluation of the total polyphenol content, total flavonoids, total catechin, total coumarins, etc. and used as a criterion of the nutritional value of food products derived from plants. These analyses can be easily performed from spectrophotometric measurements, using specific analytical reagents that have adequate selectivity to avoid obtaining false-positive values that affect the quality of the estimate. The detection of various compounds can be done through a UV–vis spectrophotometer since they efficiently absorb UV light. The advances of the "green chemistry" and the need for simultaneously carrying out a high number of samples have led to the development of innovations of the classical methods, drastically decreasing the contaminating residues of the laboratory and making the management of resources more efficient. In this way, the volumes of reagents have been adjusted for this new format, which allows the analysis of a high number of samples simultaneously in a 96-well microplate. This is how in recent years many modified methods have been developed for small volumes, resulting in being as efficient as the results obtained by classical methods (Herald et al., 2012).

8.7.6   NUCLEAR MAGNETIC RESONANCE

NMR is widely used for the elucidation of structures, mainly of an organic nature, since in principle it focuses on the study of proton NMR, subsequently expanding to other magnetically active nuclei. Therefore, this technique was almost exclusively a tool for organic chemists; however, with the development of magnets and superconducting materials, its field of application has been extended. Currently, the fields of application are biochemistry and biology, pharmacology and toxicology, medicine, environment, and food. It is noteworthy the great boom that has experienced in recent years in the field of analytical quantification. It is a spectroscopic method for nondestructive analysis, which is based on the absorption of electromagnetic radiation energy by atomic nuclei that have a magnetic moment. The nuclear spin electrons, neutrons, and protons, constituents of the nucleus of an atom, possess an intrinsic property that is given by the so-called spin term. The spin is the fourth quantum number that takes place when solving, in an approximate way, the wave equation. The spin represents a general property of the particles that can easily be explained by similarity to the properties of the electrons. The electrons of the atom circulating around its nucleus generate a magnetic field associated with an angular momentum described with the quantum number of spin $+1/2$ and $-1/2$. For NMR, the proton and neutron spins of the nucleus are very interesting. In the nucleus, the protons and neutrons are paired combining the spins $+1/2$ and $-1/2$. Following are some conditions for the purification and characterization of active principles of medicinal plants (Claridge, 2016).

8.8 CONCLUSION

In recent years, humanity has returned to the use of medicinal plants to improve their health. However, many plants lack sufficient studies to guarantee the safety of their consumption. Therefore, the researchers focus their investigations on studying medicinal plants of interest and establish techniques for the extraction, isolation and identification of active compounds. Additionally, they develop biological assays to evaluate the effectiveness of the compounds. Among the most frequently used chemical extraction techniques are decoction, infusion, maceration, soxhlet extraction, percolation, reflux, and extraction with supercritical fluids. Ethanol is the solvent that is used most often without considering the extraction technique. However, nonpolar solvents or enzymatic techniques are used to extract some compounds.

On the other hand, the most commonly used techniques for purifying medicinal plant compounds include liquid–liquid partitioning, gravitational liquid chromatography, liquid vacuum chromatography, liquid flash chromatography, TLC, GC, ion-exchange chromatography, MPLC, HPLC, and UPLC.

The selection and implementation of the technique are based on the chemical structure of each compound and the affinity it has for both the mobile phase and the stationary phase. Finally, the techniques for purifying and identifying the compounds include HPLC-MS, GC-MS, NMR, IR spectroscopy, and UV–visible spectroscopy.

TABLE 8.4 Conditions for the Purification and Characterization of Active Compounds of Medicinal Plants

Compound/Plant/ Group Compound	Purification Conditions	Caracterization Conditions	Reference
Meranzin hydrate I/*Citrus grandis*/coumarin	HPLC analyses were performed on a Waters 2695 separation module (Waters, Manchester, U.K.) equipped with a 2998 photodiode array detector (Waters, Manchester, UK) and an Alltech 3300 evaporative light scattering detector (Alltech Inc., Deerfield, IL, USA) using a Phenomenex Gemini C18 column (5 μm, φ 4.6 × 250 mm; Phenomenex Inc., Torrance, CA, USA). The semipreparative and preparative HPLC were carried out on a Waters 1515 isocratic HPLC pump (Waters, Manchester, U.K.) coupled to a 2489 UV/vis detector (Waters, Manchester, U.K.) using a Phenomenex Gemini C18 column (5 μm, φ 10 × 250 mm; Phenomenex Inc., Torrance, CA, USA).	1D- and 2D-NMR spectra were obtained with a Bruker AV 600 instrument (Bruker Co. Ltd., Bremen, Germany) using solvent signals (DMSO-d6, δH 2.50/δC 39.5; CD3OD, δH 3.31/δC 49.0) as an internal reference; IR spectra were obtained on a JASCO FT/IR-480 plus spectrometer (JASCO International Co. Ltd., Tokyo, Japan); UV/vis spectra were acquired on a JASCO V-550 UV/vis spectrometer (JASCO International Co. Ltd., Hachioji, Tokyo, Japan); and HR-ESI-MS spectra were obtained on a Waters Synapt G2 mass spectrometer (Waters, Manchester, U.K.).	(Tian et al., 2019)
Herpetosiol A, herpetosiol C/*Herpetospermum pedunculosum*/lignans	Column chromatography (CC) was performed with silica gel (200–300 mesh or 300–400 mesh, Qingdao Marine Chemical Factory) and Sephadex LH20 (GE Healthcare Bio-Sciences AB, Uppsala, Sweden); TLC was performed on silica gel 60 F254 (Merck, Germany); the spots on TLC were visualized by UV light (254 nm) and sprayed with 10% H₂SO₄, followed by heating; semi-preparative HPLC was carried out on a Shimadzu LC6AD chromatograph equipped with an ODS column (RP-18, 250 × 10 mm YMC, 5 μm) with a flow rate of 2.0 ml/min	IR spectra were recorded as KBr pellets;UV spectra were measured on a Shimadzu UV-260 spectrophotometer in absolute methanol (MeOH); λ max (log ε) in nm; NMR spectra were obtained at 400 MHz for ¹H NMR and 100 MHz for ¹³C NMR, with TMS as an internal standard, CD₃COCD₃ and CDCl₃ as solvent, δ in ppm. and *J* in Hz; High-resolution mass spectra were determined on a Xevo G2-S Q-TOF MS; and CD measurements were performed using a JASCO J-1500 spectropolarimeter	(Ma et al., 2019)

TABLE 8.4 *(Continued)*

Compound/Plant/ Group Compound	Purification Conditions	Caracterization Conditions	Reference
Eleutherins A y B/ *Eleutherine americana* L. Merr./quinones	High-resolution mass spectra were determined on a Xevo G2-S Q-TOF MS; CD measurements wereperformed using a JASCO J-1500 spectropolarimeter; CC was performed with silica gel (200–300 mesh or 300–400 mesh) and Sephadex LH-20; TLC was per-formed on silica gel 60 F254; the spots on TLC were visualized by UV light (254 nm) and sprayed with 10% H_2SO_4, followed by heating; semi-preparative HPLC was carried out on a Shimadzu LC-6AD chromatograph equipped with an ODS column (RP-18,250 × 10 mm YMC, 5 μm) with aflow rate of 2.0 ml/min	IR spectra were recorded as KBr pellets on the Avatar 360 E.S.P. spectrophotometer; UV spectra were measured on a Shimadzu UV-260 spectrophotometer in absolute methanol; λ max (log ε) in nm; NMR spectra were obtained at 400 MHz for ¹H NMR and 100 MHz for ¹³C NMR, on a Bruker DRX400 spectrometer with TMS as an internal standard and CD_3COCD_3 and $CDCl_3$ as a solvent in ppm	(Chen et al., 2019)

KEYWORDS

- **Applied Techniques**
- **Extraction**
- **Purification, Characterization**
- **Medicinal Plants**
- **Active Compounds**
- **Mass spectroscopy**

REFERENCES

Abe, R.; Ohtani, K. An ethnobotanical study of medicinal plants and traditional therapies on Batan Island, the Philippines. *J Ethnopharmacol.* 2013, 145, 554—565.

Aguilar-Stoen, M.; Moe, S. R. Medicinal plant conservation and management: Distribution of wild and cultivated species in eight countries. *Biodiver Conserv.* 2007, 16, 1973—1981.

Akihisa T; Horiuchi M; Matsumoto M; Ogihara E; Ishii K; Zhang J. Melanogenesis-inhibitory activities of isomeric C-seco limonoids and deesterified limonoids. *Chem Biodivers.* 2016, 13, 1410—1421.

Akram, M. Colch*icum autumnale*: A review. *J Med Plant Res.* 2012, 6, 1489—1491.

Aldana, J.; Téllez, N.; Gamboa, F. Antimicrobial activity of fractions and subfractions of *Elaeagia utilis* against microorganisms of importance in dental caries. *Acta Odontol Latinoam.* 2013, 6,104-111.

Al-Nahain. A.; Jahan, R.; Rahmatullah, M. Zing*iber officinale*: A potential plant against rheumatoid arthritis. *Arthritis.* 2014, 2014, 8.

Alonso-Castro, A. J.; Domínguez, F.; Maldonado-Miranda, J. J.; Castillo-Perez, L. J.; Carranza-Álvarez, C.; Solano, E.; Isiordia-Espinoza, M. A.; Juárez-Vázquez, M. D.; Zapata-Morales, J. R.; Argueta-Fuertes, M. A.; Ruiz-Padilla, A. J.; Solorio-Alvarado, C. R.; Rangel-Velázquez, J. E.; Ortiz-Andrade, R.; González-Sánchez, I.; Cruz-Jiménez, G.; Orozco-Castellanos, L. M., Use of medicinal plants by health professionals in Mexico. *J Ethnopharmacol.* 2017, 198, 81—86.

Alonso-Castro, A. J.; Maldonado-Miranda, J. J.; Zarate-Martinez, A.; Jacobo-Salcedo, M. D.; Fernandez-Galicia, C.; Figueroa-Zuniga, L. A.; Rios-Reyes, N. A.; de Leon-Rubio, M. A.; Medellin-Castillo, N. A.; Reyes-Munguia, A.; Mendez-Martinez, R.; Carranza-Alvarez, C. Medicinal plants used in the Huasteca Potosina, Mexico. *J Ethnopharmacol.* 2012, 143, 292—298.

Amakura, Y.; Yoshimura, A.; Yoshimura, M.; Yoshida, T. Isolation and characterization of phenolic antioxidants from plantago herb. *Molecules.* 2012, 17, 5459—5466.

Ávalos García, A.; Pérez-Urria Carril, E. Secondary metabolism of plants. *Reduca (Biología). Serie Fisiolog Vegetal.* 2009, 2, 119—145

Aziz, M.A.; Adnan, M.; Khan, A. H.; Shahat, A.A.; Al-Said, M.S.; Ullah, R. Traditional uses of medicinal plants practiced by the indigenous communities at Mohmand Agency, FATA, Pakistan. *J Ethnobiol Ethnomed.* 2018, 14, 2.

Baiano, A. Recovery of biomolecules from food wastes–A review. *Molecules*. 2014, 19, 14821—14842.

Balunas, M. J.; Kinghorn, A. D. Drug discovery from medicinal plants. *Life Sci*, 2005, 78, 431—441.

Bartomeu Costa-Amic, L.; Aldape Barrera F. Pharmaco*poeia of the Mexican United States*, 1st ed. Altres Costa Amic y coeditors: Mexico, 1998.

Benkeblia, N. Antimicrobial activity of essential oil extracts of various onions (*Allium cepa*) and garlic (*Allium sativum*). *LWT – Food Sci Technol*. 2004, 37, 263–268.

Benkeblia, N. Free-radical scavenging capacity and antioxidant properties of some selected onions (*Allium cepa* L.) and garlic (*Allium sativum* L.) extracts. *Braz Arch Biol Technol*. 2005, 48, 753–759.

Blanco, M.; Villaroya, I. NIR spectroscopy: A rapid-response analytical tool. *Trends Anal Chem*. 2002, 21, 240–250.

Bonomo, M. G.; Russo, D.; Cristiano, C.; Calabrone, L.; Tomaso, K. D.; Milella, L.; Salzano, G. Antimicrobial activity, antioxidant properties and phytochemical screening of *Echinacea angustifolia*, *Fraxinus excelsior* and *Crataegus oxyacantha* mother tinctures against food-borne bacteria. *EC Microbiol*. 2017, 9, 173–181.

Bourgaud, F.; Gravot, A.; Milesi, S.; Gontier, E. Production of plant secondary metabolites: A historical perspective. *Plant Sci*. 2001, 161, 839–851.

Bousta, D.; Soulimani, R.; Jarmouni, I.; Belon, P.; Falla, J.; Froment, N.; Younos, C. Neurotropic, immunological and gastric effects of low doses of *Atropa belladonna* L., *Gelsemium sempervirens* L. and *Poumon histamine* in stressed mice. *J Ethnopharmacol*. 2001, 74, 205–215.

Cam, M.E.; Hazar-Yavuz, A.N.; Yildiz, S.; Ertas, B.; Ayaz Adakul, B.; Taskin, T.; Alan, S.; Kabasakal, L. The methanolic extract of *Thymus praecox subsp. skorpilii var. skorpilii* restores glucose homeostasis, ameliorates insulin resistance and improves pancreatic β-cell function on streptozotocin/nicotinamide-induced type 2 diabetic rats. *J Ethnopharmacol*. 2019, 231, 29–38.

Camou-Guerrero, A.; Reyes-García, V.; Martínez-Ramos, M.; Casas, A. Knowledge and use value of plant species in a Rarámuri community: A gender perspective for conservation. *Hum Ecol*. 2008, 36, 259–272.

Cao, Y.J.; Xu, Y.; Liu, B.; Zheng, X.; Wu, J.; Zhang, Y.; Li, X.S.; Qi, Y.; Sun, Y.M.; Wen, W.B.; Hou, L., Wan, C.P. Dioscin, a steroidal saponin isolated from *Dioscorea nipponica*, attenuates collagen-induced arthritis by inhibiting Th17 cell response. *Am J Chin Med*. 2019, 47, 423–437.

Cardia, G.F.E.; Silva-Filho, S.E.; Silva, E.L.; Uchida, N.S.; Cavalcante, H.A.O.; Cassarotti, L.L.; Salvadego, V.E.C.; Spironello, R.A.; Bersani-Amado, C.A.; Cuman, R.K.N. Effect of Lavender (*Lavandula angustifolia*) essential Oil on Acute inflammatory response. *Evid Based Complement Alternat Med*. 2018, 2018, 1–10.

Cardoso-Ugarte, G. A.; López-Malo, A.; Sosa-Morales, M. E. Cinnamon (*Cinnamomum zeylanicum*) essential oils. *In Essential Oils in Food Preservation, Flavor and Safety*. Academic Press: San Diego, 2016, Chapter 38, pp 339–347.

Cardoso-Ugarte, G.A.; Juárez-Becerra, G.P.; Sosa-Morales. M.E.; López-Malo, A. Micro-wave-assisted extraction of essential oils from herbs. *J Microw Power Electromagn Energy*. 2013, 47, 63–72.

Castro-Vazquez, L.; Alañón, M. E.; Rodríguez-Robledo, V.; Pérez-Coello, M. S.; Hermosín-Gutierrez, I.; Díaz-Maroto, M. C.; Jordán, J.; Galindo, M.F.; Arroyo-Jiménez, M. M.

Bioactive flavonoids, antioxidant behaviour, and cytoprotective effects of dried grapefruit peels (*Citrus paradisi* Macf.). *Oxid Med Cell Longev.* 2016, 2019, 8915729.

Caudal, D.; Guinobert, I.; Lafoux, A.; Bardot, V.; Cotte, C.; Ripoche, I.; Chalard, P.; Huchet C. Skeletal muscle relaxant effect of a standardized extract of *Valeriana officinalis* L. after acute administration in mice. *J Trad Complement Med.* 2017, 8, 335–340.

Cela, R.; Lorenzo, R. A.; Casais, M.A. *Separation Techniques in Analytical Chemistry.* 1st ed. Síntesis: Spain, 2002.

Chang, M. A genoma -wide screen for methyl methanessulfonate-sensitive mutans reveals genes required for S pase progression in the presence of DNA damage. *Prod Natl Acad Sci USA.* 2002, 99, 16934–16939.

Chávez-Santoscoy, R. A.; Lazo-Vélez, M. A.; Serna-Sáldivar, S. O.; Gutiérrez-Uribe, J. A. Delivery of flavonoids and saponins from black bean (*Phaseolus vulgaris*) seed coats incorporated into whole wheat bread. *Int J Mol Sci.* 2016, 17, 222- 236.

Chen, D.; Qiao, J.; Sun, Z.; Liu, Y.; Sun, Z.; Zhu, N.; Xu, X.; Yang, J.; Ma, G. New naphtoquinones derivatives from the edible bulbs of *Eleutherine americana* and their protective effect on the injury of human umbilical vein endothelial cells. *Fitoterapia.* 2019, 132, 46–52.

Chen, M.X.; Huo, J.M.; Hu, J.; Xu, Z.P.; Zhang, X. Amaryllidaceae alkaloids from *Crinum latifolium* with cytotoxic, antimicrobial, antioxidant, and anti-inflammatory activities. *Fitoterapia.* 2018, 130, 48–53.

Cicció, J.F. A source of almost pure methyl chavicol: Volatile oil from the aerial parts of *Tagetes lucida* (Asteraceae) cultivated in Costa Rica. *Rev Biol Trop.* 2004, 52, 853–7.

Claridge, T. D.W. *High-Resolution NMR Techniques in Organic Chemistry.* 3rd ed. Elsevier Science: UK, 2016.

Craciunescu, O.; Constantin, D.; Gaspar, A.; Toma, L.; Utoiu, E.; Moldovan, L. Evaluation of antioxidant and cytoprotective activities of *Arnica montana* L. and *Artemisia absinthium* L. ethanolic extracts. *Chem Cent J.* 2012, 6, 97.

de Oliveira, J.R.; Camargo, S.E.A.; de Oliveira, L.D. *Rosmarinus officinalis* L. (rosemary) as therapeutic and prophylactic agent. *J Biomed Sci.* 2019, 26, 5.

Domínguez, X.A. *Methods of phytochemical research.* 1st ed. Limusa: Mexico, 1979.

Ebrahimpour koujan, S.; Gargari, B. P.; Mobasseri, M.; Valizadeh, H.; Asghari-Jafarabadi, M. Effects of *Silybum marianum* (L.) Gaertn. (Silymarin) extract supplementation on antioxidant status and Hs-CRP in patients with type 2 diabetes Mellitus: A randomized, triple-blind, placebo-controlled clinical trial. *Phytomedicine.* 2015, 22, 290–296.

Efferth, T.; Olbrich, A.; Sauerbrey, A.; Ross, D. D.; Gebhart, E.; Neugebauer, M. Activity of ascaridol from the anthelmintic herb *Chenopodium anthelminticum* L. against sensitive and multidrug-resistant tumor cells. *Anticancer Res.* 2002, 22, 4221–4224.

Eghbaliferiz, S.; Emami, S.A.; Tayarani-Najaran, Z.; Iranshahi, M.; Shakeri, A.; Hohmann, J.; Asili, J. Cytotoxic diterpene quinones from *Salvia tebesana* Bunge. *Fitoterapia.* 2018, 128, 97–101.

Falcão, S.; Gomes, P.; Vale, N.; Domingues, M.R. Phenolic profiling of Portuguese propolis by LC-MS Spectrometry uncommon propolis rich in flavonoid glycosides. *Phytochem Anal.* 2013, 24, 309–318.

Felixsson, E.; Persson, I. A.-L.; Eriksson, A. C.; Persson, K. Horse chestnut extract contracts bovine vessels and affects human platelet aggregation through 5-HT2A receptors: An *in vitro* study. *Phytother Res.* 2010, 24, 1297–1301.

Gao, X.; Wang, C.; Chen, Z.; Chen, Y.; Santhanam, R.K.; Xue, Z.; Ma, Q.; Guo, Q.; Liu, W.; Zhang, M.; Chen, H. Effects of N-trans-feruloyltyramine isolated from laba garlic on antioxidant, cytotoxic activities and H_2O_2-induced oxidative damage in HepG$_2$ and L0$_2$ cells. *Food Chem Toxicol.* 2019, 130, 130–141.

Gholamhoseinian, A., Moradi, M.N., Sharifi-Far, F. Screening the methanol extracts of some Iranian plants for acetylcholinesterase inhibitory activity. *Res Pharm Sci.* 2009, 4, 105–112.

Gilbert, M.T. *High Performance Liquid Chromatography.* 3rd ed. Elsevier Science: U.K., 1987. doi.org/10.1016/C2013-0-11934-4.

Gómez Caravaca, A.M.; Gómez, A.M.; Arráez Román, D.; Segura Carretero; A., Fernández Gutiérrez, A. Advances en the analysis of phenolic compounds in products derived from bees. *J Pharm Biomed Anal.* 2006, 41, 1220–34.

González-Burgos, E.; Liaudanskas, M.; Viškelis, J.; Žvikas, V.; Janulis, V.; Gómez-Serranillos, M.P. Antioxidant activity, neuroprotective properties and bioactive constituents analysis of varying polarity extracts from *Eucalyptus globulus* leaves. *J Food Drug Anal.* 2018, *26*, 1293–1302.

Hajaji, S.; Sifaoui, I.; López-Arencibia, A.; Reyes-Batlle, M.; Jiménez, I.A.; Bazzocchi, I.L.; Valladares, B.; Pinero, J.E.; Lorenzo-Morales, J.; Akkari, H. Correlation of radical-scavenging capacity and amoebicidal activity of *Matricaria recutita* L. (Asteraceae). *Exp Parasitol.* 2017, 183, 212–217.

Halberstein, R. A. Medicinal plants: Historical and cross-cultural usage patterns. *Ann Epidemiol.* 2005, 15, 686–699.

Harris, D.C. *Quantitative Chemical Analysis.* 2nd ed. Reverté: Spain. 2001.

Harvey, D. *Química Analítica Moderna. 1*st *ed.* McGraw-Hill: Colombia, 2004.

Hassan, Z.; Bosch, O.G.; Singh, D.; Narayanan, S.; Kasinather, B.V.; Seifritz, E.; Kornhuber, J.; Quednow, B.B.; Müller, C.P. Novel psychoactive substances-recent progress on neuropharmacological mechanisms of action for selected drugs. *Front Psychiatry.* 2017, 8, 152.

Hassanain, E.; Silverberg, J.I.; Norowitz, K.B.; Chice, S.; Bluth, M.H.; Brody, N.; Joks, R.; Durkin, H.G.; Smith-Norowitz, T.A. Green tea (*Camelia sinensis*) suppresses B cell production of IgE without inducing apoptosis. *Ann Clin Lab Sci.* 2010, 40, 135–43.

He, L.; Zhang, S.; Luo, C.; Sun, Y.; Lu, Q.; Huang, L.; Chen, F.; Tang, L. Functional teas from the stems of *Penthorum chinense* Pursh.: Phenolic constituents, antioxidant and hepatoprotective activity. *Plant Foods Hum Nutr.* 2019, 74, 83–90.

Herald, TJ, Gadgil, P; Tilley, M. High-throughput micro plate assays for screening flavonoid content and DPPH-scavenging activity in sorghum bran and flour. *J Sci Food Agric.* 2012, 92, 2326–2331.

Hernández-Ocura, L.; Carranza-Rosales, P.; Cobos-Puc, L.E.; López-López, L.I.; Ascasio-Valdés, J.A.; Silva-Belmares, S.Y. *Solanum elaeagnifolium* to evaluate toxicity on cellular lines and breast tumor explants. *VITAE.* 2017, *24*, 124–131.

Higashi, Y.; Smith, T.J.; Jez, J.M.; Kutchan, T.M. Crystallization and preliminary X-ray diffraction analysis of salutaridine reductase from the opium poppy *Papaver somniferum. Acta Crystallogr Sect F: Struct Biol Cryst Commun.* 2010, *66*, 163–166.

Huang, S.; Zhang, C.P.; Wang, K.; Li, G.; Hu, F.L. Recent advances in the chemical composition of propolis. *Molecules.* 2014, 19, 19610–19632.

Husain, I.; Ahmad, R.; Chandra, A.; Raza, S. T.; Shukla, Y.; Mahdi, F. Phytochemical characterization and biological activity evaluation of ethanolic Extract of *Cinnamomum zeylanicum. J Ethnopharmacol.* 2018, 219, 110–116.

Ilína, A.D. The use of enzymes and microorganisms for solving fundamental and applied problems. *Herald Acad Sci Repub Bashkortostan.* 2013, 18, 40–43.

James, P.B.; Wardle, J.; Steel, A.; Adams, J. Traditional, complementary and alternative medicine use in Sub-Saharan Africa: A systematic review. *BMJ Glob Health*. 2018, 3, e000895.

Jamila, N.; Khan, N.; Hwang, I.M.; Choi, J.Y.; Nho, E.Y.; Khan, S.N.; Atlas, A.; Kim, K.S. Determination of macro, micro, trace essential, and toxic elements in *Garcinia cambogia* fruit and its anti-obesity commercial products. *J Sci Food Agric*. 2019, 99, 2455–2462.

Jiang, Y.; Li, D.; Ma, X.; Jiang, F.; He, Q.; Qiu, S.; Li, Y.; Wang, G. Ionic Liquid ultrasound-based extraction of biflavonoids from *Selaginella helvetica* and investigation of their antioxidant activity. Molecules. 2018, 23, pii: E3284.

Johnson, T.; Boon, H. where does homeopathy fit in pharmacy practice? *Am J Pharm Educ*. 2007, 71, 7.

Karamalakova, Y.D.; Nikolova, G.D.; Georgiev, T.K.; Gadjeva, V.G.; Tolekova, A.N. Hepatoprotective properties of *Curcuma longa* L. extract in bleomycin-induced chronic hepatotoxicity. *Drug Discov Ther*. 2019, 13, 9–16.

Kerkoub, N.; Panda, S.K.; Yang, M.R.; Lu, J.G.; Jiang, Z.H.; Nasri, H.; Luyten, W. Bioassay-guided isolation of anti-candida biofilm compounds from methanol extracts of the aerial parts of *Salvia officinalis* (Annaba, Algeria). *Front Pharmacol*. 2018, 9, 1418.

Khan, H. Medicinal plants in light of history: Recognized therapeutic modality. *J Evid Based Complement Altern Med*. 2014, 19, 216–219.

Kim, H.K.; Bak, Y.O.; Choi, B.R.; Zhao, C.; Lee, H.J.; Kim, C.Y.; Lee, S.W.; Jeon, J.H.; Park, J.K. The role of the lignan constituents in the effect of *Schisandra chinensis* fruit extract on penile erection. *Phytother Res*. 2011, 25, 1776–82.

Kindscher, K. The uses of *Echinacea angustifolia* and other echinacea species by Native Americans. *In Echinacea: Herbal Medicine with a Wild History*; Springer International Publishing: Switzerland, 2016, pp. 9–20.

Kooti, W.; Servatyari, K.; Behzadifar, M.; Asadi-Samani, M.; Sadeghi, F.; Nouri, B.; Marzouni, H.Z. Effective medicinal plant in cancer treatment, Part 2: Review study. *J Evid Based Complementary Altern Med*. 2017, 22, 982–995.

Kurek, J. Cytotoxic colchicine alkaloids: From plants to drugs. *In Cytotoxicity*. InTechOpen: London, 2018, Chapter 4, pp. 45–63.

Kuś PM, Okińczyc P, Jakovljević M, Jokić S, Jerković I. Development of supercritical CO_2 extraction of bioactive phytochemicals from black poplar (*Populus nigra* L.) buds followed by GC-MS and UHPLC-DAD-QqTOF-MS. *J Pharm Biomed Anal*. 2018, 158, 15–27.

Kwakye, G. F.; Jiménez, J.; Jiménez, J. A.; Aschner, M. *Atropa belladonna* neurotoxicity: Implications to neurological disorders. *Food Chem Toxicol*. 2018, 116, 346–353.

Lemes, R.S.; Alves, C.C.F.; Estevam, E.B.B.; Santiago, M.B.; Martins, C.H.G.; Santos, T.C.L.D.; Crotti, A.E.M.; Miranda, M.L.D. Chemical composition and antibacterial activity of essential oils from *Citrus aurantifolia* leaves and fruit peel against oral pathogenic bacteria. *An Acad Bras Cienc*. 2018, 90, 1285–1292.

Li, N.; Shi, Z.; Tang, Y.; Chen, J.; Li, X. Recent progress on the total synthesis of acetogenins from Annonaceae. *Beilstein J Org Chem*. 2008, 4, 1–62.

Li, Y.; Tran, V.H.; Duke, C.C.; Roufogalis B.D. Preventive and protective properties of *Zingiber officinale* (Ginger) in diabetes mellitus, diabetic complications, and associated lipid and other metabolic disorders: A brief review. *Evid Based Complement Alternat Med*. 2012, 2012, 516870.

Liaw, C.C.; Chang, F.R.; Chen, S.L.; Wu, C.C.; Lee, K.H.; Wu, Y.C. Novel cytotoxic monotetrahydrofuranic Annonaceous acetogenins from *Annona montana*. *Bioorg Med Chem*. 2005, 13, 4767–76.

Liu, H.; Zhang, J.; Li, X.; Qi, Y.; Peng, Y.; Zhang, B.; Xiao, P. Chemical analysis of twelve lignans in the fruit of *Schisandra sphenanthera* by HPLC-PAD-MS. Phytomedicine. 2012, 19, 1234–1241.

Liu, L; Anderson, G.A.; Fernandez, T.G.; Doré, S. Efficacy and mechanism of *Panax ginseng* in experimental stroke. *Front Neurosci.* 2019, 13, 294.

Liu, R.X.; Liao, Q.; Shen, L.; Wu. J. Krishnagranatins A–I: New limonoids from the mangrove, *Xylocarpus granatum*, and NF-κB inhibitory activity. *Fitoterapia*. 2018, 131, 96–104.

Loh, Y.C.; Tan, C.S.; Ch'ng, Y.S.; Ahmad, M.; Asmawi, M.Z.; Yam, M.F. Vasodilatory effects of combined traditional chinese medicinal herbs in optimized ratio. *J Med Food*. 2017, 20, 265–278.

Long, S.Y.; Li, C.L.; Hu, J.; Zhao, Q.J.; Chen, D. Indole alkaloids from the aerial parts of *Kopsia fruticosa* and their cytotoxic, antimicrobial and antifungal activities. *Fitoterapia*. 2018, 129, 145–149.

López, Z.; Femenia, A.; Núñez-Jinez, G.; Salazar Zúñiga, M.N.; Cano, M.E.; Espino, T.; Knauth, P. In vitro immunomodulatory effect of food supplement from *Aloe vera*. *Evid Based Complement Alternat Med*. 2019, 2019, 5961742.

Lyß, G.; Knorre, A.; Schmidt, T. J.; Pahl, H. L.; Merfort, I. The anti-inflammatory sesquiterpene lactone helenalin inhibits the transcription factor NF-KB by directly targeting P65. *J Biol Chem*. 1998, 273, 33508–33516.

Ma, N.; Tang, Q.; Wu, W.T.; Huang, X.A.; Xu, Q.; Rong, G.L.; Chen, S.; Song, J.P. Three constituents of *Moringa oleifera* seeds regulate expression of Th17-relevant cytokines and ameliorate TPA-induced psoriasis-like skin lesions in mice. *Molecules*. 2018, 23, pii: E3256.

Ma, Y.; Wang, H.; Wang, R.; Meng, F.; Dong, Z.; Wang, G.; Lan, X.; Quan, H.; Liao, Z.; Chen, M. Cytotoxic lignans from the stems of *Herpetospermum pedunculosum*. Phytochemistry. 2019, 164,102-110

Mahdi, J. G. Medicinal potential of willow: A chemical perspective of aspirin discovery. *J Saudi Chem Soc*. 2010, 14, 317–322.

Maldonado, E.; Díaz-Arumir, H., Toscano, R.A.; Martínez, M. Lupane triterpenes with a δ-lactone at ring E, from *Lippia mexicana*. *J Nat Prod*. 2010, 73, 1969–72.

Malik, S.; de Mesquita, L.S.S.; Silva, C.R.; de Mesquita, J.W.C.; de Sá Rocha, E.; Bose, J.; Abiri, R.; de Maria Silva Figueiredo, P.; Costa-Júnior, L.M. Chemical profile and biological activities of essential oil from *Artemisia vulgaris* L. cultivated in Brazil. *Pharmaceuticals (Basel)*. 2019, 12, pii: E49.

Mancuso, G.; Borgonovo, G.; Scaglioni, L.; Bassoli, A. Phytochemicals from *Ruta graveolens* activate TAS2R bitter taste receptors and TRP channels involved in gustation and nociception. *Molecules*. 2015, 20, 18907–18922.

Mandalari, G.; Bennet, R. N.; Kirby, A. R.; Lo Curto, R. B.; Bisignano, G.; Waldron, K. W.; Fauls, C. B. Enzymatic hydrolysis of flavonoids and pectic oligosacchrids from Bergamot (*Citrus bergamis* Risso) peel. *J Agric Food Chem*. 2006, 54, 8307–8313.

Marrelli, M.; Amodeo, V.; Statti, G.; Conforti, F. Biological properties and bioactive components of *allium cepa* l.: Focus on potential benefits in the treatment of obesity and related comorbidities. *Molecules*. 2019, 24, 119.

Matias, D.; Nicolai, M.; Fernandes, A.S.; Saraiva, N.; Almeida J; Saraiva L; Faustino C; Díaz-Lanza AM; Reis CP; Rijo P. Comparison study of different extracts of *Plectranthus madagascariensis*, *P. neochilus* and the Rare *P. porcatus* (Lamiaceae): Chemical characterization, antioxidant, antimicrobial and cytotoxic activities. *Biomolecules*. 2019, 9, pii: E179.

Michelini, F. M.; Alché, L. E.; Bueno, C. A. Virucidal, antiviral and immunomodulatory activities of B escin and *Aesculus hippocastanum* extract. *J Pharm Pharmacol*. 2018, 70, 1561–1571.

Mirosławski, J.; Paukszto, A. Determination of the cadmium, chromium, nickel, and lead ions relays in selected polish medicinal plants and their infusion. *Biol Trace Elem Res*. 2018, 182, 147–151.

Misra, A.K.; Varma, S.K.; Kmar, R. Anti-inflammatory effect of an extract of *Agave americana* on experimental animals. *Pharmacognosy Res*. 2018, 10, 104–108.

Mutiah, R.; Widyawaruyanti, A.; Sukardiman, S. Calotroposid A: A glycosides terpenoids from *calotropis gigantea* Induces Apoptosis of Colon Cancer WiDr cells through cell cycle arrest G2/M and caspase 8 expression. *Asian Pac J Cancer Prev*. 2018, 19, 1457–1464.

Nabavi, S.M.; Šamec, D.; Tomczyk, M.; Milella, L.; Russo, D.; Habtemariam, S.; Suntar, I.; Rastrelli, L.; Daglia, M.; Xiao, J.; Giampieri, F.; Battino, M.; Sobarzo-Sanchez, E.; Nabavi, S.F.; Yousefi, B.; Jeandet, P.; Xu, S.; Shirooie, S. Flavonoid biosynthetic pathways in plants: Versatile targets for metabolic engineering. *Biotechnol Adv*. 2018, pii: S0734-9750, 30181–30182.

Narayani, M.; Srivastava, S. Elicitation: A stimulation of stress in in vitro plant cell/tissue cultures for enhancement of secondary metabolite production. *Phytochemy Rev*. 2017, 16, 1227–1252.

Noreikaitė, A.; Ayupova, R.; Satbayeva, E.; Seitaliyeva, A.; Amirkulova, M.; Pichkhadze, G.; Datkhayev, U.; Stankevičius, E. General toxicity and antifungal activity of a new dental gel with essential oil from *Abies Sibirica* L. *Med Sci Monit*. 2017, 23, 521–527.

Nuutila, A. M.; Puupponen-Pimiä, R.; Aarni, M.; Oksman-Caldentey, K.-M. Comparison of antioxidant activities of onion and garlic extracts by inhibition of lipid peroxidation and radical scavenging activity. *Food Chem*. 2003, 81, 485–493.

Ohikhena, F.U.; Wintola, O.A.; Afolayan, A.J. Quantitative phytochemical constituents and antioxidant activities of the mistletoe, *Phragmanthera capitata* (Sprengel) balle extracted with different solvents. *Pharmacog Res*. 2018, 10, 16–23.

Olioso, D.; Marzotto, M.; Bonafini, C.; Brizzi, M.; Bellavite, P. *Arnica montana* Effects on gene expression in a human macrophage cell line. Evaluation by quantitative real-time PCR. *Homeopathy* 2016, 105, 131–147.

Paarakh, P.M.; Sreeram, D.C.; Shruthi, S. D; Ganapathy, S.P. S. In vitro cytotoxic and in silico activity of piperine isolated from *Piper nigrum* fruits Linn. *In Silico Pharmacol*. 2015, 3, 9.

Paek, K. Y.; Chakrabarty, D.; Hahn, E. J. Application of bioreactor systems for large scale production of horticultural and medicinal plants. *Plant Cell Tissue Organ Culture*. 2005, 81, 287–300.

Pandey, S.S.; Singh, S.; Pandey, H.; Srivastava, M.; Ray, T.; Soni, S.; Pandey, A.; Shanker, K.; Vivek Babu, C. S.; Banerjee, S.; Gupta, M. M.; Kalra A. Endophytes of *Withania somnifera* modulate in planta content and the site of withanolide biosynthesis. *Sci Rep*. 2018, 8, 5450.

Panth, N.; Paudel, K.R.; Karki, R. Phytochemical profile and biological activity of *Juglans regia*. *J Integr Med*. 2016, 14, 359–73.

Patlolla, J. M. R.; Rao, C. V. Anti-inflammatory and anti-cancer properties of β-escin, a triterpene saponin. *Curr Pharmacol Rep*. 2015, 1, 170–178.

Pellati, F.; Pinetti, D.; Orlandini, G.; Benvenuti, S. HPLC-DAD and HPLC-ESI-MS/MS methods for metabolite profiling pf propolis extracts. *J Pharm Biomed Anal*. 2011, 55, 934–948.

Petrovska, B. B. Historical review of medicinal plants' usage. *Pharmacogn Rev*. 2012, 6, 1–5.

Pittler, M. H.; Ernst, E. Horse chestnut seed extract for chronic venous insufficiency. *Cochrane Database System Rev.* 2012, 11, 1465–1858.

Popovaa, M.P.; Bankovaa, V.S.; Bogdanovb, S.; Tsvetkovac, I.; Naydenskic, C.; Marcazzand, G.L.; Sabatinid A.G. Chemical characteristics of poplar type propolis of different geographic origin. *Apidologie.* 2007, 38, 306-306.

Posadzki, P.; Watson, L.; Ernst, E., Herb-drug interactions: An overview of systematic reviews. *Br J Clin Pharmacol.* 2013, 75, 603–618.

Prakash, D.; Singh, B. N.; Upadhyay, G. Antioxidant and free radical scavenging activities of phenols from onion (*Allium cepa*). *Food Chem.* 2007, 102, 1389–1393.

Pukdeekumjorn, P., Ruangnoo, S., Itharat, A. Anti-inflammatory activities of extracts of *Cinnamomum porrectum* (Roxb.) Kosterm. Wood (Thep-tha-ro). *J Med Assoc Thai.* 2016, 99, S138-143.

Rakelly de Oliveira, D.; Relison Tintino, S.; Morais Braga, M. F. B.; Boligon, A. A.; Linde Athayde, M.; Douglas Melo Coutinho, H.; de Menezes, I. R. A.; Fachinetto, R. *In vitro* Antimicrobial and modulatory activity of the natural products silymarin and silibinin. *BioMed Res Int.* 2015, 2015, 1–8.

Rasheed, D.M.; Porzel, A.; Frolov, A.; El Seedi, H.R.; Wessjohann, L.A.; Farag, M.A. Comparative analysis of *Hibiscus sabdariffa* (roselle) hot and cold extracts in respect to their potential for α-glucosidase inhibition. *Food Chem.* 2018, 250, 236–244.

Rodríguez-García; A.; Peixoto, I.T.; Verde-Star, M.J., De la Torre-Zavala, S.; Aviles-Arnaut, H., Ruiz, A.L. In vitro antimicrobial and antiproliferative activity of *Amphipterygium adstringens. Evid Based Complement Alternat Med.* 2015, 2015, 175497.

Sánchez-Tena, S.; Fernández-Cachón, M. L.; Carreras, A.; Mateos-Martín, M. L.; Costoya, N.; Moyer, M. P.; Nuñez, M. J.; Torres, J. L.; Cascante, M. Hamamelitannin from witch hazel (*Hamamelis virginiana*) displays specific cytotoxic activity against colon cancer cells. *J. Nat. Prod.* 2012, 75, 26–33.

Sawaya, W. N.; Khalil, J. K.; Al Mohammad, M. M. Nutritive value of prickly pear seeds, Opuntia ficus indica. *Plant Foods Hum. Nutr.* 1983, 33, 91–97.

Schifter Aceves, L., Mexican Pharmacopeias in the construction of a national identity. *Rev Mexicana Ciencias Farmacéut.* 2014, 45, 43–54.

Shaffique, S.; Rehman, A.; Ahmad, S.; Anwer, H.; Asif, H. M.; Husain, G.; Rehman, T.; Javed, S. A panoramic review on ethnobotanical, phytochemical, pharmacological and homeopathic uses of *Echinacea angustifolia. RADS J Pharm Pharm Sci.* 2018, 6, 282–286.

Sim, D.S.; Navanesan, S.; Sim, K.S.; Gurusamy, S.; Lim, S.H.; Low, Y.Y.; Kam, T.S. Conolodinines A–D, aspidosperma- aspidosperma bisindole alkaloids with antiproliferative activity from *Tabernaemontana corymbosa. J Nat Prod.* 2019, 82, 850–858.

Singh, O.; Khanam, Z.; Misra, N.; Srivastava, M.K. Chamomile (*Matricaria chamomilla* L.): An overview. *Pharmacogn Rev.* 2011, 5, 82–95.

Sirtori, C. R. Aescin: Pharmacology, Pharmacokinetics and Therapeutic Profile. *Pharmacolog Res.* 2001, 44, 183–193.

Solís-Salas, L.M.; Cobos-Puc, L.E.; Aguilar-González, C.N.; Sierra Rivera, C.A.; Iliná, A.; Silva Belmares, S.Y. *Chemistry of medicinal plants, foods, and natural products and their composites*; CRC Press; Taylor & Francis:Canada, 2019, Chapter 16, pp. 357–381

Son, Y.; An, Y.; Jung, J.; Shin, S.; Park, I.; Gwak, J.; Ju, B.G.; Chung, Y.H.; Na, M.; Oh, S. Protopine isolated from *Nandina domestica* induces apoptosis and autophagy in colon cancer cells by stabilizing p53. *Phytother Res.* 2019, 2019, 1–8.

Temrangsee, P.; Kondo, S.; Itharat, A. Antibacterial activity of extracts from five medicinal plants and their formula against bacteria that cause chronic wound infection. *J Med Assoc Thai.* 2011, 94, S166-171.

Teng, Y.; Zhang, H.; Zhou, J.; Zhan, G.; Yao, G. Hebecarposides A–K, antiproliferative lanostane-type triterpene glycosides from the leaves of *Lyonia ovalifolia* var. hebecarpa. *Phytochemistry.* 2018, 151,32-41.

Tetali, S.D. Terpenes and isoprenoids: A wealth of compounds for global use. *Planta.* 2019, 249, 1–8.

Theisen, L. L.; Erdelmeier, C. A. J.; Spoden, G. A.; Boukhallouk, F.; Sausy, A.; Florin, L.; Muller, C. P. Tannins from *Hamamelis virginiana* bark extract: Characterization and improvement of the antiviral efficacy against influenza a virus and human papillomavirus. *PLoS One.* 2014, 9, e88062.

Tian, D.; Wang, F.; Duan, M.; Cao, L.; Zhang, Y.; Yao, X.; Tang, J. Coumarin analogues from the *Citrus grandis* (L.) osbeck and their hepatoprotective activity. *J Agric Food Chem.* 2019, 67, 1937–1947.

Tomaza, I.; Maslova, L.; Stupića, D.; Preinera, D.; Ašpergerb, D.; Kontića, J. K. Recovery of flavonoids from grape skins by enzyme-assisted extraction. *Separ Sci Technol.* 2016, 51, 255–268.

Touriño, S.; Lizárraga, D.; Carreras, A.; Lorenzo, S.; Ugartondo, V.; Mitjans, M.; Vinardell, M. P.; Juliá, L.; Cascante, M.; Torres, J. L. Highly galloylated tannin fractions from witch hazel (*Hamamelis virginiana*) bark: Electron transfer capacity, *in vitro* antioxidant activity, and effects on skin-related cells. *Chem Res Toxicol.* 2008, 21, 696–704.

Tu, W.C.; Qi, Y.Y.; Ding, L.F.; Yang, H.; Liu, J.X.; Peng, L.Y.; Song, L.D.; Gong, X.; Wu, X.D.; Zhao, Q.S. Diterpenoids and sesquiterpenoids from the stem bark of *Metasequoia glyptostroboides*. *Phytochemistry.* 2019, 161, 86–96.

van Rooyen, D.; Pretorius, B.; Tembani, N.M.; ten Ham, W. Allopathic and traditional health practitioners' collaboration. *Curationis.* 2015, 38, 1495.

Vanisree, M.; Lee, C. Y.; Lo, S. F.; Nalawade, S. M.; Lin, C. Y.; Tsay, H. S. Studies on the production of some important secondary metabolites from medicinal plants by plant tissue cultures. *Botan Bull Acad Sin.* 2004, 45, 1–22.

Vašková, J.; Fejer, A.; Mojžišová, G.; Vaško, L.; Patlevi, P. Antioxidant potential of *Aesculus hippocastanum* extract and escin against reactive oxygen and nitrogen species. *Eur Rev Med Pharmacol Sci.* 2015, 19, 879–886.

Verspohl, E. J.; Bauer, K.; Neddermann, E. Antidiabetic effect of *Cinnamomum cassia* and *Cinnamomum zeylanicum in vivo* and *in vitro*. *Phytother Res.* 2005, 19, 203–206.

Wagner, H.; Bladt, S. *Plant Drug Analysis a Thin Layer Chromatography Atlas.* 2nd ed. Springer: New York, 2001.

Wang, A.; Zhang, F.; Huang, L.; Yin, X.; Li, H.; Wang, Q.; Zeng, Z.; Xie, T. New progress in biocatalysis and biotransformation of flavonoids. *J Med Plants Res.* 2010, 4, 847–856.

Weid, M.; Ziegler, J.; Kutchan, T.M. The roles of latex and the vascular bundle in morphine biosynthesis in the opium poppy, *Papaver somniferum*. *Proc Natl Acad Sci USA.* 2004, *101*, 13957–13962.

Wellington, K.; Jarvis, B. Silymarin: A review of its clinical properties in the management of hepatic disorders. *BioDrugs.* 2001, 15, 465–489.

Wilson, S. A.; Roberts, S. C. Metabolic engineering approaches for production of biochemicals in food and medicinal plants. *Curr Opin Biotechnol.* 2014, 26, 174–182.

Wójciak-Kosior, M.; Sowa, I.; Dresler, S.; Kováčik, J.; Staniak, M.; Sawicki, J.; Zielińska, S.; Świeboda, R.; Strzemski, M.; Kocjan, R. Polyaniline based material as a new SPE sorbent for pre-treatment of *Chelidonium majus* extracts before chromatographic analysis of alkaloids. *Talanta.* 2019, 194, 32–37.

Yadav, N.; Vasudeva, N.; Singh, S.; Sharma, S. K. Medicinal properties of genus *Chenopodium Linn. Nat Prod Rad.* 2007, 6, 131–134.

Yasir, M.; Das. S.; Kharya, M. D. The phytochemical and pharmacological profile of *Persea americana* Mill. *Pharmacogn Rev.* 2010, 4, 77–84.

Zeng, K.; Ren, T.; Zhang, H.; Zhuo, R.; Peng, L.; Chen, C.; Zhou, Y.; Zhao, Y.; Li, W.J.; Jin, X.; Yang, L. *Moringa oleifera* seed extract protects against brain damage in both the acute and delayed stages of ischemic stroke. *Exp Gerontol.* 2019, 122, 99–108.

Zhang, S., Lu, B., Han, X., Xu, L., Qi, Y., Yin, L., Xu, Y., Zhao, Y., Liu, K., Peng, J. Protection of the flavonoid fraction from *Rosa laevigata* Michx fruit against carbon tetrachloride-induced acute liver injury in mice. *Food Chem Toxicol.* 2013, 55, 60–69.

Zhao, W.; Zhou, T.; Fan, G.; Chai, Y.; Wu, Y. Isolation and purification of lignans from *Magnolia biondii Pamp* by isocratic reversed-phase two-dimensional liquid chromatography following microwave-assisted extraction. *J Sep Sci.* 2007, 30, 2370–81.

Zhong, L.; Yuan, Z.; Rong, L.; Zhang, Y.; Xiong, G.; Liu, Y.; Li, C. An optimized method for extraction and characterization of phenolic compounds in *Dendranthema indicum var. aromaticum* Flower. *Sci Rep.* 2019, 9, 7745.

CHAPTER 9

Panoramic View of Biological Barricades and Their Influence on Polysaccharide Nanoparticle Transport: An Updated Status in Cancer

MAYA SREERANGANATHAN, BABUKUTTAN SHEELA UNNIKRISHNAN, PREETHI GOPALAKRISHNAN USHA, and THERAKATHINAL THANKAPPAN SREELEKHA*

Regional Cancer Centre, Thiruvananthapuram, Kerala, India

Corresponding author. E-mail: ttsreelekha@gmail.com; sreelekhatt@rcctvm.gov.in

ABSTRACT

Biological barriers form a natural defensive barricade for the body, which screens the entry of foreign materials. Despite their protective role, they provide a major challenge for the transport of therapeutics across the biological barriers such as skin, oral, and intestinal mucosa, and blood–brain barrier. Pharmaceutical researchers are nowadays focusing on next-generation therapeutics based on larger biologicals (carrier-based systems). Polysaccharides, complex polymeric carbohydrates, have been studied far and wide as drug carrier systems in various forms. Polysaccharides belong to a class of drug carriers with advantageous properties such as nontoxic, biocompatible, biodegradable, and economic. The current review summarizes the studies reported on polysaccharide nanoparticles formulated for the transport across biological barricades with the aim of improving the anticancer potential.

9.1 INTRODUCTION

The advanced research studies on the technological approaches for formulating and transporting a pharmaceutically active ingredient (or a drug) in the human body to achieve its desired therapeutic effect within the safety window have evolved through decades under the field of "drug delivery." With the aim of improving patient convenience and compliance, various drug delivery systems (DDS) like controlled release systems were born in early 1950s (Park, 2016; Tiwari et al., 2012; Mirza and Siddiqui, 2014; NIH, 2016). The researchers working in the area of drug delivery in collaboration with drug discovery have contributed immensely in developing new drug molecules as well as delivery systems toward almost all types of human diseases, where "cancer" is the main focus in this review. The World Health Organization (WHO) has listed cancer as one among the deadly and devastating diseases, which is characterized by the abnormal proliferation of cells with an invasive and metastatic potential (WHO, 2014, 2017). Irrespective of all of the therapeutic modalities developed against cancer, chemotherapy still remains to be the first line of treatment option.

Chemotherapy, the major cancer therapeutic strategy, involves the use of chemical molecules that act on fast-dividing cancer cells, thereby preventing their growth and killing them (Chabner and Roberts, 2005; Huang et al., 2017). The main class of anticancer drugs include alkylating agents (cisplatin, procarbazine, carmustine, etc.), antimicrotubule agents (vinblastine, paclitaxel (PTX), etc.), antimetabolites (methotrexate, gemcitabine, etc.), topoisomerase inhibitors (etoposide, doxorubicin (DOX), etc.), and cytotoxic agents (bleomycin) (Cirenajwis et al., 2010). Anticancer drugs are generally administered orally in the form of pills or via different parenteral routes such as intramuscular, intravenous (IV), intraperitoneal, or subcutaneous in the form of injectable formulations or infusions. Drugs administered orally go through absorption steps and hence circulate for longer time in the body. Also considering patient convenience, oral formulations are preferred. However, in general, chemotherapies prefer parenteral injections or infusions since they bypass first pass metabolism as the drugs are directly given to the blood stream for its accelerated action. However, irrespective of these routes of administrations, the drug molecule has to pass through different biological barriers.

The organisms evolved by creating certain biological barriers on their own, which was differentiated upon the requirement to regulate water content, homeostasis, gas exchange, nutritional availability, waste excretion, and the protective role indeed. The functions were managed by the development of tissues for specific requirements such as skin for environment sensing (water and homeostasis regulation), lungs for oxygen exchange, intestine for nutrient passage, etc. The aforementioned play the role of structural barriers for drug delivery, and hence, DDS developers focus on the specific interaction of the developed systems with these tissues and finally deliver the cargo to the desired target site. The preliminary consideration would be on the main biological barriers such as skin, oral and nasal mucosa, gastrointestinal (GI) tract, and blood–brain barrier (BBB). Cancer represents the best example of a disease condition where the adequacy of delivering chemotherapeutics with highly potent, yet toxic, mechanisms of action means the difference between efficacious responses and severe morbidity. Pharmaceutical industries started utilizing larger biological "carrier-bound drugs" as next-generation cancer chemotherapeutics, where nanocarriers opened up new opportunities. These nanoplatforms have been materialized as suitable vectors for overcoming pharmacokinetic limitations associated with conventional drug formulations. Many of these nanoplatforms improved patient safety and morbidity, which led to their clinical approval. However, these platforms offer only marginal improvements over conventional formulations. The complex series of biological barricades reduce the site-specific bioavailability of the therapeutic agent from these platforms and thereby prevent achieving proper therapeutic outcomes, resulting in undesired toxicological profiles. Here comes the need for nontoxic biocompatible materials for developing drug carriers, where natural polysaccharides are gaining great interests.

This monograph attempted to give a brief update on biocompatible polysaccharide-based nano-DDS for the administration of anticancer drugs through different routes, highlighting the potential to cross various biological barricades and reach the diseased target site and eliciting their anticancer therapeutic potentials.

9.2 BIOLOGICAL BARRICADES TO CANCER DRUG DELIVERY

The delivery of chemotherapeutic agents to the target tumor site encounters various biological barriers depending upon the location of the diseased site and the route of administration. In general, these biological barriers act as body's first line of defense by repelling the foreign materials and thereby preventing the entry of xenogens or pathogens into the body. They also play a key role in maintaining homeostasis for the physiological process (Mitragotri, 2013; Yu et al., 2018; Schneider et al., 2013). Irrespective of their functional and protective roles, these barriers interfere with drug penetration and hence challenging the DDS affecting the therapeutic efficacy.

Therefore, it is very important to understand the structure of these biological barricades for designing and developing efficient DDS. The current review is mainly focused on skin, oral mucosa, nasal mucosa, GI mucosa, and BBB.

9.2.1 SKIN

Skin, a complex epithelial and mesenchymal tissue, accounts for about 15% of the total body weight with a surface area of 1.8–2 m² making it the largest organ of the human body. Skin forms an effective mechanical barrier preventing the entry of pathogens, unnecessary chemicals and protecting the body from physical attacks, and it is also required for thermoregulation and fluid balance (Uchechi et al., 2014; Lai-cheong, 2017). This indispensable barrier is composed of three main zones; a multilayered stratified epidermis, a dermis containing collagen and elastic fibers, and the underlying subcutaneous fat. Skin also comprises of adnexal structures such as hair follicles, sweat glands, and sebaceous glands (Figure 9.1). Epidermis, the outermost skin layer, is composed of 95% keratinocytes and 5% of melanocytes, Langerhans cells, and Merkel cells. Keratinocytes form the outermost layer of epidermis, stratum corneum (SC; corneocytes). Melanocytes synthesize melanin and are packed into melanosomes; the skin color is determined by the amount of melanosomes and the nature of melanin. Langerhans cells are found throughout the epidermis and are bone marrow-derived, antigen-presenting dendritic cells. Merkel cells transmit sensory information from the skin to the sensory nerves (poly-glutamic acid).

Skin blocks the passage of many xenobiotics including active chemotherapeutic drug molecules. In general, the drug molecules administered on the skin have to pass through the specific skin layer (penetration), pass layer by layer through different skin structures (permeation), and get absorbed into the blood vessels in the dermis and hypodermis (resorption) (Peptu et al., 2014). SC with proteo lipid structure forms the first skin barrier that is highly permeable to hydrophobic molecules, whereas impermeable to hydrophilic molecules. The granular layer composed of tightly packed cells forms the next barrier layer that stops the passage of hydrophobic molecules to the deeper skin layers. Passive diffusion of drug molecules through the transepidermal pathways and skin appendages (hair follicles, sebaceous glands, and sweat glands) resulted in percutaneous absorption (Cevc and Vierl, 2010; Bolzinger et al., 2012). To be precise, three penetration routes have been identified through skin, such as the intercellular lipid route, transcellular route, and follicular penetration (Uchechi et al., 2014).

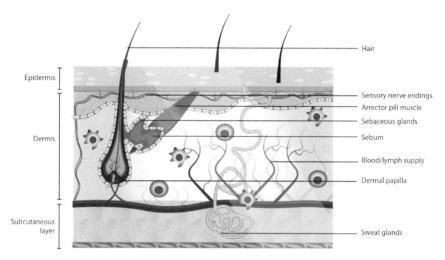

Epidermis

Dermis

Subcutaneous layer

Hair

Sensory nerve endings
Arrector pili muscle
Sebaceous glands

Sebum

Blood/lymph supply

Dermal papilla

Sweat glands

FIGURE 9.1   Schematic representation of skin with its structural characteristics.

9.2.1.1   ORAL MUCOSA

The oral cavity provides simultaneously an interesting and challenging target site for local and systemic drug delivery. Oral cavity includes lips, tongue, floor of the mouth, buccal mucosa, upper and lower gum, retromolar trigone (area behind wisdom teeth), and hard palate (bony roof of mouth) (Montero and Patel, 2015). The oral cavity is lined by a mucosal membrane termed as oral mucosa and serves to be the barrier between the mouth and oral cavity. Oral mucosa serves the role of protection of the deeper tissues in the oral cavity, sensation by providing information about the events happening within the oral cavity such as touch, pain, and taste (taste buds), salivating and reflexes such as swallowing, gagging, and retching, and finally in secretions (saliva) (Shizuko et al., 2018). Figure 9.2 represents the schematic structure of oral mucosa. The surface of the oral mucosa is covered by a stratified squamous epithelium (tightly packed epithelial cells) under which there is the basal membrane, the lamina propria (fibroblasts, connective tissue, capillaries, inflammatory cells, and extracellular matrix), and the submucosa. The oral mucosa shows considerable structural variation in different areas of the oral cavity, but three main types of mucosa can be identified, according to their primary function: masticatory mucosa (keratinized for mechanical compression of food), lining mucosa (nonkeratinized; relaxes during mastication), and specialized mucosa (could be both keratinized or nonkeratinized and its covers the tongue and functions with taste buds) (Patel et al., 2011; Kinikoglu et al., 2015; Qin et al., 2017).

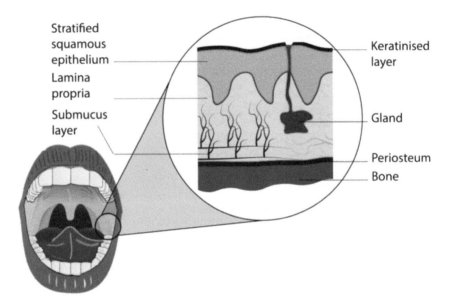

FIGURE 9.2   Schematic representation of oral mucosa.

Oral mucosa serves to be a physiological barrier for the passage of active drug molecules due to certain factors such as the thickness and permeability of oral mucosa, saliva film covering oral mucosa, pH, and enzymatic activity. The active therapeutic molecules follow either the lipophilic/transcellular pathway through the lipid bilayer of the individual cells of mucosa or the hydrophilic/paracellular pathway through aqueous pores along the polar heads of the bilayer (Peptu et al., 2014; Blanchette et al., 2004).

9.2.1.2   NASAL MUCOSA

Nasal drug delivery offers a well-tolerated noninvasive route of drug administration that has been used as an alternative route to IV administration for systemic availability of the chemotherapeutic agents. The primary function of nose is olfaction, filtration of the inhaled air from the airborne particles, and protection from foreign pathogens (Singh et al., 2013; Türker et al., 2004). The nasal cavity comprises a large surface area (150–160 cm^2) covered by mucosa rich in vessels and nerves, and the cavity opens out through apertures or nostrils. The large mucosal surface area for improved drug absorption, porous endothelial membranes with high blood flow, absence of drug degradation, overcoming the first pass metabolism, ease of administration, and patient compliance makes the nasal route an attractive target for drug administration (Singh et al., 2013; Türker et al., 2004; Kumar et al., 2016; Sachan and Singh, 2014).

From an anatomical perspective (Figure 9.3), the mucosal layer lining the nasal cavity is termed as respiratory mucosa or nasal mucosa from the nostrils to the pharynx. One-third of the nasal mucosa will be smooth epithelium composed of flat surfaced cells (stratified squamous epithelium). Beneath the squamous cell layer comes the proliferative cell layer, which is attached to a network of tough fibers, basement membrane, which supports the epithelium. The rest two-thirds of the nasal mucosa is composed of epithelial cells arranged in columns with hair (pseudo-stratified columnar ciliated epithelium)-containing mucus-producing goblet cells, which overlies a basement membrane (network of cross-linked collagen fibers). Beneath the basement membrane, there is lamina propria (chorionic tissue with connective-elastic properties, which is made up of glands secreting water and mucus, nerves, an extensive network of blood vessels, and cellular elements such as blood plasma. The entire mucosa is rich in blood vessels and contains large venous-like spaces, bodies that have a vein-like appearance and swell and congest in response to allergy or infection.

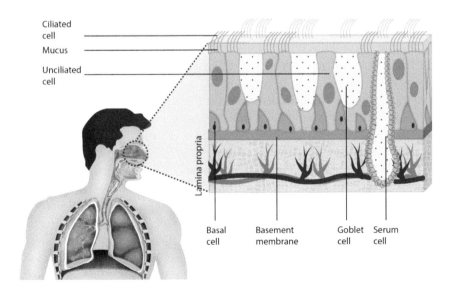

FIGURE 9.3 Schematic representation of the cellular organization of nasal mucosa.

Based on the structure of this nasal mucosa, the mucus and the ciliated epithelium form the physical barrier, the mucociliary clearance forms the temporary barrier, and the enzymatic activity forms the chemical barrier (Peptu et al., 2014). The viscous mucus prevents the interpenetration of drugs, whereas the low-viscous mucus prevents the proper adhesion of drug molecules. Thus, the epithelial mucus limits the movement of therapeutic agents. Particles trapped in the mucus layer are transported with it and, thereby, eventually cleared from the nasal cavity. The collective action of the mucus layer and cilia is called mucociliary clearance. Nasal mucociliary clearance prevents the penetration of particles through the nasal mucosa and allows for their absorption, which makes the nasal mucus a temporary barrier to drug substance transnasal absorption. The presence of oxidative and conjugative proteases and peptidases makes mucus an enzymatic barrier (Singh et al., 2013; Kumar et al., 2016; Sachan and Singh, 2014; Upadhyay et al., 2011; Alagusundaram et al., 2010).

9.2.1.3   *GASTROINTESTINAL MUCOSA*

Human gut or alimentary canal or GI tract is the continuous passage of about 9 m long between the mouth and anus, comprising the organs of digestion: stomach, small intestine, and large intestine. The GI tract consists of enormous surface area, which forms a barrier between the human body and luminal environment, playing an important role in the efficient absorption of nutrients and electrolytes from food and water (Bowen, 2002). The exclusionary properties of the gastric and intestinal mucosa are referred to as the "gastrointestinal barrier." The GI barrier is categorized into an intrinsic barrier, epithelial cells lining the digestive tract with tight junctions (TJs), and an extrinsic barrier, secretions that act on these epithelial cells and maintain the barrier functions. A layer of mucus and a layer of epithelial cells, TJ proteins connecting the intestinal epithelial cells, and the underlying lamina propria form the intestinal barrier (Bowen, 2002; König et al., 2016).

The GI tract mucosa (Figure 9.4) consists of a single layer of columnar epithelial cells covered by a layer of secreted mucus, which specifically consist of enterocytes, goblet cells, M cells, and Paneth cells. Enterocytes form a tightly packed layer at the apical surface by brush-like structures termed as microvilli (Snoeck et al., 2005). These cells form intestinal crypts to increase the intestinal surface area, facilitating absorption. Goblet cells secrete mucus, forming a jelly viscous film at the epithelial surface and contributing to the intestinal barrier function (Kim and Khan, 2013).

Paneth cells are specialized epithelial cells at the base of the crypt that secrete microbicidal proteins like α-defensins, lysozyme, and cathelicidins, preventing the entry of microorganisms to the intestinal mucosa (Schenk and Mueller, 2008). Paneth cells, located at the base of the intestinal crypts, are able to secrete antimicrobial proteins responsible for catching and killing bacteria. The M cells, located in Peyer's patches, are able to bind antigens and microorganisms and to deliver them to the local immune system (Peptu et al., 2014; Kiyono and Fukuyama, 2004; Sato and Iwasaki, 2005).

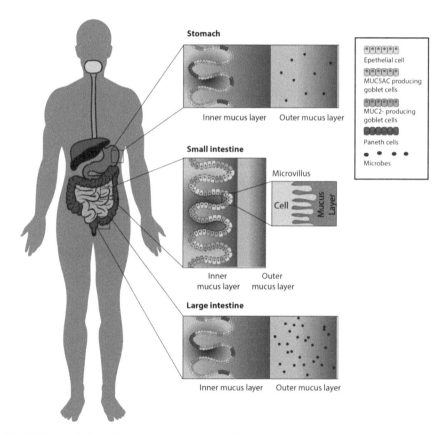

FIGURE 9.4 Schematic representation of the GI tract.

9.2.1.4 BLOOD–BRAIN BARRIER

The primary requirement for the healthy brain function and activity is to maintain stable homeostasis in the neuronal environment of the central nervous system (CNS) by evading the invading pathogens, neurotoxic molecules, and circulating blood cells. For this purpose of protection, CNS has developed certain physiological barriers such as the blood–cerebrospinal fluid barrier, the BBB, the blood–retinal barrier, and the blood–spinal cord barrier (Obermeier et al., 2016). BBB is studied to be the most selective and dynamic protective barrier composed of tightly organized endothelial cells (ECs) and a discontinuous pericyte (PC) layer. In 1885, Ehrlich observed that intravenously injected trypan blue dye stained all organs except brain, which led to the identification of the BBB existence, which blocks the entry of molecules into the brain (Saunders et al., 2014).

The diffusion barrier BBB (Figure 9.5) is composed of a brain capillary EC layer, astrocytes, PCs, and neuronal cells. The neurovascular unit is the dynamic functional unit, which exists by the intimate contact between neurons, astrocytes, microglia, PCs, and blood vessels and the functional interactions and signaling between them. The innermost lumina is the EC layer lining the brain capillaries rich in mitochondria and is adjoined to the basement membrane by tight cell-to-cell junction proteins. The basement membrane (30–40 nm) is a thick lamina, which is composed of collagen type IV, heparin sulfate proteoglycans, laminin, fibronectin, and other extracellular matrix proteins (Serlin et al., 2015; Daneman and Prat, 2015). Transmembrane proteins (junctional adhesion molecule-1, occludin, and claudins1/3, 5, and possibly 12) and cytoplasmic accessory proteins (zonulaoccludens, cingulin, AF-6, and 7H6) establish the TJs between adjacent ECs (Stamatovic et al., 2016). This dynamic TJ forms the diffusion barrier, which selectively excludes most blood-borne substances from entering the brain. These TJs also result in extremely high transendothelial electrical resistance between the blood and brain, and the passive diffusion of compounds is considerably restricted (Daneman and Prat, 2015). The endothelium is surrounded by a continuous stratum of cellular elements including PCs and astroglial foot processes. PCs envelop brain microvessels and capillaries and play a major role in the formation and maturation of BBB and regulation of tissue survival. They also regulate the cerebral blood flow, and hence, their absence results in reduced cerebral blood flow and loss of BBB integrity. The astrocytes interact with the PCs and the ECs via end feet protrusions enclosing the vessel walls tightly and thus maintaining the TJ barrier. Astrocytes play a role in homeostasis maintenance and also provide templates for migratory neuronal streams. Synaptic transmission, clearance of neurotransmitters, plasticity, and blood flow are determined by the interaction between astrocytes and neurons. Thus, the entire neurovascular unit functions as the security system to the brain (Serlin et al., 2015; Daneman and Prat, 2015).

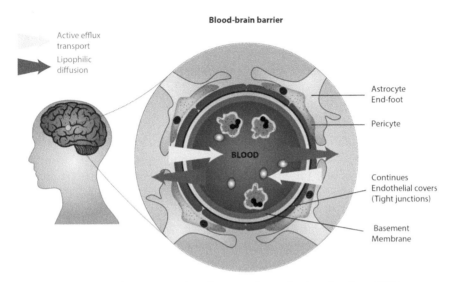

FIGURE 9.5 Schematic representation of structural organizations forming a BBB.

The BBB, therefore, is universally considered as the most important barrier in preventing molecules from reaching the brain parenchyma via extensive branches of blood capillary networks. The BBB containing the brain capillary endothelium (BCE) excludes ~100% of large-molecule neurotherapeutics and more than 98% of all small-molecule drugs from the brain. In addition to physical barriers, several functional barriers contribute to the restrictive nature of BBB, creating major obstacles to efficient drug delivery into the CNS. The molecules utilize the paracellular route of passage between ECs and the transcellular route for passing through ECs. The balance between paracellular and transcellular transports decides the degree of BBB permeability. Generally, BCEC's TJ prevents paracellular transport of small and large water-soluble compounds from blood capillaries to the brain, except for some very small or gaseous molecules such as water and carbon dioxide. To be precise, the biological characteristics that provide selective permeability to the BBB includes (a) the presence of TJs between the ECs formed by the complex transmembrane proteins and which are strengthen by the astrocytes and PCs, (b) mitochondria-rich ECs without fenestrations, (c) expression of various transporters that regulate the entry of molecules into the brain such as GLUT1 (glucose carrier), LAT1 (amino acid carrier), transferring receptors, insulin receptors, lipoprotein receptors, and ATP family of efflux transporters such as P-glycoprotein (P-gp) and multidrug-resistance-related proteins (MRPs), and (d) a lack of lymphatic drainage and absence of major histocompatibility complex antigens in CNS with immune reactivity inducible on temporary demand to provide maximum protection to neuronal function. These properties attributed BBB to function as a physical barrier (TJ), a transport barrier (P-gp), a metabolic or enzymatic barrier (specialized enzyme systems), and an immunological barrier (Bernacki et al., 2008; Weiss et al., 2009; Suciu et al., 2016).

The organization and functioning of the BBB can be altered under pathological conditions, such as in the case of tumors (Figure 9.6). In such a case, the barrier is called the blood–brain tumor barrier (BBTB) (Tellingen et al., 2015; Zhan and Lu, 2012). The pathological aspect of BBB involves cumulative effects of multiple factors released from glia cells and ECs. These factors disrupt BBB, loosening the TJs, resulting in fluid leakage into the brain, and causing edema and hemorrhage. The abnormal vessel permeability accounts for the excessive accumulation of interstitial fluid. Tumor cells break the EC layer and spread through the vessels, resulting in metastasis. They lack differentiated transport properties of normal BBB vessels. In brain tumors, especially in advanced brain tumors, the BBB is compromised in the core but is integral in the surrounding area. It is assumed that the BBTB restricts the distribution of drugs from circulation to brain tumors. Compared with blood tumor barriers in peripheral tumors, the BBTB exhibits smaller pore size and expresses a higher level of drug efflux pumps, such as P-gp and MRPs (Tellingen et al., 2015; Zhan and Lu, 2012). The cellular architecture of the BBB and key alterations occurring in a pathological context are depicted in Figure 9.6. Dysfunction of the BBB is the cause of brain tumor expansion and metastasis, and the presence of BBB limits the delivery of therapeutic agents, which necessitates alternative drug delivery strategies to cross or to bypass BBB because BBB serves as an exclusive drug target for researchers.

9.3   POLYSACCHARIDE NANOPARTICLES CROSSING BIOLOGICAL BARRIERS

Biopolymers, the polymers originated from natural sources, offer surplus advantage over conventional polymers and hence emerged as popular candidates for drug delivery carriers. In general, biopolymers are nontoxic, biocompatible, biodegradable, user friendly, and relatively cheap and hence have been found to be very promising for industrial applications in various forms. The main class of biopolymers exploited for drug delivery applications include polysaccharides (cellulose, chitin, chitosan (CS), alginate, etc.) and proteins (albumin, gelatin, collagen, silk, etc.) (Yadav et al., 2015; Raveendran et al., 2017).

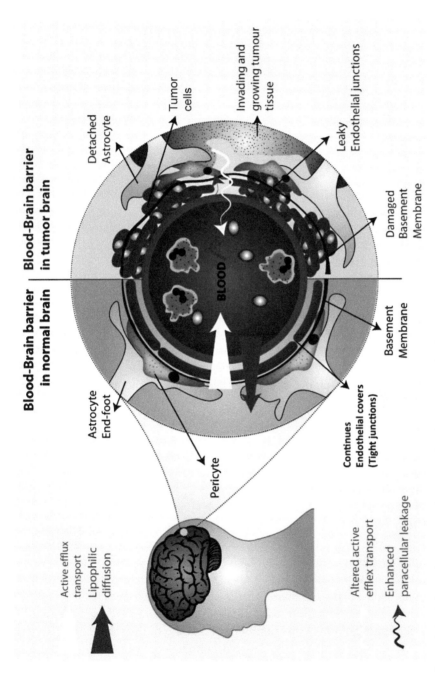

FIGURE 9.6 Schematic representation comparing the BBB organization in healthy and tumor conditions.

Polysaccharides, which are ubiquitous such as polyhydroxylated biopolymer containing sugar moieties, serve as a virtuous template for the synthesis of highly biocompatible nanoparticles (NPs) (Liu et al., 2008, 2015). Polysaccharide NPs can favor the high rate of solubility, biocompatibility, and prolonged drug delivery at target sites for the theranostic management of cancer. Mucoadhesion properties of polysaccharides enhance the bioavailability of drugs for higher therapeutic effect by binding to the mucus layer of the cell surface in tissues. Certain polysaccharide also tends to bind to cell surface receptors that lead to receptor-mediated endocytosis. For instance, biodegradable hyaluronic acid (HA)-coated CS NPs were developed to encapsulate a chemotherapeutic drug to improve the drug's antitumor efficiency by achieving targeted drug delivery via CD44 receptor overexpressed in A549 cell lines (Wang et al., 2017).

Polysaccharide NPs offer adaptable characteristics compared to synthetic or semi-synthetic polymeric NPs. First, most of the polysaccharides are biopolymer derived from microbial, plant, and animal sources, which are less expensive and easy to purify. Second, these polymers are highly biocompatible and tend to be retained in the circulating body fluids in the animal body. Third, polysaccharide NPs tend to escape from certain phagocytic cells, which mainly disturbs other polymeric NPs for their effective drug delivery system. Finally, most of the abundantly found polysaccharides possess tunable functional groups for their surface modification as both therapeutic agents as well as imaging agents, which highlights the efficiency of using polysaccharide NPs for biomedical applications (Peptu et al., 2014; Liu et al., 2008, 2015; Saravanakumar et al., 2012).

Polysaccharides used for nanoparticle synthesis can be classified into ionic polyelectrolytes and nonionic polyelectrolytes based on the functional groups (Salatin and Jelvehgari, 2017). Ionic polyelectrolytes classified further into cationic (CS) and anionic (alginate, heparin, HA, pectin) polysaccharides, while neutral polysaccharides include dextran, pullulan, xyloglucan, and galactomannan (Liu et al., 2008; Salatin and Jelvehgari, 2017) List of common polysaccharides used for the preparation of nanoparticles with it structure, charge, and source is tabulated in Table 9.1.

Based on the presence of functional groups and side chain of the polymer, the synthesis of polysaccharide NPs is categorized into self-assembled and cross-linked NPs. Self-assembled polysaccharide NPs are formed by polymer aggregation via their amphiphilic nature of the side chains (Myrick et al., 2014; Raja et al., 2016). The hydrophobic core of the polysaccharides tends to form aggregates within its central core, and the hydrophilic side chains remain in its periphery, offering higher rate of solubility, for example, amphiphilic carboxymethyl dextran with lithocholic acid labeled with Cy5.5 and loaded with the anticancer drug, DOX, for both cancer imaging and therapy (Thambi et al., 2014). These NPs exhibit controlled drug release in response to intracellular glutathione in cancer cells (Thambi et al., 2014). Cross-linked polysaccharide NPs are synthesized based on the ionic interactions between the charged moieties in polymer and the cross-linking agents (Patil and Jadge, 2008). Mostly, peptides and charged molecules tend to form an intermolecular interaction to form nanoaggregates, while certain chemical linkers such as tripolyphosphate (TPP) and calcium salts can produce polymeric NPs via ionic interactions. Previous studies revealed the formation of CS NPs in the presence of TPP due to the interaction of positively charged CS with negatively charged salt (Salatin and Jelvehgari, 2017). The encapsulation of the positively charged small-molecule drug (DOX) into CS (Janes et al., 2001; Soares et al., 2016) and xyloglucan (PST001) (Joseph et al., 2014; Joseph et al., 2014) followed by TPP cross-linking resulted in NPs with higher rate of encapsulation efficiency and successful endocytosis in cancer cells.

TABLE 9.1 List of Common Polysaccharides Used for Synthesizing Nanoparticles for Various Applications

Polysaccharide	Structure	Charge	Source
Chitosan	β-(1–4)-Linked-glucosamine and *N*-acetyl-D-glucosamine	Positive	Arthropods
Alginate	(1–4)-Linked β-D-mannuronic acid and α-l-guluronic acid	Negative	Plant
Hyaluronic acid	D-glucuronic acid and *N*-acetyl-D-glucosamine	Negative	Microbial
Pectin	D-galactouronic acid	Negative	Plant
Dextran	α-(1–6) glycosidic linkages in its main chain and a variable number of α-(1–2), α-(1–3) and α-(1–4) branched linkages	Neutral	Microbial
Pullulan	α-(1–6)-D-glucopyranose and α-(1–4)-D-glucopyranose	Neutral	Fungi
Xyloglucan	D-Galacto-D-xylo-D-glucans	Neutral	Plant

Various polysaccharide-based NPs were developed so far for the targeted drug delivery system based on the enhanced permeation and retention (EPR) effect of NPs at the tumor sites. CS NPs were developed with various surface modifications for drug encapsulation based on the ionic interactions for the development of successful controlled drug releasing nanovectors. CS NPs have been conjugated with several peptide molecules or ligands, like folic acid (FA), antibodies, and so on, for targeting the cancer cells at the site of action. These ligand-conjugated CS NPs with encapsulated drug were taken up by the cancer cells and released the intracellular drug in a controlled manner. Wang et al. (2017) exhibited the CS NPs decorated with HA encapsulated 5-fluoruracil effectively and were taken up by CD44-overexpressed tumor cells, leading to mitochondrial damage by the production of reactive oxygen species. PTX-loaded CS-coated poly(D, L-lactide-co-glycolide) NPs were found to be effective against lung tumors with increased cytotoxicity in the acidic pH. Multifunctional iron-oxide-loaded BSA NPs conjugated with dextran–FA moieties exhibited a suitable system for both tumor diagnosis and therapy in terms of MRI imaging and enhanced tumor reduction in H22 tumor-bearing mice (Hao et al., 2014).

Polysaccharide-based polymersomes are nowadays considered to be a fruitful drug carrier as these particles can encapsulate both hydrophobic and hydrophilic drugs with improved stability. Lecommandoux et al. have shown the efficient synthesis of block polymer between dextran and poly (γ-benzyl L-glutamate) that could deliver the chemotherapeutic drug, DOX. Hollow nanospheres based on polysaccharides also form a full pledged area of research in nanomedicine, as reported based on CS-poly(acrylic acid) nanospheres encapsulated with DOX drug with sustained drug release both in vitro and in vivo. Increased hydrophobicity on the surface of polysaccharides was achieved using liposomal-based polysaccharide NPs for the proficient cellular uptake of NPs. Docetaxel (DTX)-encapsulated N-palmitoyl CS-anchored liposomes exhibited higher in vitro stability with the release of the drug at the site of action. Drug-conjugated pectin NPs with targeting potential to specific cancer cells have been developed with improved drug delivery efficiency. Pectin–methotrexate NPs synthesized via the ionotropic gelation method with a hydrodynamic size of 390 nm provide sustained methotrexate delivery to hepatocellular carcinoma (Chittasupho et al., 2013). Sodium-alginate-based polysaccharide NPs enhance the effectiveness of the drug carrier system with controlled and targeted drug release, which is the utmost concern in pharmaceutical and medical allies. Manatunga et al. developed a novel pH-sensitive sodium alginate, hydroxyapatite bilayer-coated iron oxide NP composite (IONP/HAp-NaAlg) via the coprecipitation method that can be recognized as a potential drug delivery system for the purpose of curcumin and 6-gingerol to treat diseases such as cancer. The study focused on the designing of NPs with pH-triggered drug release with high loading and encapsulation efficiency (Manatunga et al., 2017). Pullulan produced from starch is used nowadays in the development of chemotherapeutic-loaded NPs as well as metal-conjugated NPs for various applications in the biomedical field. A PTX-loaded core-cross-linked nanoplatform was successfully engineered for targeted liver cancer treatment. In this study, reversibly cross-linked pullulan NPs with FA (FA-Pull-LA CLNPs) were fabricated for reduction-responsive liver drug delivery based on the specific affinity of pullulan and FA to overexpress asialoglycoprotein receptors and folate receptors. In vivo therapeutic efficacy studies confirmed that FA-Pull-LAPTX CLNPs achieved an enhanced antitumor effect and reduced systemic toxicity compared to free drugs (Huang et al., 2018). As evident from several studies, polysaccharide NP synthesis is found to be ecofriendly and easy for the development of stable NPs. Similarly, encapsulation of several drugs by rapid entrapment methods favors the advancement in nanosized formulations with high drug payload for the efficient management of cancer.

The development of nanomedicines as fascinating tools for enhancing the transport of drugs across the above-discussed biological barricades and treating cancer remains to be utmost challenge faced by the pharmaceutical companies. The literature shows that the use of NPs dramatically changed the future of therapeutic modalities involved in cancer screening, diagnosis, and treatment. Colloidal carriers based on natural or synthetic origin have been formulated as nanosystems within approximately 1–200 nm size range for carrying different therapeutic payloads for drug delivery. The main factors influencing the transport of NPs across biological barriers include the size (smaller NPs, <100 nm is favored), shape (spherical, cubic, rod, etc.; spherical NPs are most studied and recently rod-shaped NPs were reported to show adhesion propensity compared to their spherical counterparts), and zeta potential (positively charged NPs can cause brain toxicity, so negatively charged stable NPs are preferred for brain delivery). The nanosystems developed for drug delivery can open the TJs and enable the drugs to penetrate the barrier membranes or tissue layers, can be endocytosed by the cells and release the drugs intracellularly, and can also inhibit the transmembrane efflux systems. An ideal nanodrug delivery system designed for cancer management should have the following characteristics: (a) it should be controlled; (b) it should not damage the barrier; (c) the carrier system should be biodegradable and nontoxic; (d) systemic delivery should be targeted to the barrier tissues and the site of intended action; (e) the drug load transported through the barriers should be adequate for reaching therapeutic concentrations in the diseased tissue; and (f) therapeutic concentrations should be maintained for a sufficient duration of time for the desired efficacy.

9.3.1   *POLYSACCHARIDE NANOPARTICLES FOR ORAL DRUG DELIVERY*

Based on the above-discussed ambit, natural and biodegradable polysaccharides could be a preferred option for oral drug delivery due to their biocompatibility, nontoxicity, and adjustable controlled release properties. In 2014, Feng et al. designed coacervate microcapsules immobilized with multilayer sodium alginate beads (CMs-M-ALG beads) for oral drug delivery. The CMs-M-ALG beads were prepared by immobilization of doxorubicin hydrochloride-loaded CS/carboxymethyl coacervate microcapsules (DOX:CS/CMCS-CMs) in the core and layers of sodium alginate beads. The obtained CMs-M-ALG beads exhibited a layer-by-layer structure and a rough surface with many nanoscale particles. These carrier systems showed good gastric tolerance, and the ex vivo studies demonstrated that CMs-M-ALG beads were able to enhance the absorption of DOX by controlled release and prolong the contact time between the DOX:CS/CMCS-CMs and small intestinal mucosa (Feng et al., 2014). Later, the same group constructed sodium alginate beads with a porous core (internal cross-linking) and without a porous core (external cross-linking) based on CS and carboxymethyl chitosan (CMC) NPs immobilized in multilayer of sodium alginate beads (Li et al., 2016). The study showed that the internally cross-linked beads could efficiently pass through acidic gastric fluid and provide sustained release of a potent anticancer agent, doxorubicin hydrochloride, and hence presumed to have excellent potentials as carriers for the sustained release of oral drugs (Li et al., 2016). In these carrier systems, the prolonged contact time of these nanosystems with the intestinal mucosa could be due to the slow and sustained release of DOX-loaded CS NPs from the alginate beads and the hydrogen bonding between the carboxylate ions in the carrier systems and mucin; the high binding capacity to Ca^{2+} of deprotonated carboxyl groups on ALG could further enhance the absorption of DOX (Feng et al., 2014; Li et al., 2016). CS-coated polycaprolactone (PCL) NPs loaded with curcumin demonstrated their strong ability to electrostatically interact with the mucin glycoprotein and hence found to be mucoadhesive. These NPs showed in vitro anticancer profile on oral cancer cell line SCC-9 and hence were observed to be used as a formulation that could deliver curcumin directly to the oral cavity and reduce oral cancer (Mazzarino et al., 2015). Thiolated chitosan (TCS)-coated poly(methyl methacrylate) core was reported to be useful for the oral delivery of hydrophobic drug, DTX. In vitro drug release studies showed an initial burst release, but after that, a sustained release for 10 days was seen. When formulated as NPs, the transportation of DTX from the intestinal membrane showed a significant increase in ex vivo studies (Shahrooz et al., 2011). TCS-modified PCL NPs were studied for the oral delivery of taxol for lung cancer. Preliminary studies reported that TCS increased the mucoadhesiveness and permeation properties, resulting in enhanced NP uptake by the GI mucosa and improving drug absorption. These mucoadhesive NPs increased PTX transport by opening TJs and bypassing the efflux pump of P-gp of lung cancer cell lines (A549). Further in vivo evaluation could make these NPs eligible for applications in oral chemotherapy (Jiang et al., 2013). An oral delivery system was devised for entrapping PTX into *N*-((2-hydroxy-3-trimethylammonium)propyl)chitosan chloride (HTCC) NPs of about 130 nm. In vitro and in vivo transport investigations had proved that the presence of positive charges on the surface enhanced the intestinal permeability of these NPs. Additional exploration into side effects revealed that HTCC-NP:PTX caused lower Cremophor EL-associated toxicities compared with taxol. These results strongly suggested that HTCC NPs could be exploited as an oral carrier of PTX for cancer therapy (Lv et al., 2011). CS was also exploited as a coating for improving the stability, controlled delivery, mucoadhesion, and eventually cellular uptake by cancer cell lines. CS capping was shown to further improve the stability of solid lipid NPs (SLNPs) in simulated gastric condition by forming a distinct and thick layer around the SLNPs against their aggregation, as a result of which they can be developed as the oral drug delivery system for hydrophobic drugs (Luo et al., 2014). A blend of phospholipid (lecithin) and CS was used for the nanoformulation of tamoxifen citrate (TAM) (Barbieri et al., 2015), where CS played the role of bridging the negative polar heads in the phosphatidyl choline-based liposomes and strengthened the vesicle structure. Ex vivo studies showed that the penetration of TAM drug solution across the rat intestinal walls was enhanced by the lecitihin/CS NPs. The effect of enzymes on intestinal permeation of tamoxifen was shown only when tamoxifen-loaded NPs were in intimate contact with the mucosal surface. The encapsulation of tamoxifen in lecithin/CS NPs improved the nonmetabolized drug passage through the rat intestinal tissues via paracellular transport (Barbieri et al., 2015). Oxaliplatin-loaded HA-coupled CS NPs encapsulated in Eudragit pellets were reported to be effective for colonic tumors. In tumor-bearing Balb/C mice, orally administered formulations showed significant accumulation in the colon and tumors post 12 h of administration. These DDS show relatively high drug concentration in the colonic milieu and colonic tumors with prolonged exposure time, which indicates enhanced antitumor efficacy with low systemic toxicity for the treatment of colon cancer (Jain et al., 2010). In another study, Tsai et al. designed cisplatin-HA nanoconjugate (HCNPs), which were further coated with Eudragit for colon-targeted anticancer effect toward colon cancers (Tsai et al., 2013). HCNPs were entrapped in Eudragit S100-coated pectinate/alginate microbeads (PAMs) by using an electrospray method and a polyelectrolyte multilayer-coating technique in aqueous solution. Orally administered microbeads in rats facilitated colon-specific delivery of cisplatin ligand–receptor relationships and pH-dependent degradation, which limited the release in acidic conditions and only under simulated colonic conditions. Eudragit-coated HCNP-PAMs also reduced cisplatin-affiliated nephrotoxicity in vivo (Tsai et al., 2013). Similarly preliminary studies on 5-fluorouracil (5-FU)-loaded CS NPs with enteric coating (Eudragit) reported them to be effective against colon cancers (Tummala et al., 2014). Gemcitabine-loaded CS NPs formed by the ionic gelation using pluronic F-127 resulted in less than 200-nm-sized NPs. The mucoadhesion study indicated that the positively charged amino groups of CS interact ionocially and forms hydrogen bonding with the mucin chains, thereby enhancing the mucoadhesivity. These CS NPs could be further exploited as a potential candidate for the oral delivery of Gemcitabine (Hosseinzadeh et al., 2012). pH-sensitive CS/Fucoidan NPs were formed at pH 1.2 with a size distribution of 200–300 nm. These NPs became larger at higher pH and release the loaded drug (Curcumin). This property will help in developing these NPs as an efficient oral delivery system (Huang and Lam, 2011).

Natural gums play a dual role in protecting the drug in the gastric as well as intestinal conditions. An improved delivery system was developed for 5-FU using polysaccharides such as guar gum (GG) and xanthan gum (XG) NPs. These NPs reported to show dual advantage of targeted release of 5-FU specifically in colon and replenishing the colonic microflora, which was damaged by the chemical attack of the drug. Accordingly, the study focused on overcoming the issue of toxicity toward the natural microbial flora and offered a solution to the problem in terms of concomitant administration of probiotics along with colon-targeted 5-FU (Singh et al., 2015).

Dextran sulfate-PLGA NPs were studied as a carrier system for vincristine sulfate (VCR). In vivo pharmacokinetics in rats after oral administration of vincristine solution (VCR-Sol) and vincristine sulfate-loaded dextran sulfate-PLGA hybrid NPs (VCR-DPNPs) indicated that the apparent bioavailability of VCR-DPNPs was increased to approximately 3.3-fold compared to that of VCR-Sol. The relative cellular uptake of VCR-DPNPs was 12.4-fold higher than that of VCR-Sol in MCF-7/Adr cells, implying that P-gp-mediated drug efflux was diminished by the introduction of DPNs. Observations from this study indicated that it could be an effective strategy for oral delivery of VCR with improved encapsulation efficiency and oral bioavailability (Ling et al., 2010). The role of polysaccharides in GI residence of the NPs and bioenhancement of loaded therapeutic agents was studied by Anisha et al. using curcumin as a model drug (Souza and Devarajan, 2016). They extensively analyzed the NPs adsorbed with polysaccharides: arabinogalactan (AG) and kappa carrageenan (KC); galactose-based polysaccharides and pullulan (Pul); and glucose-based polysaccharides. Following oral administration of these nanoformulations, it was observed that AG- and KC-based formulation enhanced the gastroretention and improved the oral bioavailability of curcumin. The study pointed toward the advantage of using galactose-based polysaccharide nanoformulations, which enhance the drug absorption through galectin-mediated absorption. Bioenhancement could therefore be manifold for drugs that exhibit good stability throughout the GI tract (Souza and Devarajan, 2016).

Among various polysaccharides, pectin, which is enzyme-resistant (protease and amylase, which are active in the upper GI tract), has been shown to be a better matrix for the delivery of orally administered drugs and for colon-specific drug delivery with controlled release properties. Pectin/ethyl cellulose film-coated 5-FU tablets given orally to rats exhibited an accumulation specifically in the cecum and colon, whereas uncoated pellets remained distributed in the upper GI tract (Wei et al., 2008). Eudragit S-100-coated citrus pectin NPs were developed by Subudhi et al. for 5-FU (Subudhi et al., 2015). In vivo data clearly depicted that Eudragit S100 successfully guarded NPs to reach the colon region, where NPs were taken up and showed drug release for a prolonged period. Citrus pectin also acts as a ligand for galectin-3 receptors that are overexpressed on colorectal cancer cells and was found to enhance the targeting potential of E-CPNs to cancer cells (Subudhi et al., 2015). In 2016, Izadi et al. reported β-lacto globulin pectin NPs as a carrier for anticancer platinum complex to colon. The drug release profile in simulated GI conditions demonstrated that β-lacto globulin with a secondary coating is stable in acidic conditions and able to release its cargo at pH 7, which could serve as an effective vehicle for oral drug delivery preparations (Izadi et al., 2016). Earlier study from our laboratory reported that a natural polysaccharide PST001, isolated from *Tamarindus indica*, was conjugated with Imatinib mesylate, a tyrosine kinase inhibitor and an antitumor drug that inhibits the BCR-ABL kinase caused by chromosomal rearrangements in chronic myeloid leukemia and acute lymphoblastic leukemia, and found to be effective against K562-resistant cell lines in vitro and showed no toxicity to mice when administered orally (James et al., 2017).

9.3.2 POLYSACCHARIDE NANOPARTICLES FOR TOPICAL DRUG DELIVERY

The principal function of skin is to defend our body from external pathogens as a topical barrier and is exploited for cutaneous and percutaneous delivery of therapeutics considering the advantageous bypassing of first pass metabolism and patient compliance. Based on the structure of SC, it could be highlighted that the skin integrity, dimensions of orifices, aqueous pores, lipidic fluid paths, and the density of appendages affect the absorption of topically applied therapeutic agents (Baroli, 2010). The major routes of penetration through the skin barrier include the intercellular lipid route and transcellular or transappendageal (follicular) pathways (Paliwal, 2017). Nanocarriers interact with skin based on size, surface charge, lamellarity, and mode of application and can translocate intact into the skin without being degraded or they can be degraded near the skin surface and the incorporated therapeutic molecule can penetrate into skin layers. However, most particles do not cross the SC and the transappendageal route appears to be the dominant pathway of NP entry into skin irrespective of the nanomaterial.

Chitin is a natural polysaccharide like CS that has been exploited for various biomedical applications, and its positive charge aids in its use as a skin penetration enhancer. Sabitha et al. reported the use of chitin in the form of nanogels for the transdermal delivery of anticancer molecules like curcumin (Mangalathillam et al., 2012) and 5-FU (Sabitha et al., 2013) for melanoma. They developed 70–80 nm curcumin-loaded chitin nanogels (CCNGs) and 120–140-nm-sized 5-FU-loaded chitin nanogels (FCNGs), which were found to be effective against melanoma cell lines (A-375) in vitro. Ex vivo penetration studies performed in porcine skin showed a fourfold increase in steady-state transdermal flux of curcumin from CCNGs as compared to that of the control curcumin solution. FCNGs showed more or less similar steady-state flux as that of control 5-FU, but the retention in the deeper layers of skin was found to be 4–5 times more. Histopathological evaluation revealed loosening of the outer most layer of epidermis, SC, by interaction of catatonically charged chitin, with no observed signs of inflammation; hence, nanogels could be a good option for treatment of skin cancers (Mangalathillam et al., 2012; Sabitha et al., 2013). Many studies are reported for the use of CS NPs for transdermal drug delivery. Recently, transdermal drug delivery mechanisms of CS NPs with the synergistic action of microwave in skin modification was investigated by Nawaz and Wong (2017). The transdermal drug delivery profiles of 5-FU-loaded CS NPs across untreated and microwave-treated skin were examined. The drug transport was mediated via NPs carrying drug across the skin and/or diffusion of earlier-released drug molecules from skin surfaces. The CS NPs largely affected the palmitic acid and keratin domains. Combined microwave and nanotechnologies synergize polysaccharide-mediated transdermal drug delivery (Nawaz and Wong, 2017).

The skin comprising a linear polysaccharide HA is being investigated as a topical delivery carrier in recent times. The properties of HA that aid skin permeability include the following: (1) it is a hygroscopic polyanion, (2) it has the ability to hydrate the skin, (3) the hydrophobic patch of HA interacts with the lipid components in SC during penetration and disrupt the skin barriers, and (4) the presence of HA receptors on the skin resident cells (keratinocytes in epidermis and fibroblasts in the dermis over expressed HA receptors) (Kim, 2014). A nanographene oxide HA conjugate (NGO-HA) was synthesized for photothermal ablation therapy of melanoma skin cancer using a near-infrared (NIR) laser. The transdermal delivery of NGO-HA was enhanced by the EPR effect and the overexpressed HA receptors around the tumor tissues. The NIR irradiation resulted in complete ablation of tumor with no recurrence of tumorigenesis in tissues (Jung et al., 2013). Lee et al. (2016) reported the targeted delivery of death receptor 5 antibody-conjugated hyaluronate-gold nanorod (HA-AuNR/DR5 Ab) for the transdermal theranosis of skin cancer. Wang et al. (2018) worked on polysaccharide-based microneedles for the sustained delivery of anti-(PD1) for immunotherapy of cancer. They used pH-sensitive dextran NPs integrated with biocompatible HA that encapsulated αPD1 and glucose oxidase (GOx), which converts blood glucose to gluconic acid. Altering the pH of the environment to acidic promotes the self-dissociation of NPs and subsequently results in the substantial release of αPD1. They reported immediate immune responses in a B16F10 mouse melanoma model with a single administration of the microneedle patch (Wang et al., 2018).

9.3.3 POLYSACCHARIDE NANOPARTICLES CROSSING THE BLOOD–BRAIN BARRIER

Polysaccharide-based colloidal drug carriers in the form of emulsions, micelles, liposomes dendrimers, and NPs were developed as nanomedicines, which provide a potential platform to be developed as therapeutic strategies to pass through BBB. However, only a few investigations have been done in the case of polysaccharide anticancer strategies against BBB. Plenty of CS-based transport systems have been reported for overcoming BBB. β-cyclodextrin (β-CD) is a potential polysaccharide studied for BBB crossing. Aktas et al. (2005) synthesized transferrin receptor (TfR)-targeted CS nanospheres conjugated with poly(ethylene glycol) (PEG) bearing the OX26 monoclonal antibody that triggered receptor-mediated transport across the BBB. They designed these NPs for the delivery of the caspase inhibitor peptide, and the systemically administered systems were found to provide significant concentration of the peptide in the brain tissues. The electrostatic interaction between these cationic particles and the negatively charged brain endothelium could be responsible for the targeted-induced transport across the BBB (Aktas et al., 2005). β-CD was modified using poly(β-amino ester) and formulated into nanogels coloaded with DOX and insulin. The hydrophobic cavity provided space for hydrophobic drugs and the outer polysaccharide layer gives opportunity for loading hydrophilic moieties and showed enhanced permeability over blood–brain microendothelial cells. The same group prepared NPs with the same material and found that they had higher permeability across in vitro BBB models (Gil et al., 2012). Hydroxypropyl-b-cyclodextrin enhanced the solubility of puerarin and gelatin, which in turn improved the viscosity of theinner water phase, resulting in enhanced entrapment of puerarin. It was also proposed that nanocarriers or cyclodextrin inclusion complex enhanced the penetration of drug across the BBB (Tao et al., 2013).

9.3.4   POLYSACCHARIDE NANOPARTICLES CROSSING OTHER BARRIERS

In addition to the above-discussed barriers, drug delivery candidates were exposed to nasal mucosa, rectal mucosa, and vaginal and ophthalmic barriers. However, not many researchers have exploited these routes for delivering therapeutic agents for cancer treatment.

Nasal mucosa forms a barrier for delivering therapeutic agents to lungs as well as to brain. Nasal administrations of drugs provide an alternative administration route for brain disorders, where BBB forms the main challenging barricade. Mucoadhesive polymer CS was reported to be beneficial in binding to nasal mucosa and improving the drug concentration in the lungs as well as CS aerosols directly reaching brain from the nose bypassing BBB (Kaur et al., 2012). Majority of the studies focused on the use of CS NPs for the nasal delivery of proteins, peptides, or nucleic acids as antigens, vaccines, or genetherapy (Kaur et al., 2012; van Woensel et al., 2013). DNA vaccine loaded into mannosylated chitosan (MCS) NPs was intranasally administered to prostate cancer (subcutaneous)-bearing mice by Yao et al. (2013). They reported that these MCS targeted antigen-presenting cells, enhanced mannose-mediated cancer cell uptake, and enhanced the intranasal administration, resulting in high level of anti-GRP antibody and thereby inhibiting tumor growth. Thus, MCS offered efficient tumor immunotherapy (Yao et al., 2013). PTX-loaded alginate microparticles have been developed for site-specific pulmonary delivery of drugs to mucosal tissues. Results show that the exposure of cells to pure PTX and PTX-loaded microparticles effectively inhibited the growth of A549 and Calu-6 cells, similarly in concentration- and time-dependent manners (Alipour et al., 2010). Similarly HA, carboxymethyl cellulose, and cyclodextrins have been reported to be studied for nasal drug delivery (Kaur et al., 2012). Feng et al. reported DOX-loaded CS and CMC nanogels with high mucin binding and inhibiting paracellular transport to the colon. This enhanced mucoadhesion and limited permeability prolonged the contact time with the intestinal mucosa, resulting in local drug concentration. These nanogels can be further exploited for local colorectal cancer therapy (Feng et al., 2015). In 2012, Pandekal et al. developed a CS–polycarbophil interpolyelectrolyte (IPEC) complex, which was later compressed with tablets of anticancer drug 5-FU. This 5FU-IPEC was evaluated for drug release at different pH values simulating buccal cavity, vagina, and rectum. The formulations showed controlled release of 5-FU with the highest bioadhesive property and satisfactory residence in both buccal and vaginal cavities of rabbit. Formulation with 3% of sodium deoxycholate exhibited maximum permeation of 5-FU. The study thus proved that a suitable combination of IPEC, CS, and polycarbophil demonstrated a potential candidate for controlled release of 5-FU in buccal, vaginal, and rectal pH (Pendekal and Tegginamat, 2012).

Extensive research is needed where biocompatible natural polysaccharides can be utilized to form cancer nanomedicines to exploit other noninvasive administration routes so that lives of cancer patients can be saved worldwide.

9.4 CONCLUSIONS

The interaction between the nanocarriers and the target cancer tissues by overcoming the biobarriers plays a major role in the efficient delivery of the therapeutic molecules and hence their cancer therapeutic efficacy. Biological barriers are vital for the development, function, and integrity of many organs. Several layers of cells help in the formation of these barricades, and the cell type may vary from organ to organ or site to site. Mainly, these barriers are characterized by the formation of TJs, adherens junctions, desmosomes, and gap junctions. The protective role of the biological barricades has been greatly studied, but the manipulation of these barricades for the delivery of therapeutics needs more research and a major challenge is the transport of drugs through these barriers. Currently, many of the anticancer therapeutic nanosystems rely on the transendothelial system, which is inefficient, slow, and requires high drug concentrations. The DDS requires specific properties that enable them to cross these biological barricades. For instance, neutral/hydrophilic NPs are preferred for crossing skin barriers, whereas transferrin receptor-targeted NPs will be beneficial for BBB crossing. Biological polymers such as polysaccharides can be used as carrier molecules to breach the barriers and enhance the functionality, thereby serving the "purpose." Multivalent features of NPs make them cross the barriers and competently deliver drug moieties to the tumors. Multifunctionality may increase the complexity of the systems, and regulatory problems may also arise. Therefore, it is necessary to keep the preparations simple, cost effective, and patient friendly. In this scenario, nanocarriers, exploiting various synchronized modalities responsible for overcoming several sequential biological barricades, could highly improve the therapeutic efficacy of the drug. Even more efforts have to be made for exploiting these natural biodegradable, biocompatible, and flexible carrier materials, polysaccharides, with standard quality evaluations in preclinical and clinical perspectives to proactively address rapid advances in developing cancer treatment strategies.

ACKNOWLEDGMENT

We greatly acknowledge the Council of Scientific and Industrial Research (60(112)/16/EMR-11, dated April 06, 2016), Govt. of India, for providing funding to carry out the work associated with the current review and also a fellowship to author Maya. S and the University Grants Commission for the research fellowship provided to PGU and UKBS.

KEYWORDS

- **polysaccharides**
- **nanoparticles**
- **biological barrier**
- **oral mucosa**
- **intestinal mucosa**
- **blood–brain barrier**

REFERENCES

Aktaş Y., Yemisci M., Karine Andrieux, R., Neslihan Gürsoy, Maria Jose Alonso, Eduardo Fernandez-Megia, Ramón Novoa-Carballal, Emilio Quiñoá, Ricardo Riguera, Mustafa F. Sargon, H. Hamdi Celik, Ayhan S. Demir, A. Atilla Hincal, Turgay Dalkara, Yilmaz Capan, Patrick Couvreur. Development and brain delivery of chitosan PEG nanoparticles functionalized with the monoclonal antibody OX26. *Bioconjug Chem* (2005); 16(6): 1503–1511.

Alagusundaram, M., Chengaiah B., Gnanaprakash K., Ramkanth S., Madhusudhana C., Chetty, & D. Dhachinamoorthi. Nasal drug delivery system—an overview. *Int J Res Pharm Sci* (2010); 1(4): 454–465.

Alipour, S., Montaseri, H. & Tafaghodi, M. Preparation and characterization of biodegradable paclitaxel loaded alginate microparticles for pulmonary delivery. *Colloids Surf B Biointerfaces* (2010); 81(2): 521–529.

Barbieri S., Buttini F., Rossi A., Bettini R., Colombo P., Ponchel G., Sonvico F., Colombo G. Ex vivo permeation of tamoxifen and its 4-OH metabolite through rat intestine from lecithin/chitosan nanoparticles. *Int J Pharm* (2015); 491(1–2): 99–104.

Baroli, B. Penetration of nanoparticles and nanomaterials in the skin: fiction or reality? *J Pharm Sci* (2010); 99(1): 21–50.

Bernacki, J., Dobrowolska, A. & Nierwińska, K. Physiology and pharmacological role of the blood-brain barrier. *Pharmacol Rep* (2008); 60(5): 600–22.

Blanchette, J., Kavimandan, N. and Peppas, N. A. Principles of transmucosal delivery of therapeutic agents. *Biomed Pharmacother* (2004); 58(3): 142–151.

Bolzinger, M. A., Briançon, S., Pelletier, J. and Chevalier, Y. Penetration of drugs through skin, a complex rate-controlling membrane. *Curr Opin Colloid Interface Sci* (2012); 17(3): 156–165.

Bowen, R. A. The gastrointestinal barrier. *The Veterinary ICU Book,* CRC press, Taylor and Francis, USA, (2002), pp. 40–46.

Cevc, G. and Vierl, U. Nanotechnology and the transdermal route. A state of the art review and critical appraisal. *J Control Release* (2010); 141(3): 277–99.

Chabner, B. A. and Roberts, T. G. Timeline: chemotherapy and the war on cancer. *Nat Rev Cancer* (2005); 5(1), 65–72.

Chittasupho, C., Jaturanpinyo, M. and Mangmool, S. Pectin nanoparticle enhances cytotoxicity of methotrexate against hepG2 cells. *Drug Deliv.* (2013); 20(1): 1–9.

Cirenajwis, Helena, Smiljanic, Sandra, Honeth, Gabriella, Hegardt, Cecilia, Marton, Laurence J, Oredsson, and Stina M. Reduction of the putative CD44⁺CD24⁻ breast cancer stem cell population by targeting the polyamine metabolic pathway with PG11047. *Anti-Cancer Drugs* (2010); 21(10) 897–906.

Daneman, R. and Prat, A. The blood–brain barrier. *Cold Spring Harb Perspect Biol* (2015); 7(1) a020412.

Chao Feng, Ruixi Song, Guohui Sun, Ming Kong, Zixian Bao, Yang Li, Xiaojie Cheng, Dongsu Cha, Hyunjin Park, Xiguang Chen. Immobilization of coacervate microcapsules in multilayer sodium alginate beads for efficient oral anticancer drug delivery. *Biomacromolecules* (2014); 15(3): 985–96.

Chao Feng, Jing Li, Ming Kong, Ya Liu 2, Xiao Jie Cheng, Yang Li, Hyun Jin Park, Xi Guang Chen. Surface charge effect on mucoadhesion of chitosan based nanogels for local anti-colorectal cancer drug delivery. *Colloids Surf B: Biointerfaces* (2015); 1(128): 439–447.

Gil, E. S., Wu, L., Xu, L. and Lowe, T. L. β-Cyclodextrin-poly (β-amino ester) nanoparticles for sustained drug delivery across the blood–brain barrier. *Biomacromolecules* (2012). 13 (11): 3533–3541.

Hao, H., Ma, Q., He, F. and Yao, P. Doxorubicin and Fe_3O_4 loaded albumin nanoparticles with folic acid modified dextran surface for tumor diagnosis and therapy. *J Mater Chem B* (2014) 2(45): 7978–7987.

Hosniyeh Hosseinzadeh, Fatemeh Atyabi, Rassoul Dinarvand, Seyed Naser Ostad. Chitosan–pluronic nanoparticles as oral delivery of anticancer gemcitabine: preparation and in vitro study. *Int J Nanomed* (2012). 7, 1851–1863.

Huang, C.-Y., Ju, D.-T., Chang, C.-F., Muralidhar Reddy, P. and Velmurugan, B. K. A review on the effects of current chemotherapy drugs and natural agents in treating non-small cell lung cancer. *BioMedicine* (2017); 7(4): 23.

Liping Huang, Birendra Chaurasiya, Dawei Wu, Huimin Wang, Yunai Du, Jiasheng Tu, Thomas J Webster, Chunmeng Sun. Versatile redox-sensitive pullulan nanoparticles for enhanced liver targeting and efficient cancer therapy. *Nanomedcine* (2018). 14(3): 1005–1017.

Huang, Y. and Lam, U. Chitosan/fucoidan pH sensitive nanoparticles for oral delivery system. *J Chin Chem Soc*. (2011). 58(6) 779–785.

Izadi, Z., Divsalar, A., Building, R. L. and Brown, A. C. b-lactoglobulin–pectin nanoparticle-based oral drug delivery system for potential treatment of colon cancer. *Chem Biol Drug Des* (2016); 88(2): 209–216.

Jain, A., Jain, S. K., Ganesh, N., Barve, J. and Beg, A. M. Design and development of ligand-appended polysaccharidic nanoparticles for the delivery of oxaliplatin in colorectal cancer. *Nanomed Nanotechnol Biol Med* (2010). 6(1): 179–190.

James A. R., Unnikrishnan B. S., Priya R., Joseph M. M., Manojkumar T.K., Raveendran Pillai K., Shiji R., Preethi G. U., Kusumakumary P., Sreelekha T. T. Computational and mechanistic studies on the effect of galactoxyloglucan: Imatinib nanoconjugate in imatinib resistant K562 cells. *Tumor Biol* (2017); 39(3): 1010428317695946.

Janes, K. A., Fresneau, M. P., Marazuela, A., Fabra, A. and Alonso, M. J. Chitosan nanoparticles as delivery systems for doxorubicin. *J Control Release* (2001); 73(2–3): 255–67.

Jiang, L., Li, X., Liu, L. and Zhang, Q. Thiolated chitosan-modified PLA-PCL-TPGS nanoparticles for oral chemotherapy of lung cancer. *Nanoscale Res Lett* (2013); 8(1): 66.

Manu M. Joseph, Aravind S. R., Suraj K. George, Raveendran K. Pillai, Mini S., Sreelekha T. T. Co-encapsulation. of doxorubicin with galactoxyloglucan nanoparticles for intracellular tumor-targeted delivery in murine ascites and solid tumors. *Transl Oncol* (2014); 7(5): 525–36.

Ho Sang Jung, Won Ho Kong, Dong Kyung Sung, Min-Young Lee, Song Eun Beack, Do Hee Keum, Ki Su Kim, Seok Hyun Yun, and Sei Kwang Hahn. Nanographene oxide–hyaluronic acid conjugate for photothermal ablation therapy of skin cancer. *ACS Nano* (2014); 8(1): 260–268.

Kaur, G., Narang, R. K., Rath, G. and Goyal, A. K. Advances in pulmonary delivery of nanoparticles. *Artif Cells Blood SubstitImmobil Biotechnol* (2012). 40(1–2): 75–96.

Kim, J. and Khan, W. Goblet cells and mucins: role in innate defense in enteric infections. *Pathogens* (2013); 2(1): 55–70.

Kim, K. S. Enhancing the transdermal penetration of nanoconstructs: could hyaluronic acid be the key? *Nanomedicine* (2014); 9(6): 743–745.

Kinikoglu, B., Damour, O. and Hasirci, V. Tissue engineering of oral mucosa: a shared concept with skin. *J Artif Organs* (2015); 18(1): 8–19.

Kiyono, H. and Fukuyama, S. Nalt versus Peyer's-patch-mediated mucosal immunity. *Nat Rev Immunol* (2004); 4(9): 699–710.

Julia König, Jerry Wells, Patrice D Cani, Clara L García-Ródenas, Tom MacDonald, Annick Mercenier, Jacqueline Whyte, Freddy Troost 7, Robert-Jan Brummer. Human intestinal barrier function in health and disease. *Clin Transl Gastroenterol* (2016); 7(10): e196.

Kumar, A., Pandey, A. N. and Jain, S. K. Nasal-nanotechnology: revolution for efficient therapeutics delivery. *Drug Deliv* (2016); 23(3): 681–93.

Lai-Cheong, J. E. Structure and function of skin, hair and nails: key points. *Medicine (Baltimore)* (2017); 45: 347–351.

Hwiwon Lee, Jung Ho Lee, Jeesu Kim, Jong Hwan Mun, Junho Chung, Heebeom Kooll, Chulhong Kim, Seok Hyun Yun, and Sei Kwang Hahn. Hyaluronate–gold nanorod/DR5 antibody complex for noninvasive theranosis of skin cancer. *CS Appl Mater Interfaces* (2016); 8(47): 32202–32210.

Jing Li, Changqing Jiang, Xuqian Lang, Ming Kong, Xiaojie Cheng, Ya Liu, Chao Feng, Xiguang Chen. Multilayer sodium alginate beads with porous core containing chitosan based nanoparticles for oral delivery of anticancer drug. *Int J Biol Macromol* (2016); 85: 1–8.

Guixia Ling, Peng Zhang, Wenping Zhang, Jin Sun, Xiaoxue Meng, Yimeng Qin, Yihui Deng, Zhonggui He. Development of novel self-assembled DS-PLGA hybrid nanoparticles for improving oral bioavailability of vincristine sulfate by P-gp inhibition. *J Control Release* (2010); 148(2): 241–248.

Liu, J., Willför, S. and Xu, C. A review of bioactive plant polysaccharides: biological activities, functionalization, and biomedical applications. *Bioact Carbohydr Diet Fibre* (2015); 5(1): 31–61.

Liu, Z., Jiao, Y., Wang, Y., Zhou, C. and Zhang, Z. Polysaccharides-based nanoparticles as drug delivery systems. *Adv Drug Deliv Rev.* (2008); 60(15): 1650–1662.

Luo, Y., Teng, Z., Li, Y. and Wang, Q. Solid lipid nanoparticles for oral drug delivery: Chitosan coating improves stability, controlled delivery, mucoadhesion and cellular uptake. *Carbohydr Polym.* (2014); 122: 221–229.

Lv, P., Wei, W., Yue, H., Yang, T. and Wang, L. Porous quaternized chitosan nanoparticles containing paclitaxel nanocrystals improved therapeutic efficacy in non-small-cell lung cancer after oral administration. *Biomacromolecules* (2011); 12(12): 4230–4239.

Danushika C Manatunga, Rohini M de Silva, K M Nalin de Silva, Nuwan de Silva, Shiva Bhandari, Yoke Khin Yap, N Pabakara Costha. pH responsive controlled release of anti-cancer hydrophobic drugs from sodium alginate and hydroxyapatite bi-coated iron oxide nanoparticles. *Eur J Pharm Biopharm.* (2017); 117: 29–38.

Mazzarino L., Loch-neckel G., Bubniak S., Mazzucco S., Santos-silva M.C., Borsali R., Lemos-senna E. Curcumin-loaded chitosan-coated nanoparticles as a new approach for the local treatment of oral cavity cancer. *J Nanosci Nanotechnol.* (2015); 15(1): 781–791.

Paliwal, S. Transdermal drug delivery: opportunities and challenges for controlled delivery of therapeutic agents using nanocarriers. *Curr Drug Metab.* (2017); 18(5): 481–495.

Mirza, A. Z. and Siddiqui, F. A. Nanomedicine and drug delivery: a mini review. *Int Nano Lett* (2014); 4: 94.

Montero, P. H. and Patel, S. G. Cancer of the oral cavity. *Surg Oncol Clin N Am.* (2015); 24(3): 491–508.

Mitragotri, S. Devices for overcoming biological barriers: the use of physical forces to disrupt the barriers. *Adv Drug Deliv Rev.* (2013); 65(1): 100–103.

Myrick, J. M., Vendra, V. K. and Krishnan, S. Self-assembled polysaccharide nanostructures for controlled-release applications. *Nanotechnol Rev.* (2014); 3(4): 319–346.

Mangalathillam, S., Rejinold, N. S., Nair, A. and Lakshmanan, V. Curcumin loaded chitin nanogels for skin cancer treatment via the transdermal route. *Nanoscale* (2012); 4(1): 239–250.

Nawaz, A. and Wong, T. W. Microwave as skin permeation enhancer for transdermal drug delivery of chitosan-5-fluorouracil nanoparticles. *Carbohydr. Polym.* (2017); 157: 906–919.

NIH. *Drug Delivery Systems: Getting Drugs to Their Targets in a Controlled Manner*, National Institute of Biomedical Imaging and Bioengineering. (October 1, 2016).

Obermeier, B., Verma, A. and Ransohoff, R. M. The blood–brain barrier. *Handb Clin Neurol* (2016); 133: 39–59.

Park, K. Drug delivery research: the invention cycle. *Mol Pharm.* (2016); 13 (7), 2143–2147.

Patel, V. F., Liu, F. and Brown, M. B. Advances in oral transmucosal drug delivery. *J Control Release* (2011); 153(2): 106–116.

Patil, S. and Jadge, D. Crosslinking of polysaccharides: methods and applications. *Pharm Rev* (2008); 6; 2. ISSN 1918-5561.

Pendekal, M. S. and Tegginamat, P. K. Development and characterization of chitosan-polycarbophil interpolyelectrolyte complex-based 5-fluorouracil formulations for buccal, vaginal and rectal application. *DARU J Facult Pharm.* (2012); 20(1): 67.

Peptu, C. A., Ochiuz, L., Alupei, L., Peptu, C. and Popa, M. Carbohydrate based nanoparticles for drug delivery across biological barriers. *J Biomed Nanotechnol.* (2014); 10(9): 2107–2148.

Najar, I. N. and Das S. Poly-glutamic acid (PGA)—structure, synthesis, genomic organization and its application: a review. *In. J Pharm Sci Res.* (2015); 6(6): 2320–5148.

Qin, R., Steel, A. and Fazel, N. Oral mucosa biology and salivary biomarkers. *Clin Dermatol.* (2017); 35(5): 477–483.

Raja, M. A., Zeenat, S., Arif, M. and Liu, C. Self-assembled nanoparticles based on amphiphilic chitosan derivative and arginine for oral curcumin delivery. *Int J Nanomed.* (2016); 11: 4397–4412.

Sreejith Raveendran, Ankit K Rochani, Toru Maekawa, D Sakthi Kumar. Smart carriers and nanohealers: a nanomedical insight on natural polymers. *Materials (Basel).* (2017); 10(8): 929.

Sabitha M, Sanoj Rejinold N., Amrita Nair, Vinoth-Kumar Lakshmanan, Shantikumar V Nair, Jayakumar, R. Development and evaluation of 5-fluorouracil loaded chitin nanogels for treatment of skin cancer. *Carbohydr Polym.* (2013); 91(1): 48–57.

Sachan, A. K. and Singh, S. Nanoparticles: nasal delivery of drugs. *Int J Pharm Res Sch.* (2014); 3(3): 33–44.

Salatin, S. and Jelvehgari, M. Natural polysaccharide based nanoparticles for drug/gene delivery. *Pharm Sci* (2017); 23: 84–94.

Saravanakumar, G., Jo, D.-G. and H. Park, J. Polysaccharide-based nanoparticles: a versatile platform for drug delivery and biomedical imaging. *Curr Med Chem* (2012); 19(19): 3212–3229.

Sato, A. and Iwasaki, A. Peyer's patch dendritic cells as regulators of mucosal adaptive immunity. *Cellul Mol Life Sci.* (2005); 62(12): 1333–1338.

Norman R Saunders, Jean-Jacques Dreifuss, Katarzyna M. Dziegielewska, Pia A. Johansson, Mark D. Habgood, Kjeld Møllgård, Hans-Christian Bauer. The rights and wrongs of blood–brain barrier permeability studies: a walk through 100 years of history. *Front Neurosci* (2014); 8: 404.

Schenk, M. and Mueller, C. The mucosal immune system at the gastrointestinal barrier. *Best Pract Res Clin Gastroenterol* (2008); 22(3): 391–409.

Marc Schneider, Maike Windbergs, Nicole Daum, Brigitta Loretz, Eva-Maria Collnot, Steffi Hansen, Ulrich F Schaefer, Claus-Michael Lehr. Crossing biological barriers for advanced drug delivery. *Eur J Pharm Biopharm.* (2013); 84(2): 239–241.

Serlin Y., Shelef I., Knyazer B., Friedman A. Anatomy and physiology of the blood–brain barrier. *Semin Cell Dev Biol.* (2015); 38: 2–6.

Shahrooz Saremi, Fatemeh Atyabi, Seyedeh Parinaz Akhlaghi, Seyed Nasser Ostad, Rassoul Dinarvand. Thiolated chitosan nanoparticles for enhancing oral absorption of docetaxel: preparation, in vitro and ex vivo evaluation. *Int J Nanomed.* (2011); 6: 119–128.

Shizuko Satoh-Kuriwada, Noriaki Shoji, Hiroyuki Miyake, Chiyo Watanabe, and Takashi Sasano. Effects and mechanisms of tastants on the gustatory-salivary reflex in human minor salivary glands. *BioMed Res Int.* (2018); 2018: 3847075.

Singh, R. M. P., Kumar, A. and Pathak, K. Mucoadhesive in situ nasal gelling drug delivery systems for modulated drug delivery. *Expert Opin Drug Deliv.* (2013); 10(1): 115–130.

Sima Singh, Niranjan G Kotla, and Omprakash Sunnapu. A nanomedicine-promising approach to provide an appropriate colon-targeted drug delivery system for 5-fluorouracil. *Int J Nanomed.* (2015); 10: 7175–7182.

Snoeck, V., Goddeeris, B. and Cox, E. The role of enterocytes in the intestinal barrier function and antigen uptake. *Microbes Infect.* (2005); 7(7–8): 997–1004.

Soares P IP, Sousa AI, Silva JC, Ferreira IMM, Novo CMM, Borges JP. Chitosan-based nanoparticles as drug delivery systems for doxorubicin: optimization and modelling. *Carbohydr Polym.* (2016); 147: 304–312.

Souza, A. A. D. and Devarajan, P. V. Bioenhanced oral curcumin nanoparticles: role of carbo-hydrates. *Carbohydr Polym.* (2016); 136: 1251–1258.

Stamatovic, S. M., Johnson, A. M., Keep, R. F. and Andjelkovic, A. V. Junctional proteins of the blood–brain barrier: new insights into function and dysfunction. *Tissue Barriers.* (2016) 4(1): e1154641.

Biswaranjan Subudhi M., Ankit Jain, Ashish Jain, Pooja Hurkat, Satish Shilpi, Arvind Gulbake, Sanjay K Jain. Eudragit S100 coated citrus pectin nanoparticles for colon targeting of 5-fluorouracil. *Materials (Basel).* (2015); 8(3): 832–849.

Suciu, M., Hermenean, A. and Wilhelm, I. Heterogeneity of the blood–brain barrier. *Tissue Barriers.* (2016); 4(1): e1143544.

Tao HQ, Meng Q, Li M, Yu H, Liu M, Du D, Sun S, Yang H, Wang Y, Ye W, Yang L, Zhu D, Jiang C, Peng H. HP-β-CD-PLGA nanoparticles improve the penetration and bioavailability of puerarin and enhance the therapeutic effects on brain ischemia-reperfusion injury in rats. *Naunyn Schmiedebergs Arch Pharmacol.* (2013); 386(1): 61–70.

Tellingen, O., Van, Gooijer, M. C. De, Wesseling, P., Wurdinger, T. and Vries, H. E. De. Over-coming the blood–brain tumor barrier for effective glioblastoma treatment. *Drug Resist Updat.* (2015); 19: 1–12.

Thavasyappan Thambi, Dong Gil You, Hwa Seung Han, V G Deepagan, Sang Min Jeon, Yung Doug Suh, Ki Young Choi, Kwangmeyung Kim, Ick Chan Kwon, Gi-Ra Yi, Jun Young Lee, Doo Sung Lee, Jae Hyung Park. Bioreducible carboxymethyl dextran nanoparticles for tumor-targeted drug delivery. *Adv Healthcare Mater.* (2014); 3(11): 1829–1838.

Tiwari G, Tiwari R, Sriwastawa B, Bhati L, Pandey S, Pandey P, Bannerjee SK. Drug delivery systems: an updated review. *Int J Pharm Investig.* (2012); 2(1): 2–11.

Tsai S.-W., Yu D.-S., Tsao S.-W., Hsu F.-Y. Hyaluronan–cisplatin conjugate nanoparticles embedded in Eudragit S100-coated pectin/alginate microbeads for colon drug delivery. *Int J Nanomed.* (2013); 8: 2399–2407.

Tummala, S., Satish Kumar, M.N. and Prakash, A. Formulation and characterization of 5-fluorouracil enteric coated nanoparticles for sustained and localized release in treating colorectal cancer. *SAUDI Pharm J.* (2014); 23: 308–314.

Türker, S., Onur, E. and Özer, Y. Nasal route and drug delivery systems. *Pharm World Sci* (2004); 26: 137–142.

Uchechi, O., Ogbonna, J. D. N. and Attama, A. A. Nanoparticles for dermal and transdermal drug delivery. In *Applications of Nanotechnology in Drug Delivery* (2014); Intech Open, London.

Upadhyay, S., Parikh, A., Joshi, P., Upadhyay, U. M. and Chotai, N. P. Intranasal drug delivery system—a glimpse to become maestro. *J Appl Pharm Sci.* (2011); 1(03): 34–44.

van Woensel, M., Wauthoz, N., Rosière, R., Amighi, K., Mathieu, V., Lefranc, F., van Gool, S. W., & de Vleeschouwer, S. Formulations for intranasal delivery of pharmacological agents to combat brain disease: a new opportunity to tackle GBM?. *Cancers* (2013); 5(3): 1020–1048.

Wang, C., Ye, Y., Hochu, G. M., Sadeghifar, H. and Gu, Z. Enhanced cancer immunotherapy by microneedle patch-assisted delivery of anti-PD1 antibody partners. *Nano Lett.* (2016); 16(4): 2334–2340.

Wang, T., Hou, J., Su, C., Zhao, L. and Shi, Y. Hyaluronic acid - coated chitosan nanoparticles induce ROS-mediated tumor cell apoptosis and enhance antitumor efficiency by targeted drug delivery via CD44. *J. Nanobiotechnol.* (2017); 15: 7.

Wei, H., Qing, D., De-ying, C., Bai, X. and Li-Fang, F. Study on colon-specific pectin/ethylcellulose film-coated 5-fluorouracil pellets in rats. *Int J Pharm.* (2008); 348(1–2): 35–45.

Weiss, N., Miller, F., Cazaubon, S. and Couraud, P. The blood–brain barrier in brain homeostasis and neurological diseases. *Biochim. Biophys. Acta Biomembr.* (2009); 1788(4): 842–857.

WHO, 2014. The top 10 causes of death. Global burden of disease: Fact Sheet No. 310 (Updated May 2014) (2014). doi:/entity/mediacentre/factsheets/fs310/en/index.html.

WHO. WHO_Cancer (2017).

Yadav, P., Yadav, H., Shah, V. G., Shah, G. and Dhaka, G. Biomedical biopolymers, their origin and evolution in biomedical sciences: a systematic review. *J Clin Diagn Res.* (2015); 9(9): 21–25.

Yao, W., Peng, Y., Du, M., Luo, J. and Zong, L. Preventative vaccine-loaded mannosylated chitosan nanoparticles intended for nasal mucosal delivery enhance immune responses and potent tumor immunity. *Mol Pharm.* (2013); 10(8): 2904–2914.

Yu, F., Selva Kumar, N. D. O., Choudhury, D., Foo, L. C. and Ng, S. H. Microfluidic platforms for modeling biological barriers in the circulatory system. *Drug Discov Today* (2018); 23(4): 815–829.

Zhan, C. and Lu, W. The blood–brain/tumor barriers: challenges and chances for malignant gliomas targeted drug delivery. *Curr Pharm Biotechnol* (2012); 13(12): 2380–2387.

CHAPTER 10

Selective Targeting of Human Colon Cancer HCT116 Cells by Phytomediated Silver Nanoparticles

V. S. SHANIBA*, AHLAM ABDUL AZIZ, JOBISH JOSEPH, P. R. JAYASREE, and P. R. MANISH KUMAR

Department of Biotechnology, University of Calicut, Malappuram 673635, Kerala, India

Corresponding author. E-mail: Shanibavsn@gmail.com, manishramakrishnan@rediffmail.com

ABSTRACT

Biosynthesis of nanoparticles has emerged as a key area of research in nanotechnology. Plant extracts offer innumerable, yet to reveal, bioresources that are cost effective, nonhazardous, reducing, and stabilizing compounds that are being employed for the synthesis of silver nanoparticles (AgNPs). This is a first report on the *in vitro* cytotoxic effects of *Annona muricata* fruit extract-mediated AgNPs (AMFAgNPs) against the HCT116 colon cancer cell line. Physicochemical parameters for the effective synthesis of AgNPs were optimized for the reactants—silver nitrate and fruit extract. The resultant AMFAgNPs generated were characterized structurally. UV–vis spectrum of AMFAgNPs revealed an absorption maximum at 430 nm; XRD analysis showed an average particle size of 19 nm, predominantly spherical, as evidenced by field-emission scanning electron microscopy imaging. Fourier transform infrared spectroscopy confirmed the presence of high amounts of biomolecules responsible for the efficient stabilization of nanoparticles. Their cytotoxicity was analyzed against HCT116 colon cancer cells by MTT, dichloro-dihydro-flourescien diacetate (DCFH-DA), rhodamine 123, and clonogenic assays. For visual detection of apoptotic cells, Hoechst/acridine orange–ethidium bromide (AO-EtBr) staining and annexin V-fluorescein isothiocyanate (FITC) binding assays were performed. Cell cycle distribution was assessed by flow cytometry, while relative gene expressions were evaluated by the reverse transcription-quantitative polymerase chain reaction technique. Apoptosis induction was observed by fluorescence and scanning electron microscopy. AMFAgNPs caused G_0/G_1 cell cycle phase arrest. Apoptosis-regulating genes—PUMA, cas-3, -8, -9, and BAX—were found to be upregulated in the nanoparticle-treated cells compared to those treated with the chemotherapeutic drug cisplatin used as a positive control. The therapeutic potential of AMFAgNPs was evident as they effectively targeted colon cancer cells without being toxic to normal human lymphocytes and erythrocytes.

10.1 INTRODUCTION

10.1.1 CANCER AND CONVENTIONAL THERAPEUTICS

Cancer is the second leading cause of human death outnumbered only by heart disease-related mortality worldwide. Of the 17.5 million cancer cases reported globally in 2015, 8.7 million succumbed to the disease. According to WHO (2018) fact sheets, lung cancer occurred in 1.8 million people and resulted in 1.69 million deaths, followed by liver (788,000 deaths), colorectal (774,000 deaths), stomach (754,000 deaths), and breast (571,000 deaths). Conventional cancer treatment strategies include surgery, radiotherapy, and the use of chemotherapeutic drugs, all of which lack specificity. Surgical treatment does not ensure complete elimination of cancer cells, can be traumatic with large tumors, and becomes invalid in the case of subclinical metastasis. Damage to healthy cells and nearby tissues near the treated area, fatigue, and development of secondary cancers are the risks associated with radiation therapy. The many harmful side effects of chemotherapeutic drugs, including the toxicity of their byproducts, adversely affect the nontargeted host cells. Further, the development of intrinsic and acquired resistance against these drugs curtails the success of chemotherapy (Peer et al. 2007). Targeted therapies involve the use of inhibitors of angiogenesis, signal transduction pathways including immunotherapy with monoclonal antibodies that interfere with molecular pathways regulating tumor growth and progression.

10.1.2   NANOTECHNOLOGY AND NANOMEDICINE FOR CANCER CURE

Despite the development and availability of several aggressive treatment modes, cancer-related death rates continue to rise (Jemal et al., 2011; Miller et al., 2016). Thus, the development of efficacious, biocompatible, and cost-effective methods for cancer cure becomes indispensable. Nanotechnology has emerged as a fast-growing branch of science and engineering with several applications in various fields like electronics, food storage, medicine, fuel /solar cells, batteries, chemical sensors, fabrics, and molecular manufacturing (Prabhu and Poulose, 2012). Nanoparticles (NPs) comprise clusters of atoms in the size range of 1–100 nm. Nanomaterials have great importance as the physicochemical properties of a metal are changed as it reaches the nanosize capable of displaying properties that are different as compared to those of the bulk metal. Gold, silver, and platinum have been used mostly for the synthesis of stable dispersions of NPs, which are useful in areas such as photography, catalysis, biological labeling, photonics, and optoelectronics (Jacob et al., 2012). Chemical, physical, and biological methods have been adopted for their synthesis. The fabrication of NPs through chemical and physical methods is most popular but requires the use of toxic and hazardous chemicals and large amounts of energy, and generates harmful wastes (Song and Kim, 2009; Rajasekharreddy et al., 2010). Both chemical and physical methods adversely affect the environment in addition to being technically laborious and expensive. As an alternative, a "greener" biogenic synthesis of NPs employing natural products such as sugars, biopolymers, microorganisms, and plant extracts containing phytochemicals is believed to play a key role as reductants in the conversion of metal ions to NPs as well as provide capping agents to stabilize them. The use of plant extract has been found to be relatively more advantageous over microbial systems for large-scale production of biogenic NPs since the maintenance of aseptic conditions for microbial cultures is not industrially feasible (Ahmed et al., 2015).

A plethora of *in vitro* studies on different cancer cell lines has confirmed anticancer activities of biogenic gold, silver, copper, titanium, zinc, and iron prepared from different biosources (Salunke et al., 2016; Barabadi et al., 2017). Among these, silver NPshave gained more popularity and acceptability due to their antibacterial, antifungal, larvicidal, and antiparasitic properties and industrial applications (Khatoon et al., 2017). AgNPs are cytotoxic to cancer cells and show excellent potential as an antitumor agent.

The exact mechanism of AgNPs anticancer activity is not fully understood yet. However, the proposed mechanisms involve the generation of reactive oxygen species (ROS), causing mitochondrial damage, inducing sub-G$_1$ arrest in cells (Ovais et al., 2016), upregulation of p53 protein and caspase-3 expression (Kovacs et al., 2016), and inhibiting vascular endothelial growth factor (VEGF)-induced activities (Hackenberg et al., 2011). Cytotoxicity is dependent on parameters such as particle size, surface area/reactivity, distribution pattern, as well as cell-type specificity (Feng et al., 2016).

The physicochemical and biological traits confer several advantages to these particles, making them excellent drug delivery vehicles that have increasingly become popular in clinical use in the recent past. The use of surface-modified nanoscale liposomes (nanosomes)/nanocapsules/lipid micelles to deliver hydrophilic and hydrophobic anticancer drugs as well as short interfering RNA (siRNA) formulations is a promising development in cancer therapy. Further, "active" and "passive" targeted drug delivery are also being achieved by covalent coupling of NPs with appropriate ligands such as lectins, antibodies, aptamers, folate, and peptides (Bannerjee and Sengupta, 2011; Renganathan et al., 2012). Thus, the knowledge and tools of nanotechnology offer tremendous potential for efficacious drug delivery, enhanced diagnosis imaging for treatment of cancers, and is sometimes referred to as nano-oncology within the broader discipline termed as nanomedicine.

10.1.3 PLANTS SELECTED FOR EXTRACT PREPARATION

Annonaceae is a very large plant family consisting of about 120 genera and more than 2000 species. The medicinal uses of the family members have been known since more than a century ago (Billon, 1869), and this species has attracted human attention due to its bioactivity and toxicity. *Annona muricata*, also known as Soursop or Graviola, is a member of the family. It is an evergreen plant distributed in the tropical and subtropical regions of the world, attaining heights of mostly up to 4 m. It produces heart-shaped edible fruits of about 5–20 cm in diameter, green in color, with white flesh inside used extensively to prepare syrups, candies, beverages, ice creams, and shakes (Patel and Patel, 2016). In traditional medicine, the *A. muricata* plant has been used for the treatment of cancers, especially of the liver and breast. It also shows antiarthritic, antiparasitic, antimalarial, hepatoprotective, antidiabetic, analgesic hypotensive, anti-inflammatory, and immune-enhancing effects (Gajalakshmi et al., 2012).

10.1.4 BIOSYNTHESIS AND CHARACTERIZATION OF SILVER NANOPARTICLES

Shade-dried, cut pieces of the mature *A. muricata* were powdered and boiled in distilled water to obtain an aqueous extract, which was then used for the synthesis of silver NPs. Various parameters such as the extract volume, silver nitrate concentration, temperature, pH, and the time period of incubation were optimized for the generation of biogenic NPs. The reaction was then periodically monitored at wavelengths ranging between 200 and 700 nm to detect the formation of AMFAgNPs as indicated by the development of brown coloration in the reaction mixture. The NPs were then subjected to characterization by Fourier transform infrared (FTIR) spectroscopy, X-ray diffraction (XRD) spectroscopy, field-emission scanning electron microscopy (FESEM), and energy dispersive X-ray spectroscopy (EDX).

10.1.4.1 EVALUATION OF CELLULAR EFFECTS OF AMFAGNPS ON HCT116 COLON CANCER CELLS

Human colorectal carcinoma adherent HCT116 cells with epithelial morphology were employed to study the cellular effects of biogenic silver NPs. With a doubling time of about 18 h, these cells facilitate *in vitro* experimental studies to better understand the defective pathways of cancerous colon cells as well as their responses to potential anticancer drugs. According to a recent estimate, as many as 1.4 million cases of colon cancer have been reported globally (Andres et al., 2018). The disease is highly heterogeneous due to numerous alterations in signaling pathways and is known to affect 20–25% of younger individuals with a family history of colon cancer and heritable susceptibility (Kuppusamy et al., 2016).

The cytotoxicity of AMFAgNPs was evaluated on HCT116 cells by MTT and clonogenic assays, while the cytomorphological changes were monitored by light, fluorescence, and scanning electron microscopy. The determination of cellular ROS and changes in mitochondrial membrane potential in addition to the quantification of apoptosis-related gene expression by reverse transcription-quantitative polymerase chain reaction (RTq-PCR) was also carried out to assess the apoptogenic potential of the biogenic NPs. Cisplatin, a standard anticancer drug, was taken as a positive control to gauge the effect of NPs. Further, changes in cell cycle distribution of HCT116, exposed to AMFAgNPs, were analyzed using flow cytometry. Parallel experiments were also carried out on normal control cells such as human peripheral blood lymphocytes and erythrocytes to study the cellular effects of AMAgNPs.

10.2 MATERIALS

10.2.1 EQUIPMENT

The equipment used are sterile 50 mL conical flasks, UV–vis spectrophotometer, lyophilizer, X-ray diffractometer, scanning electron microscope, 96-well flat-bottom culture plates, 12-well culture plates, six-well plates, ELISA plate reader, Illumina Eco™ Real-Time system, inverted fluorescence microscope, slides, and coverslips.

10.2.2 REAGENTS

The reagents used are silver nitrate, *A. muricata* fruits, HCT116 cell lines, Dulbecco's modified Eagle's medium (HiMedia), fetal bovine serum (FBS), streptomycin and penicillin, 3-(4,5-dimethylthiazol-2-yl)-2,5-diphenyltetrazolium bromide (MTT), dimethyl sulfoxide (DMSO), phosphate-buffered saline (PBS), fluorescence stains (AO, EtBr, Hoechst 33258, rhodamine 123, and dichloro-dihydro-fluorescein), glutaraldehyde, HiKaryo XL™ RPMI medium, crystal violet, ethanol, trypsin, and SYBR Green PCR Master Mix.

10.3 METHODS

10.3.1 SYNTHESIS OF BIOSYNTHESIZED AMFAGNPS

1. Fresh mature fruits were cut into pieces, shade-dried, and powdered finely.
2. Ten grams of fruit powder was weighed, mixed with 200 mL of Milli-Q water, and boiled for 15 min.
3. After cooling, the extract was filtered through muslin cloth and Whatman no. 1 filter paper and stored at 4 °C in a refrigerator (Mubarakali et al., 2011).
4. For effective synthesis, different parameters such as the volume of fruit extract, concentration of metal ions, temperature, pH, and time were optimized (Shaniba et al., 2017).

5. The reaction mixture was periodically monitored in the range of 200–700 nm using a UV–vis spectrophotometer (PerkinElmer Lambda25) to detect the formation of silver NPs as indicated by the appearance of brown color.

The optimum conditions for the effective green synthesis of AMFAgNPs were 8.0 mL of AMF extract mixed with 92.0 mL of 1.0 mM AgNO$_3$ (pH 7.0), at 90 °C for 30 min. Absorption spectra of AMFAgNPs formed in the reaction media, as evidenced by the formation of brown color (Figure 10.1), had an absorption maximum at 421 nm (Figure 10.2) due to surface plasmon resonance of AgNPs (Krishnaraj et al., 2010).

FIGURE 10.1   Aqueous solution of 1mM AgNO$_3$ with AMF extract: (a) before the addition of the leaf extract and (b) after the addition of leaf extract at 30 min.

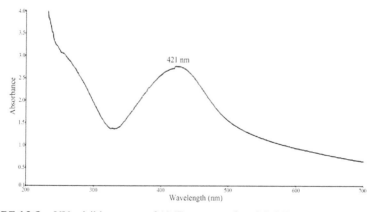

FIGURE 10.2   UV–visible spectra of AMF extract-reduced AgNPs.

10.3.2   *CHARACTERIZATION OF AMFAGNPS*

10.3.2.1   *FTIR ANALYSIS*

1. The reaction mixture was subjected to centrifugation at 12,000 rpm for 20 min.
2. The resultant pellet was resuspended in Milli-Q water and lyophilized (ScanVac CoolSafe).
3. The freeze-dried powder was pelletized with potassium bromide (KBr) powder and subjected to FTIR analysis.
4. The spectra were recorded using a JASCO 4100 at a wavelength ranging from 4000 to 400 cm^{-1}.

FTIR analysis of AMFAgNPs showed an intensive peak at 3650 cm^{-1} corresponding to the O–H stretching of alcohols and phenols (Figure 10.3). The peaks at 3422 cm^{-1} indicated N–H groups. The peaks observed at 2923 cm^{-1}, 2853 cm^{-1}, 1457 cm^{-1}, and 672 cm^{-1} represented C–H functional groups. Further, the peaks at 1747 cm^{-1} represented carbonyl groups (C=O) and the peaks at 1023 cm^{-1} indicated C–O bonds. The results obtained depicted that the presence of these functional groups influenced the synthesis and stabilization of NPs. These results confirmed the encapsulation of AgNPs with the phytoconstituents from plant extracts.

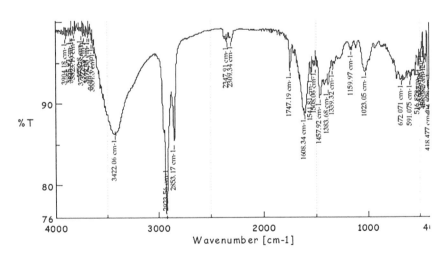

FIGURE 10.3   FTIR spectrum of AMFAgNPs.

10.3.2.2 X-RAY DIFFRACTION ANALYSIS

The dried AMFAgNPs were coated on an XRD grid, and the spectrum was recorded using a Rigaku MiniFlex X-ray diffractometer with CuKα radiation (40 kV, 15 mA). The crystalline structure of silver NPs was identified by XRD analysis (Figure 10.4). XRD showed intense peaks in the whole spectrum of 2θ values ranging from 20° to 70°. Four strong peaks observed at the 2θ values 27.6°, 38°, 46°, and 63.6° corresponded to the (110), (111), (200), and (220) planes of face-centered cubic phase, obtained from JCPDS card 89-3722. Some of the unassigned peaks were observed at 2θ values of 28°, 42.8°, and 53.2°, which are attributable to the biomolecules present in the plant extract impregnated on the surface of AgNPs. The average crystal size was found to be 19 nm, as calculated by the Debye–Scherrer equation $D = 0.94\lambda/\beta \cos\theta$, where D is the average crystallite domain size perpendicular to the reflecting planes, λ is the X-ray wavelength, β is the full width at half-maximum, and θ is the diffraction angle (Vivek et al., 2012).

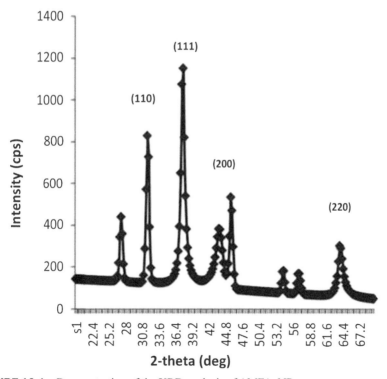

FIGURE 10.4   Demonstration of the XRD analysis of AMFAgNPs.

10.3.2.3   *FIELD-EMISSION SCANNING ELECTRON MICROSCOPY*

The average particle size was determined using FESEM, Horiba S46600, equipped with EDX. The morphology and size determination of the synthesized AMFAgNPs were based on FESEM images (Figure 10.5). The SEM image showed a spherical and relatively uniform shape of NPs with diameters varying from 30 to 100 nm. EDX analysis revealed the presence of elemental silver signal emanating from the silver NPs (Figure 10.6).

FIGURE 10.5   FESEM micrograph of AMFAgNPs.

10.3.3   *EXPERIMENTAL EVIDENCE ON THE CELLULAR EFFECTS OF AMFAGNPS*

The colorectal cancer cell line HCT116 was obtained from the National Centre for Cell Sciences, Pune. The cells were grown as a monolayer, supplemented with 10% FBS and 100 U/mL each of penicillin and streptomycin, and incubated at 37 °C in 5% CO_2.

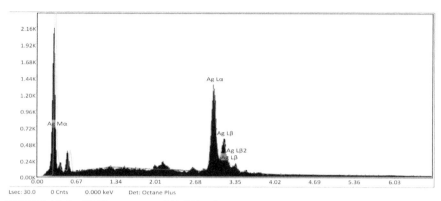

FIGURE 10.6 EDX spectrum of AMFAgNPs.

10.3.3.1 MTT ASSAY FOR CYTOTOXICITY EVALUATION OF AMFAGNPS ON HCT116 CELLS

MTT is a colorimetric, nonradioactive assay for measuring cell viability through increased metabolism of tetrazolium salt (Mosmann, 1983).

1. HCT116 cells (1×10^4 cells/well) were seeded into 96-well tissue culture plates and incubated for 24 h (see Note 1).
2. The cells were then exposed to different concentrations of AMFAgNPs/cisplatin (2.5–70 µg/mL) and incubated for 24, 48, and 72 h.
3. Following the addition of MTT, the cells were incubated further for another 3 h at 37 °C (Figure 10.7).
4. Thereafter, the formazan crystals were dissolved in 150 µL of DMSO.
5. The absorbance was recorded at 620 nm using an ELISA plate reader (Multiscan EX, Thermo Scientific, USA).
6. The obtained optical density (OD) value was used to compute the percentage viability using the following formula:

$$\text{Percentage viability} = \frac{\text{OD value of experimental samples}}{\text{OD value of experimental control}} \times 100$$

This is the first report on the cytotoxicity of biogenic AgNPs obtained from aqueous *A. muricata* fruit extract against HCT116 colon cancer cell line. AgNP-treated cells showed decreased metabolic activity, which in turn is dependent on the cell type and size of NPs (Lima et al., 2012). The MTT results showed that the AMFAgNPs decreased cell viability in a time- and dose-dependent manner. AMFAgNPs showed potent cytotoxicity as evidenced by the IC_{50} values of 25 µg/mL and 7.5 µg/mL obtained following 48 and 72 h of treatment, respectively. Similar results of time- and dose-dependent cytotoxicity of biogenic silver NPs have been reported previously (Jeyaraj et al., 2013; Awasthi et al., 2015).

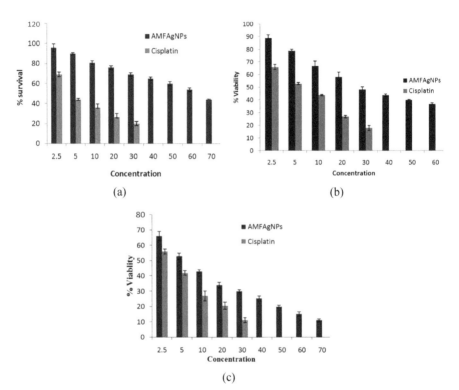

FIGURE 10.7 Cell viability of HCT116 cells by the MTT method. (a) 24 h treatment, (b) 48 h treatment, and (c) 72 h treatment.

10.3.3.2 *MORPHOLOGICAL OBSERVATION*

1. HCT116 cells (1.5×10^5) were treated with different concentrations of AMFAgNPs for 48 h.
2. The control and treated cells were harvested and washed with ice-cold PBS.
3. These were then observed under a bright-field light microscope to assess the cytomorphological changes.

The treated cells showed distinct cellular morphological changes compared to their normal healthy counterparts (Figure 10.8a). Control cells appeared as irregular, confluent aggregates with rounded and polygonal cells. AMFAgNP-treated cells appeared shrunken and spherical with restricted cell-spreading patterns compared to controls.

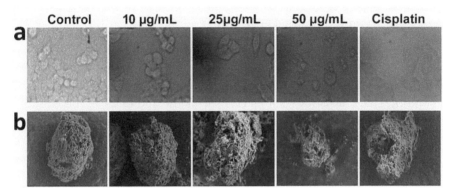

FIGURE 10.8   Morphological changes in HCT116 cells after 48 h treatment with AMFAgNPs: (a) light microscopy image and (b) FESEM image.

10.3.3.3   SCANNING ELECTRON MICROSCOPY

Morphological changes in cells treated with AMFAgNPs in comparison to untreated control cells were observed using field-emission scanning electron microscopy (Figure 10.10b).

1.   Cells (1.5×10^5) were seeded in 12-well plates and treated with various concentrations (10, 25, and 50 µg/mL) of AMFAgNPs.
2.   The treated and control cells were trypsinized, followed by centrifugation.
3.   Cells were then fixed in 4% glutaraldehyde at room temperature for 3 min, followed by two washes with PBS.
4.   Cells were then dehydrated by passing through ascending graded acetone series (75–85–100%) and dried.
5.   The dried cells were then coated with gold and observed using a Hitachi SU6600 scanning electron microscope.

The treated cells displayed distinct morphological changes. A dose-dependent mild shrinkage of cells was evident at lower concentrations with severe shrinkage and blebbing observable at higher concentrations tested. The control cells showed a normal organized structure.

10.3.3.4   HOECHST 33258 STAINING

1. Cells (1.5×10^5) were seeded in 12-well plates and treated with various concentrations (10, 25, and 50 µg/mL) of AMFAgNPs for 48 h (see Note 2).
2. After collection, cells were washed with PBS and resuspended in Hoechst 33258 solution (1 mg/mL), followed by incubation at 37 °C in the dark (Baghbani et al., 2017).

Images of the treated and control cells were then captured using a fluorescence microscope (Figure 10.9). Cells exposed to AMFAgNP concentrations below, at, and above IC_{50} concentrations for 48 h displayed apoptotic characteristics such as cell shrinkage and nuclear condensation and fragmentation in a dose-dependent manner compared to the normal cell morphology observed in control cells.

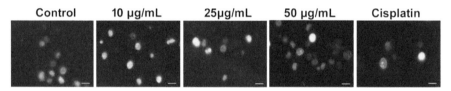

| Control | 10 µg/mL | 25µg/mL | 50 µg/mL | Cisplatin |

FIGURE 10.9   Fluorescence microscopy images of Hoechst 33258-stained HCT116 cells.

10.3.3.5   ACRIDINE ORANGE–ETHIDIUM BROMIDE DUAL STAINING

The morphological evidence of apoptosis in AMFAgNP-treated cells was detected by dual staining with AO and EtBr. AO penetrates into living cells, emitting green fluorescence after intercalation into deoxyribonucleic acid (DNA). The second dye EtBr emits red fluorescence in cells with an altered cell membrane. Viable cells have bright green nuclei with an organized structure. Apoptotic cells have orange-to-red nuclei with condensed or fragmented chromatin, while necrotic cells display uniform orange-to-red nuclei with a condensed structure.

1. Following treatment with AMFAgNPs/cisplatin, the cells were subjected to trypsinization.
2. Centrifugation at 5000 rpm for 5 min was followed by a wash with PBS.
3. Approximately 10 µL of the dye mixture (50 µg/mL each of AO and EtBr) was mixed with 200 µL of cell suspension (1.5×10^5 cell/mL).

4. Following a brief incubation for 2–3 min at room temperature, 20 µL of cell suspension was then placed on a glass slide and observed under a fluorescence microscope at 40 × magnification.

The percentage of apoptotic cells was determined by the following formula:

$$\% \text{ of apoptotic cells} = \frac{\text{Total number of apoptotic cells}}{\text{Total number of normal and apoptotic cells}} \times 100$$

The induction of apoptosis, following treatment with the different concentrations of AMFAgNPs (10, 25, and 50 µg/mL), was assessed by fluorescence microscopy. The untreated control cells appeared green with intact nuclei and normal cell morphology. Cells treated with biogenic NPs were stained orange to red and showed characteristics of apoptosis such as chromatin condensation, nuclear fragmentation, and alterations in the size and shape of cells (Figure 10.10a).

Necrotic cells were observed with red fluorescence due to the loss of membrane integrity (Thangam et al., 2012). A quantitative analysis based on cell counts showed a dose-dependent induction of apoptosis, leading to 80% cell death at the highest test concentration of 50 µg/mL of AMFAgNPs (Figure 10.10b)

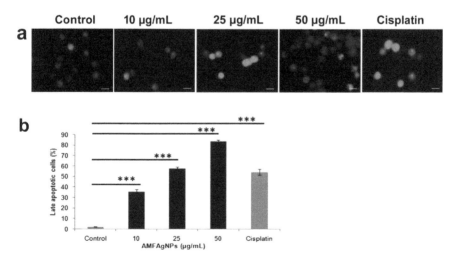

FIGURE 10.10 (a) Fluorescence microscopy images of AO/EtBr-stained HCT116 cells and (b) quantitative analysis of late apoptotic cells.

10.3.3.6 QUANTIFICATION OF INTRACELLULAR ROS

ROS generation in cells undergoing chemical or environmental stress results in oxidative damage to macromolecules such as lipids, proteins, and DNA. Intracellular ROS levels were measured using DCFH-DA, which measures hydrogen peroxide, an indirect procedure for estimating ROS. The dye passively enters the cell, where it reacts with ROS to form the highly fluorescent compound, dichlorofluorescein.

1. Cells were seeded in a 12-well plate (1.5×10^5 cells/well).
2. AMFAgNP treatment was given at various concentrations (10, 25, and 50 µg/mL) for 48 h.
3. Following exposure to the NPs, the cells were washed with PBS and then incubated in 20 µL of working solution of DCFH-DA (1 mg/mL stock solution was diluted to yield a 20 µg/mL working solution) at 37 °C for 30 min (see Note 6).

The presence of ROS was visually examined as evidenced by the fluorescence observed using a fluorescence microscope. NP-induced ROS generation and oxidative stress have been previously proposed as mediators of apoptotic cell death (Zhu et al., 2016). Figure 10.11a shows a dose-dependent increase of intracellular ROS generation after 48 h exposure to the NPs. The results indicated the involvement of ROS in the observed cytotoxicity of AgNPs.

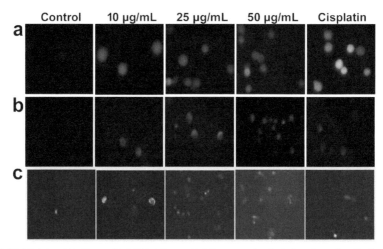

FIGURE 10.11 (a) Effect of AMFAgNPs on ROS generation in HCT116 cells, (b) detection of the mitochondrial membrane potential by fluorescence microscopy, and (c) fluorescence microscopy images of Annexin V-FITC-stained HCT116 cells.

10.3.3.7   *ASSESSMENT OF MITOCHONDRIAL MEMBRANE POTENTIAL*

Mitochondria are an integral part of apoptotic machinery. The depletion of mitochondrial membrane potential, an early marker of apoptosis, leads to the release of its intermembrane space proteins, such as cytochrome-c, into the cytosol, which then activates caspases.

1. HCT116 cells were seeded (1×10^5 cells/well) in 12-well plates.
2. The cells were cultured with or without AMFAgNPs (10, 25, 50 µg/mL)/cisplatin for 48 h.
3. After treatment, the cells were incubated with the lipophilic cationic dye rhodamine 123 (10 µg/mL) for 30 min.
4. Cells were then harvested and washed with PBS.
5. The green fluorescence from the depolarized mitochondria was observed using a fluorescence microscope.

A dose-dependent decrease in the mitochondrial membrane potential was observed in both AMFAgNPs and cisplatin-treated cells (Figure 11b).

10.3.3.8   *ANNEXIN V/PI DOUBLE STAINING ASSAY*

There was translocation of phosphatidyl serine from the inner to the outer phase of plasma membrane during the early stage of apoptosis. This was determined by staining cells with Annexin-FITC. The inclusion of the otherwise impermeable propidium iodide (PI) stain assisted further in detecting necrotic cell death, which is characterized by the loss of membrane integrity.

Apoptosis-mediated cell death of cancer cells was examined using FITC-labeled Annexin V/PI, according to the manufacturer's protocol (Sigma, USA).

1. HCT116 cells were seeded (1.5×10^5 cells/well) in 12-well plates and incubated at 37 °C for 48 h.
2. Treatment with different concentrations of AMFAgNPs (10, 25, and 50 µg/mL)/cisplatin was given for 48 h.
3. Cells were then pelleted by trpsinization followed by two washes with ice-cold PBS.
4. The cells were resuspended in 500 µL of 1X binding buffer and 5.0 µL of Annexin V-FITC solution.
5. PI (10 µL) was added, and the cells were then incubated for 10 min at room temperature.
6. The stained cells were observed using a fluorescence microscope.

AMFAgNPs induced a concentration-dependent increase in the number of Annexin V-positive HCT116, which is indicative of the occurrence of apoptosis in the treated cells (Figure 10.11c).

10.3.3.9   CLONOGENIC ASSAY

Clonogenic assay or colony formation assay is an *in vitro* cell survival assay based on the ability of a single cell to grow into a colony. The colony is defined to consist of at least 50 cells. The assay essentially evaluates cell's reproductive ability (Varun et al., 2014).

1. HCT116 cells were plated at a density of 5×10^2 cells per well in six-well plates.
2. Treatment with various concentrations of AMFAgNPs (10, 25, and 50 µg/mL) was carried out; the untreated cells were maintained as control.
3. After 48 h of incubation, the media was changed and the cells were subjected to a further incubation for seven days.
4. The growth medium was then decanted, and the cells were carefully rinsed with PBS.
5. The PBS was carefully removed, and then, 2–3 mL of a mixture of 6% glutaraldehyde and 0.5% crystal violet prepared in sterile distilled water was added.
6. The cells were then allowed to stand for at least 30 min for fixation/staining.
7. Cells were then rinsed with tap water (see Note 3).
8. The plates with colonies were then allowed to dry at room temperature.

AMFAgNP-treated HCT116 cells displayed a concentration-dependent reduction in the colony-forming capability compared to that of control cells, indicating that biogenic AgNPs effectively suppressed the growth and proliferation of these cells (Figure 10.12a and b).

10.3.4   CELL CYCLE ANALYSIS

Cell cycle distribution and the percentage of apoptotic cells were determined by flow cytometry. Extensive DNA degradation occurs as part of the apoptotic process; flow cytometry analysis usually involves the observation of a "hypodiploid" or "sub-G1" peak in the DNA histogram.

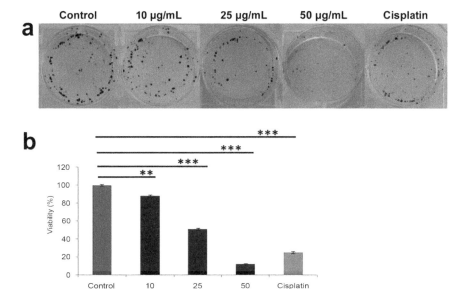

FIGURE 10.12 Clonogenic capacity in AMFAgNP-treated HCT116 cells as measured by the colony-forming assay. (a) Cells stained with crystal violet and (b) histogram of the percentage of viable colonies.

1. Cells were seeded at 1×10^5 cells/well in a six-well plate.
2. After treatment with AMFAgNPs for 48 h, the cells were trypsinized.
3. Following centrifugation at 5000 rpm for 5 min, the cells were washed with cold PBS buffer and fixed with 70% ethanol overnight at $-20\ °C$.
4. Following recentrifugation to discard ethanol, the cells were rinsed twice with cold PBS.
5. The recovered cells were then resuspended in 300 µL of PBS and treated by adding 5–7 µL of RNase (10 mg/mL).
6. Following incubation for 30 min at 37 °C with intermittent shaking, the cells were stained with propidium iodide (working solution 1 mg/mL).
7. The cells were then analyzed using a flow cytometer (BD FACS Diva software, version 5.0.2).

The proportion of cells in G_0/G_1, S, and G_2/M phases has been represented as DNA histograms. For each experiment, 10,000 events per sample were recorded. FACS analysis revealed that AMFAgNP treatment led to an accumulation of up to 84% of cell population in the G_0G_1 phase in a concentration-dependent manner within the range tested (10–50 µg/mL) as compared to controls (Figure 10.13a and b). In contrast, treatment with the standard anticancer drug cisplatin, which was taken as the drug control, leads to G_2 arrest (not G_1) in 25% of the cells compared to controls. The induction of apoptotic cell death or cell cycle arrest in cancer cells is thought to be the principal strategy for tumor elimination. In addition to apoptosis, the inhibition of cell cycle progression also plays a key role in regulating the growth of cancer cells (Young et al. 2017). The induction of cell cycle arrest by AMFAgNPs thus demonstrates their therapeutic potential from this viewpoint.

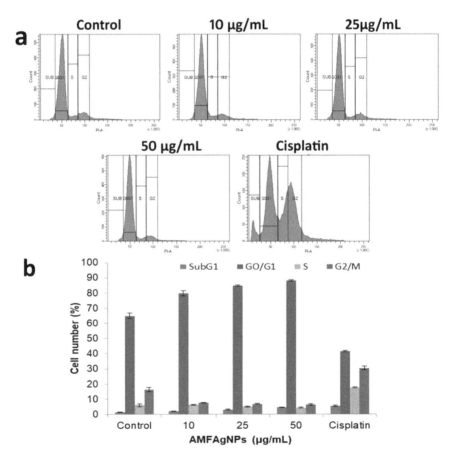

FIGURE 10.13   Cell cycle analysis by FACS. (a) AMFAgNPs induced G_0/G_1 cell cycle arrest in HCT116 cells and (b) representative histogram shows the percentage of cells in G_1, S, and G_2/M.

10.3.5 REVERSE TRANSCRIPTION QUANTITATIVE POLYMERASE CHAIN REACTION ANALYSIS

The relative levels of apoptosis-related gene expression analysis were evaluated by RTq-PCR in the presence of five sets of gene specific (PUMA, caspase-3, -8, -9, and Bax) primers listed in Table 10.1. Total RNA was isolated from treated and control cells using TRI reagent, according to the manufacturer's instructions, and employed for RTq-PCR.

TABLE 10.1 Primers Used in the RTq-PCR Assay

Gene	Primer Sequence
PUMA	Forward: 5'GACCTCAACGACAGTACGA3'
	Reverse: 5'GAGATTGTACAGGACCCTCCA3'
Caspase 3	Forward: 5'TGGCATACTCCACAGCACCTGGTTA3'
	Reverse: 5'CATGGCACACAAAGCGACTGGATGAA3'
Caspase 8	Forward: 5'CATCCAGTCACTTTGCCAGA3'
	Reverse: 5'GCATCTGTTTCCCCATGTTT3'
Caspase 9	Forward: 5'TTCCCAGGTTTTGTTTCCTG3'
	Reverse: 5'CCTTTCACCGAAACAGCATT3'
Bax	Forward: 5'GCCACCAGCCTGTTTGAG3'
	Reverse: 5'CTGCCACCCAGCCACCC3'
GAPDH	Forward: 5'AATCCCATCACCATCTTCCA3'
	Reverse: 5'CCTGCTTCACCACCTTCTTG3'

1. In a 5.0 μL reaction volume, cDNA was synthesized by taking 1.0 μL of 0.5 μg/μL RNA and 0.5 μL of oligo dT (100 ng/μL).
2. The mixture was then incubated at 65 °C for 10 min and placed on ice.
3. To this, 1 μL of 10 mM dNTP mix, 2 μL of 10× M-MLV reverse transcriptase buffer, 1 μL of M-MLV reverse transcriptase, 0.5 μL of RNasin, 0.5 μL of DTT (20 mM) was added and made up to 10 μL with sterile nuclease-free water (see Note 5).
4. The reaction mixture was incubated at 37 °C for 1 h, then heated to 95 °C for 10 min, and stored at −20 °C until use.
5. The transcript levels of individual mRNA types from control and treated cells were determined by SYBR Green PCR Master Mix using an Illumina Eco™ Real Time system. The relative expression levels were calculated using the $\Delta\Delta C_t$ method with GAPDH as an internal reference gene.

The expression levels of the apoptotic related genes were analyzed to determine their involvement, if any, in the cytotoxicity of AMAgNPs. The activation of caspase proteases is fundamental in triggering apoptotic cell death. Among the different subgroups, caspase-3 plays a decisive role in the apoptosis-associated processes, such as loss of membrane integrity and DNA damage, thereby enhancing apoptotic cell death in a caspase-3-dependant pathway. Caspase-3 expression was found to significantly increase up to 4.8-fold in a concentration-dependent manner in cells treated with AMFAgNPs (Figure 10.14a–d). Likewise, PUMA, caspase-8, and caspase-9 also showed increased gene expression. The proapototic Bax promoted apoptosis and Bcl-2 promoted survival are responsible for the induction of mitochondria-mediated apoptosis. Upregulation of Bax expression was also evident in AMFAgNP-treated cells. However, the proapoptotic Bax gene expression was found to increase by 28-fold in cells treated with AMFAgNPs at the IC_{50} concentration. AgNPs have been reported to stimulate p^{53} responsive gene expression. PUMA (p^{53}, upregulated modulator of apoptosis) induces apoptosis in colon cancer through a Bax- and mitochondria-dependent manner (Mustata et al., 2013; Kovacs et al., 2016). A-549 NSCLC cells treated with silver NPs showed increased gene expression of caspase-3 and -8 (Karthik et al., 2014). Taken together, AMFAgNPs were found to induce apoptosis by upregulating genes such as PUMA, caspase-3, -8, -9, and Bax.

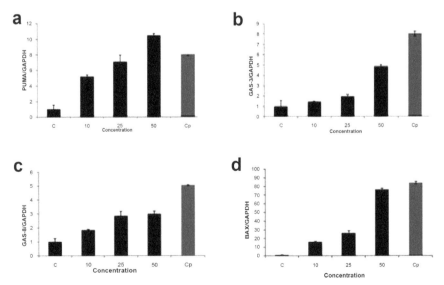

FIGURE 10.14 Relative quantification of mRNA expression of apoptosis-related genes using RTq-PCR in HCT116 cells treated with AMFAgNPs and cisplatin. (a) PUMA, (b) caspase-3, (c) caspase-8, and (d) Bax.

10.3.6 ASSESSMENT OF AMFAGNPS CYTOTOXICITY ON NORMAL HUMAN CELLS

10.3.6.1 PERIPHERAL BLOOD LYMPHOCYTES

Cellular toxicity of AMFAgNPs was also studied on human peripheral blood lymphocyte cultures (hPBLs) since many conventional anticancer treatments kill cells nonspecifically irrespective of whether they are normal or cancerous.

1. Isolated lymphocytes (10^5 cells/mL) were added to 5 mL of lymphocyte culture medium, HiKaryo XL™ RPMI, supplemented with phytohemagglutinin.
2. The medium was then incubated at 37 °C for 48 h (see Note 7).
3. Replicate lymphocyte cultures were then treated with different concentrations (10, 30, and 60 µg/mL) of AMFAgNPs and the metal precursor, silver nitrate, separately.
4. The lymphocytes were harvested by centrifugation at the 24th h (see Note 4).
5. The spent medium was discarded, and the cell pellet was resuspended in PBS and centrifuged at 1000 rpm.
6. To the pellet, 200 µL of MTT solution was added in PBS (500 µg/mL).
7. The tubes were incubated in the dark for 3–4 h.
8. The MTT solution was discarded and then 200 µL of DMSO was added to dissolve the formazan crystals.
9. The absorbance was measure at 620 nm to determine cell viability.

The results clearly indicated that AgNPs at IC_{50} concentration were not cytotoxic and did not affect their proliferation (Figure 10.15). However, the precursor metal, silver nitrate, was found to be toxic to lymphocytes.

10.3.6.2 RED BLOOD CELLS

The cellular toxicity of AMFAgNPs was also tested against human erythrocytes.

1. Heparinized fresh human blood was taken and centrifuged at 1500 rpm for 10 min at 4 °C.
2. Plasma and buffy coat were removed by aspiration.
3. The supernatant was discarded; the erythrocyte pellet was washed thrice with PBS.

4. The erythrocytes were then incubated with different concentrations (50, 100, 150, 200, and 500 µg/mL) of both AMFAgNPs and AgNO$_3$ separately for 90 min at 37 °C.
5. The samples were them centrifuged at 3000 rpm for 10 min.
6. The absorbance of the supernatants was measured at 543 nm using a spectrophotometer to monitor the hemolysis.
7. Percent hemolysis (*H*) was calculated by the following formula:

$$H = \frac{(ODS\ OD0)}{(OD100\ OD0)} \times 100$$

where OD$_S$, OD$_0$, OD$_{100}$ denote the optical densities of the experimental, negative (PBS), and positive (water) controls, respectively (Khan et al., 2015).

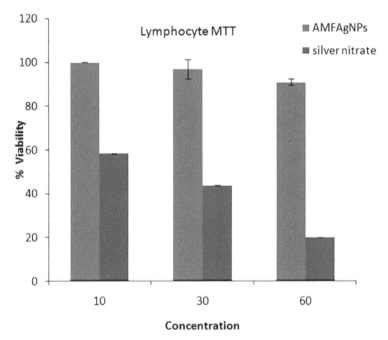

FIGURE 10.15 Effect of AMFAgNPs on hPBLs. Cell viability of normal hPBLs was assessed by the MTT method.

The percentage hemolysis values of erythrocytes exposed to AMFAgNPs were estimated as 2.5, 4.2, 5, 6, and 19 in comparison to 8, 23, 38, 44, and

56% obtained with the precursor compound $AgNO_3$. This showed that hemolysis occurred in a dose-dependent manner (Figure 10.16). Thus, it is clear that NPs apparently are far less toxic compared to its metal salt and that at the IC_{50} concentration both categories of normal cells, lymphocytes and erythrocytes, remained unaffected.

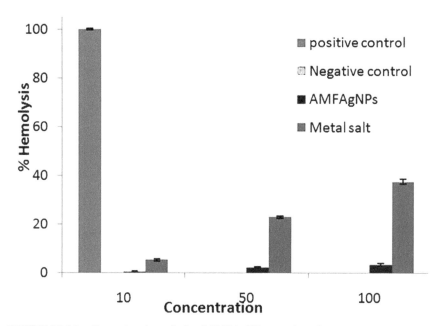

FIGURE 10.16   Percentage hemolysis of AMFAgNP-treated erythrocytes.

10.4   NOTES

1. Before all experiments, ensure that all required materials from sterile pipettes, sterile test tubes, culture dishes, and plates are ready for use. Prewarm the medium, PBS, and trypsin at 37 °C. Work out the cell dilutions and label the dishes or plates. The experiment has to continue smoothly to limit the total time, preventing the adverse effects of pH and temperature changes.
2. The dilutions have to be performed accurately to seed the correct number of cells.
3. Do not place the dishes or plates under a running tap, but fill the sink with water and immerse the dishes or plates carefully.

4. The incubation period should be at least 48 h to ensure obtaining the desired cell number; low-speed centrifugation should be to preferred to avoid cell damage.

5. The method described here for RNA isolation provides high yield of RNA, in some cases RNA prepared may not be entirely free of contaminating proteins. In this case, two options may be pursued. After ethanol precipitation, the sample can be resuspended in 400 µL of water and extracted once with phenol:chloroform:isoamyl alcohol (25:24:1) and once with chloroform:isoamyl alcohol (24:1), followed by a repeat of ethanol precipitation. Alternatively, the pellet can be resuspended in PK buffer (50 mM Tris, pH 7.5, 5 mM EDTA, and 0.5% w/v SDS) and incubated with 200 µg/ mL Proteinase K for 30 min, followed by extraction and ethanol precipitation.

6. The concentration of the stain is a critical factor for analyzing different phases of cells. All staining methods should be done in the dark condition.

7. To improve the growth condition, a gentle mixing of culture tubes is essential.

ACKNOWLEDGMENTS

The authors would like to thank the National Institute of Technology, Calicut, for use of SEM and EDX facility. Thanks are also due to our sister Departments of Chemistry and Physics, University of Calicut, for allowing the use of facilities to carry out FTIR and XRD analyses, respectively.

KEYWORDS

- *Annona muricata*
- **HCT116 cell lines**
- **apoptosis**
- **AgNPs**
- **RTq-PCR**

REFERENCES

Ahmed S, Ahmad M, Swami, BL, Ikram S (2015) Plants extract mediated synthesis of silver nanoparticles for antimicrobial applications: a green expertise. J Adv Res. http://dx.doi.org/10.1016/j.jare.2015.02.007.

Andres SF, Williams KN, Rustgi AK (2018) The molecular basis of metastatic colorectal cancer. Current Colorectal Cancer Rep (14)2: 69–79.

Awasthi KK, Awasthi AR, Verma NKumar, Roy PK, Awasthi PJ (2015) Cytotoxicity, genotoxicity and alteration of cellular antioxidant enzymes in silver nanoparticles exposed CHO cells. RSC Adv 5: 34927–34935.

Baghbani FA, Movagharniaa R, Sharifiana A, Salehib S, Ataollah SSS (2017) Photo-catalytic,anti-bacterial,and anti-cancer properties of phyto-mediated synthesis of silver nanoparticles from *Artemisia tournefortiana* Rchb extract. J Photochem. Photobiol B: Biol 173: 640–649.

Banerjee D, Sengupta S (2011) Nanoparticles in cancer chemotherapy. Prog Mol Biol Transl Sci 104: 489–507.

Barabadi H, Ovais M, Khan ZH, Saravanan M (2017) Anti-cancer green bionanomaterials: Present status and future prospects, Green Chem Lett Rev 10 (4): 285–314.

Billon H (1869) *Histoire des plantes*. Librairie Hachette, Paris, pp. 275–276.

Gajalalakshmi S, Vijayalakshmi S, Rajeswari DV (2012) Phytochemical and pharmacological properties of *Annona muricata*: A review. Int. J. Pharm. Pharm Sci 4(2): 3–6.

Hackenberg S, Scherzed A, Kessler M, Hummel S, Technau A, Froelich K, Ginzkey C, Koehler C, Hagen R, Kleinsasser N (2011) Silver nanoparticles: Evaluation of DNA damage, toxicity and functional impairment in human mesenchymal stem cells. Toxicol Lett 201(1): 27–33.

Jacob SJP, Finub JS, Narayanan A (2012) Synthesis of silver nanoparticles using *Piper longum* leaf extracts and its cytotoxic activity against Hep-2 cell line. Colloids Surf, B Biointerfaces 91: 212–214.

Jemal A, Bray F, Center MM, Ferlay J, Ward E, Forman D (2011) Gloabal cancer statistics. CA: Cancer J Clin 61(2): 69–90.

Karthik S, Sankar R, Varunkumar K, Ravikumar V (2014) Romidepsin induces cell cycle arrest, apoptosis, histone hyperacetylation and reduces matrix metalloproteinases 2 and 9expression in bortezomib sensitized non-small cell lung cancer cells. Biomed Pharmacother 68: 327–334.

Khan M, Naqvi AH, Ahmad M (2015) Comparative study of the cytotoxic and genotoxic potentials of zinc oxide and titanium dioxide nanoparticles. Toxicol Rep 2: 765–774.

Khatoon H, Singh A, Ahmad F, Kamal A (2017) Brassinosteroids an essential steroidal regulator: Its Structure, synthesis and signaling in plant growth and development—A review. Int J. Curr Res Biosci Plant Biol 4(7): 88–96.

Kovacs D, Igaz N, Keskeny C, Bélteky P, Tóth T, Gáspár R, Madarász D, Rázga Z, Kónya Z, Boros IM, Kiricsi M (2016) Silver nanoparticles defeat p53-positive and p53-negative osteosarcoma cells by triggering mitochondrial stress and apoptosis. Sci Rep 6:27902 | DOI: 10.1038/srep27902.

Krishnaraj C, Jagan EG, Rajasekar S, Selvakumar P, Kalaichelvan PT, Mohan N (2010) Synthesis of silver nanoparticles using *Acalypha indica* leaf extracts and its antibacterial activity against water borne pathogens. Colloids Surf, B Biointerfaces 76: 50–56.

Kuppusamy P, Ichwan SJA, Nur PHA, Hidayati WS, Soundharrajan I, Govindan N, Pragas GM, Mashitah MY (2016) Invtiro anticancer activity of Au, Ag nanoparticles synthesized using *Commelina nudiflora* L. Aqeous extracts against HCT116 colon cancer cells. Biol Trace Elem Res 173: 297–305.

Lima R, Seabra AB, Duran N (2012) Silver nanoparticles: A brief review of cytotoxicity and genotoxicity of chemically and biogenically synthesized nanoparticles. J Appl Toxicol 32(11): 867–879.

Miller DK, Rebecca L, Siegel L, Chieh CL, Mariotto AB, Kramer JL, Julia H, Rowland Kevin DS, Alteri R, Jemal A (2016) cancer treatment and survivorship statistics, 2016. CA: Cancer J Clin 66: 271–289.

Mosmann T (1983) Rapid colorimetric assay for cellular growth and survival: application to proliferation and cytotoxicity assays. Immunol. Methods 65(1): 55–63.

Mubarak Ali D, Thajuddin N, Jeganathan K, Gunasekaran M (2011) Plant extract mediated synthesis of silver and gold nanoparticles and its antibacterial activity against clinically isolated pathogens. Colloids Surf. B Biointerfaces 85: 360–365.

Mustata RC, Vasile G, Vallone FV, Strollo S, Lefort A, Libert F (2013) Identification of lgr5-independent spheroid-generating progenitors of the mouse fetal intestinal epithelium. Cell Rep 5: 421–432.

Ovais M, Khalil AT, Raza A, Khan MA, Ahmad I, Islam NU, Saravanan M, Ubaid MF, Ali M, Shinwari ZK (2016) Synthesis of silver nanoparticles via plant extracts: Beginning a new era in cancer theranostics. Nanomedicine 11(23): 3157–3177.

Patel S, Jayvadan KP (2016) A review on a miracle fruits of *Annona muricata*. J Pharmacogn Phytochem 5(1): 137–148.

Peer D, Karp JM, Hong S Farokhzad OC, Margalit RR, Langer (2007) Nanocarriers as an emerging platform for cancer therapy. Nat Nanotechnol 2: 751–760.

Prabhu S, Poulose EK (2012) Silver nanoparticles: Mechanism of antimicrobial action, synthesis, medical applications, and toxicity effects. Int Nano Lett 2 (32): 2–10.

Rajasekharreddy P, Rani PU, Sreedhar B (2010) Qualitative assessment of silver and gold nanoparticle synthesis in various plants: A photobiological approach. J Nanopart Res 12: 1711–1721.

Ranganathan R, Madanmohan S, Kesavan A, Baskar G, Ramia Y, Krishnamoorthy, Santosham R, Ponraju D, Kumar SR, Venkatraman G (2012) Nanomedicine: towards development of patient-friendly drug-delivery systems for oncological applications. Int J Nanomed 7: 1043–1060.

Salunke BK, Sawant SS, Lee S-I, Kim BS (2016) Microorganisms as efficient biosystem for the synthesis of metal nanoparticles. current ccenario and future possibilites. World J Microbiol Biotechnol 32(5): 1–16.

Shaniba VS, Ahlam AA, Kumar PRM (2017) Phyto-mediated synthesis of silver nanoparticles from *Annona muricata* fruit extract, assessment of their biomedical and photocatalytic potential 8(1): 170–80.

Song JY, Kim BS (2009) Rapid biological synthesis of silver nanoparticles using plant leaf extracts. Bioprocess Biosyst Eng 32: 79–84.

Thangam R, Gunasekaran P, Kaveri K, Sridevi G, Sundarraj S, Paulpandi M (2012). A novel disintegrin protein from *Naja naja* venom induces cytotoxicity and apoptosis in human cancer cell lines in vitro. Process Biochem 47: 1243–1249.

Varun S, Sellappa S (2014) Enhanced apoptosis in MCF-7 human breast cancer cells by biogenic gold nanoparticles synthesized from *Argemone mexicana* leaf extract. Int J Pharm Pharm Sci 6(8): 528–531.

Vivek R, Thangam R, Muthuchelian K, Gunasekaran P, Kaveri K, Kannan S (2012) Green biosynthesis of silver nanoparticles from *Annona squamosa* leaf extract and its *in vitro* cytotoxic effect on MCF-7 cells. Process Biochem 47: 2405–2410.

WHO (2018) Fact sheet on cancer. http.//www.who.int/en/news-fact sheets/detail

Xi Feng Z, Wei S, Sangiliyandi G (2016) Silver nanoparticles-mediatedcellular responses in various cell lines: An *in vitro* model. Int J Mol Sci17(10): 1603–1629.

Zhu B, Li Y, Lin Z, Zhao M, Xu, T, Wang C, Deng N (2016) Silver nanoparticles induce HePG-2 cells apoptosis through ROS mediatedsignaling pathways. Nanoscale Res Lett 11: 198−206.

CHAPTER 11

Promising Anticancer Potentials of Natural Chalcones as Inhibitors of Angiogenesis

DEBARSHI KAR MAHAPATRA[1*], VIVEK ASATI[2], and SANJAY KUMAR BHARTI[3]

[1]*Department of Pharmaceutical Chemistry, Dadasaheb Balpande College of Pharmacy, Nagpur 440037, Maharashtra, India*

[2]*Department of Pharmaceutical Chemistry, NRI Institute of Pharmacy, Bhopal 462021, Madhya Pradesh, India*

[3]*Institute of Pharmaceutical Sciences, Guru Ghasidas Vishwavidyalaya (A Central University), Bilaspur 495009, Chhattisgarh, India*

Corresponding author. E-mail: dkmbsp@gmail.com

ABSTRACT

This chapter focuses on the budding perspective of chalcone (prop-2-ene-1-one)-based natural inhibitors (isoliquiretigenin, butein, garcinol, hydroxysafflor yellow A, broussochalcone A, 2,4-dihydroxy-6-methoxy-3,5-dimethylchalcone, 4′-hydroxy chalcone, and Parasiticin-A, -B, and -C) and synthetic inhibitors (4-(*p*-toluenesulfonylamino)-4′-hydroxy chalcone, 4-maleamide peptidyl chalcone, and quinolyl-thienyl chalcone) that will prevent angiogenic switching (fibroblast growth factor angiogenin, TGF-β) by directly inhibiting the three vital therapeutic targets: vascular endothelial growth factor, vascular endothelial growth factor receptor-2, and matrix metalloproteinases-2/9, which will block neovascularization, vascular formation, and network formation by completely depriving the cells from the required nutrients, fluid, signaling molecules, and oxygen. The highlighted studies will positively motivate young minds, medicinal chemists, future researchers, and allied scientists for developing or exploring potential angiogenic inhibitors for the treatment of cancer with better pharmacodynamics attributes and less side or adverse effects.

11.1   INTRODUCTION

Angiogenesis is a vital process in the human body that leads to the formation and maintenance of blood vessel network (Breier, 2000). This imperative process, however, leads to the precipitation of several diseases like cancer and inflammation. With the advancement in medical research, numerous signaling molecules, such as the Delta–Notch system, ephrin–Eph receptors, vascular endothelial growth factors–vascular endothelial growth factor receptors (VEGF–VEGFRs), and angiopoietin–Tie, have been known to involve in the angiogenesis process (Weidner, 1993). The VEGF–VEGFRs play a pivotal role in the process of new blood vessel formation from the precursors in early life stages, vasculogenesis, and angiogenesis (Sridhar and Shepherd, 2003). Blood vessels play a crucial role in the supply of oxygen, nutrients, signaling molecules, fluids, etc. along with access to immune supervision. The endothelial cells, which constitute the inner surface of the blood vessels, are repaired by several angiogenic processes (Wilting and Christ, 1996). The capillary walls and vessels of fine dimensions consist of pericytes (a single cell layer), which largely contrast with the constitution of arteries and veins that are formed largely by smooth muscle cells. However, in adults, the formation of the blood vessels is mediated under a very strict process because both disproportionately high amount of neovascularization and their scarce formation will lead to several pathogenic conditions (Battegay, 1995).

During the process of vasculogenesis along with the primary vascular plexus, the formation of the vascular net (capillaries fuse into vessels, arteries, and veins) results from the already existing condition/system (Auerbach and Auerbach, 1994). During the angiogenesis progression, further transformations occur, primarily the vascular plexus expands notably due to acute branching of the capillary and formation of the highly organized network (Ribatti, 2009). The whole procedure involves commencement of the process from the nearby destructed pre-existing blood vessels, followed by the initiation of endothelial cell proliferation, and finally the migration process. The situated endothelial cells are bringing together in tubular structures followed by the formation of blood vessels around them (Otrock et al., 2007).

In the pathogenesis of tumor development, the activation of angiogenesis is a crucial phenomenon as it provides developmental support to the cancer cells by promoting easy access to food, nutrients, oxygen, and several other growth-promoting factors (Ribatti, 2008). Without these factors, a tumor nodule cannot grow more than 4 cm^3 due to prevailing hypoxia and in the due course will lead to death of cells (Shahi and Pineda, 2008). Various factors, such as fibroblast growth factor (FGF), VEGF–VEGFRs, angiogenin, TGF-β, etc., promote tumor growth by enhancing blood vessel networking in the cancer cells. The inhibition of such factors will lead to containment of "angiogenic switching" and therefore will lead to inactivation of pro-angiogenic factors, and ultimately, the progression/proliferation will halt (Tahergorabi and Khazaei, 2012).

Several natural products, such as chalcones, aurones, flavonoids, flavones, flavonols, flavanones, isoflavones, anthocyanidine, flavan-3-ols, proanthocyanidins, etc., are regarded as potential antiangiogenic compounds (Mahapatra et al., 2015a). Chalcones (prop-2-ene-1-one) are the class of natural products having a composition of two aromatic rings bound by an α,β unsaturated carbonyl bridge (Mahapatra et al., 2015b). They are the chromophoric compounds first synthesized in the 19th century by Tambor and Kostanecki (Mahapatra et al., 2015c). They are regarded as open-chain intermediates in the flavonoid synthesis in the aurone pathway (Mahapatra and Bharti, 2016). The chalcone scaffold-bearing compounds are known to have antibacterial, anti-inflammatory, anticancer, antifungal, hypnotic, antiviral, sedative, anticonvulsant, antimalarial, antileishmanial, antitrypanosomal, antiretroviral, antiplatelet, antihypertensive, antihyperlipidemic, antiarrhythmic, antigout, analgesic, antidiabetic, antiobesity, etc. properties (Mahapatra et al., 2017; Mahapatra et al., 2019a, 2019b). This chapter highlights the antiangiogenic perspectives of natural chalcones.

11.2 INHIBITION OF VASCULAR ENDOTHELIAL GROWTH FACTORS BY CHALCONES

While rationally developing some potential angiogenic inhibitors, either from nature or (semi-)synthetic, VEGF-1 and VEGF-2 remained the prime target for the pharmacotherapeutics (Figure 11.1) (Ferrara et al., 2003). These are solitarily formed in the cells and promote neovascularization. During the embryonic stages, these are vastly expressed and play a critical role in fetal vasculogenesis (Ferrara, 2002). In adults, these play a central role in the formation of blood vessels under injury conditions and cardiac blockade. Under tumor conditions, these are largely overexpressed and lead to swift growth of tumor and its spreading across the body due to the enormous availability of nutrients and oxygen in the metabolizing cells (McColl et al., 2004).

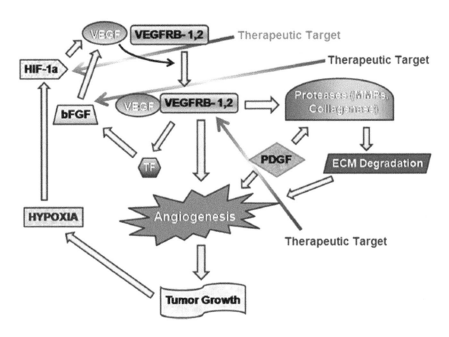

FIGURE 11.1   Illustration of the VEGF pathway.

The family of VEGF genes comprises of seven members, which include VEGF-E, a viral genome-derived component. VEGFs play an essential role in the cardiovascular system, atherosclerosis, cardiac myofibroblasts, central nervous system, nonendothelial cells, bones, hematopoietic cells, hematological malignancies, autocrine signaling, tumor cells, etc. (Przybylski, 2009). VEGF-A, VEGF-B, VEGF-C, and VEGF-D have been glimpsed to regulate the process of angiogenesis at early embryogenesis, show vascular permeability activity, have proangiogenic activity, stimulate cell migration in macrophage, and are considered as significant regulators of lymphangiogenesis (Nagy et al., 2007). VEGFs of diverse origin have proangiogenic characteristics for the development and maintenance of blood vessel physiological levels to overcome ischemic diseases. Based on the conditions, inhibitors of angiogenesis have been developed and preclinically evaluated to determine, but none of them proved very promising for a complete cure (Baka et al., 2006). Multikinase inhibitors and anti-VEGF-A neutralizing antibody have been produced as an alternative way of treatment; keeping possible side effects aside, however, the clinical efficacy is under question (Ferrara and Kerbel, 2005). Chalcone-based compounds have been reported very recently as promising inhibitors of VEGF and may be applied in the near future as angiogenesis inhibitors with several advantageous pharmacodynamics and pharmacokinetics attributes.

2-Hydroxy-4-methoxychalcone (1) has been found to exhibit in vivo anti-angiogenic activity by complete inhibition of COX-2 enzyme. The inhibition of COX-2 enzyme directly or indirectly produced a marked suppression of the proliferative nature of the cells (Lee et al., 2006).

A natural chalcone obtained from the dried flower of *Cleistocalyx operculatus*, 2,4-dihydroxy-6-methoxy-3,5-dimethylchalcone (2) (Figure 11.2), produced reversible inhibition of kinase domain receptor (KDR) tyrosine kinase phosphorylation (VEGF). VEGF is a key factor responsible for the growth of human vascular endothelial human dermal microvascular endothelial cells (HDMECs) and ultimately results in angiogenesis. The inhibition of VEGF leads to an acute reduction in the tumor vessel density and eventually causes in vivo antiangiogenic activity (Zhu et al., 2005).

Two novel methoxylated chalcones, 2,2′,4′-trimethoxychalcone (3) and 3′-bromo-2,4-dimethoxychalcone (4), demonstrated significant antiendothelial characteristics in immortalized human microvascular endothelial cells (HMEC-1) and thwarted the angiogenic growth by preventing the tumor-induced neovascularization (Bertl et al., 2004).

Chalcone (5) has been observed as a potential antiangiogenic candidate by producing considerable inhibition of VEGF-induced migration and invasion of Hep3B and human umbilical vein endothelial cells (HUVECs) (under nontoxic concentrations) as well as the inhibition of HIF-1 by down-regulating the expression of HIF-1a (under hypoxic conditions) (Wang et al., 2015).

The potential of 4′-hydroxy chalcone (6) in suppressing multiple-step angiogenesis was studied by Varinska and co-workers. The natural product showed no cytotoxic effects and effectively inhibited VEGF and fibroblast growth factor (FGF)-induced phosphorylation of extracellular signal-regulated kinase (ERK)-1/-2 and Akt kinase, which noticeably restricted the endothelial cell proliferation, their migration, and formation of tubules. The suppression of growth factor pathways translated into in vitro angiogenic attributes of endothelial cells (Varinska et al., 2012).

With antiangiogenic prospective, hydrophilic chalcones (7 and 8) have been screened against human mammary adenocarcinoma (MCF-7) cell line

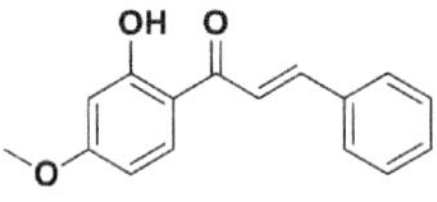

FIGURE 11.2   Chalcones as VEGF inhibitors.

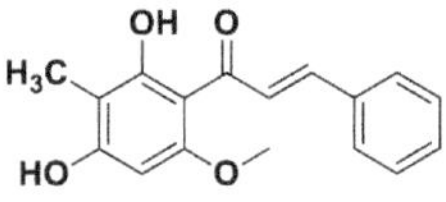

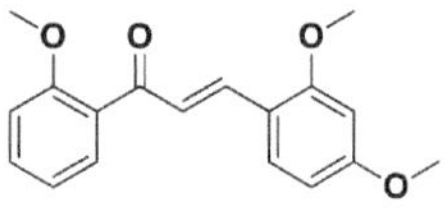

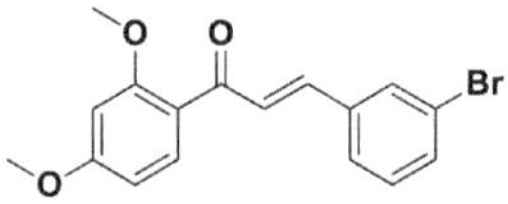

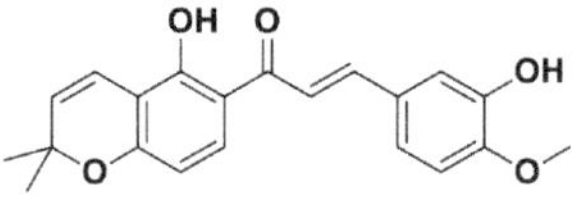

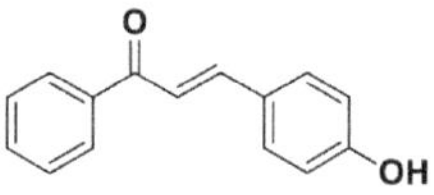

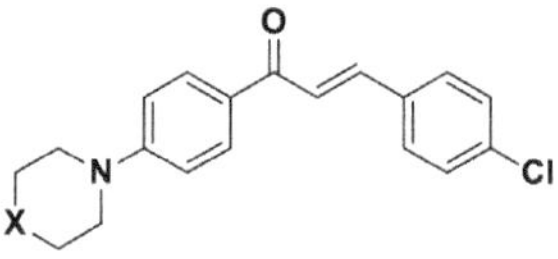

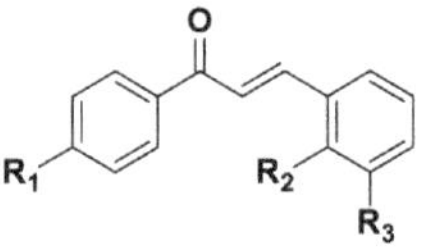

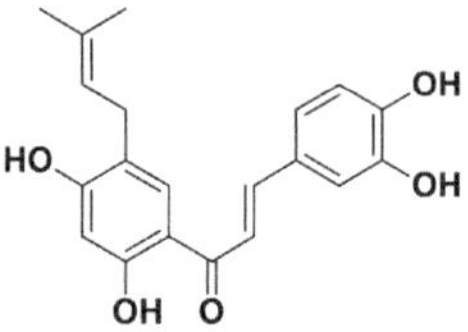

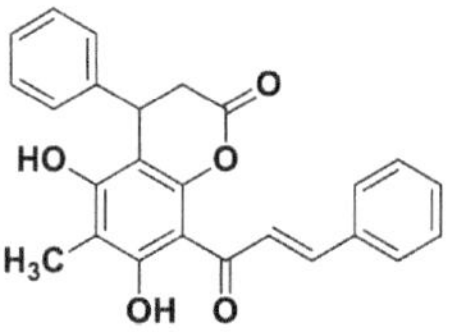

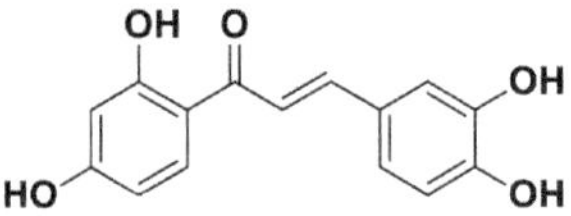

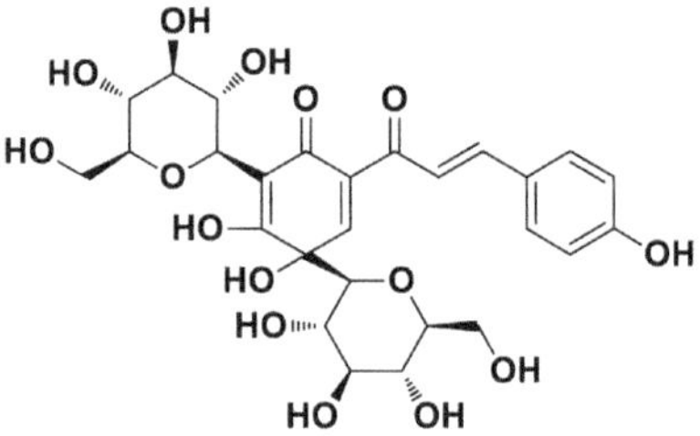

and multiple myeloma (RPMI-8226) cell line with an analogous motif. The study expressed tremendous antiangiogenic activity through the regulation of various pathways at low micromolar concentrations (Syam et al., 2012).

Quite similarly, chalcone molecules **(9–12)** demonstrated significant activities (in micromolar concentrations) against human adenocarcinoma cell line HT-29, human breast adenocarcinoma cell line MCF-7, human prostate cancer cell line PC3, human normal liver cell line WRL-68, and human lung adenocarcinoma cell line A549 (Nelson et al., 2013).

From the ethyl acetate extract of bark of *Broussonetia papyrifera*, broussochalcone A **(13)** was isolated and in vitro investigated for antiangiogenic ability against ER-positive breast cancer MCF-7 cells, where excellent results were perceived due to target modulatory effects (Guo et al., 2013).

Parasiticin-A, -B, and -C, three novel chalcone derivatives along with other four chalcone compounds, have been isolated from *Cyclosorus parasiticus* leaves and screened against six human cancer cell lines (A549, MDA-MB-231, ALL-SIL, HepG2, and MCF-7). The in vitro exploration demonstrated extensive cytotoxicity (micromolar IC_{50} values) along with the significant antiangiogenic activity. Parasiticin-C **(14)** and 2,4-dihydroxy-6-methoxy-3,5-dimethylchalcone **(2)** showed the highest antiangiogenic activity (Wei et al., 2013).

In an interesting study, a natural chalcone, butein **(15)** has been scrutinized as potent *in vivo* suppressor of microvessel formation. The chalcone inhibited the major factor VEGF, which prevents sprouting of the minor vessels from the aortic rings (Chung et al., 2013).

Chalcone-based pigment, hydroxysafflor yellow A **(16)**, obtained from the flower petals of *Carthamus tinctorius*, presented antiangiogenic activity by downregulating the βFGF mRNA expression and VEGF in transplanted gastric adenocarcinoma (Xi et al., 2009).

11.3 INHIBITION OF VASCULAR ENDOTHELIAL GROWTH FACTOR RECEPTOR-2 BY CHALCONES

Depending on the type of vertebrate species, the genes of VEGFR comprises of 3–4 members. It comprises of three loops produced by the intramolecular disulfide bonding, and loop 1 cooperates with loop 3 to exhibit necessary binding and activation cascade for angiogenic switching (Sawamiphak et al., 2010). VEGFR-1 (VEGF receptor 1 also known as Flt-1, fms-like tyrosine kinase) and VEGFR-2 (KDR/Flk-1, fetal liver kinase 1 is the murine homolog of human kinase insert domain-containing receptor) are the most prominent receptors involved in vasculogenesis, neovascularization, and proangiogenesis processes (Figure 11.3) (D'Haene et al., 2013). VEGFR-2 is a receptor of VEGF-A and mostly mediates endothelial growth and signaling. The angiogenic promoters are foremost expressed by VEGF-E (Orf-virus-derived VEGF) (Li et al., 2005). The inhibition of VEGFR-2 may be regarded as the most fitting biological target for the suppression of solid tumors (Shibuya, 2006). Researchers working on chalcone scaffold have put forward some recently tested VEGFR-2 inhibitors as perspective antiangiogenic agents.

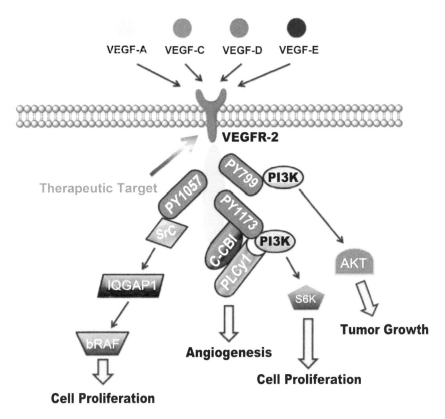

FIGURE 11.3   Illustration of the VEGFR-2 pathway.

Based on the structure-based virtual screening, quinolyl-thienyl chalcone series was in silico screened for their ability to inhibit VEGFR-2 kinase. Further, in vitro research on HUVECs produced noteworthy inhibition by the chalcone hybrids. The highest activity was recognized by inhibitor **17** (Rizvi et al., 2012).

Isoliquiritigenin **(18)**, a natural product, has been identified as antiangiogenic candidate that mediates the activity inhibiting the VEGF expression. The natural chalcone exhibits HIF-1α (hypoxia inducible factor-1α) proteasome degradation in breast cancer cells and blocks the kinase activity via interaction with VEGFR-2 (Wang et al., 2013b).

The strong VEGFR-2 inhibitory activity of 2,4-dihydroxy-6-methoxy-3,5-dimethylchalcone **(2)** (Figure 11.4) remained the key highlights of Patent CN1454895A, which further depicts its utility in chemoprevention due to antiangiogenesis activity (Sun Yat-Sen Cancer Center, 2003).

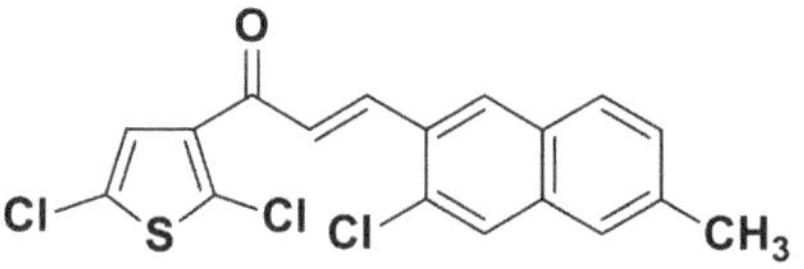

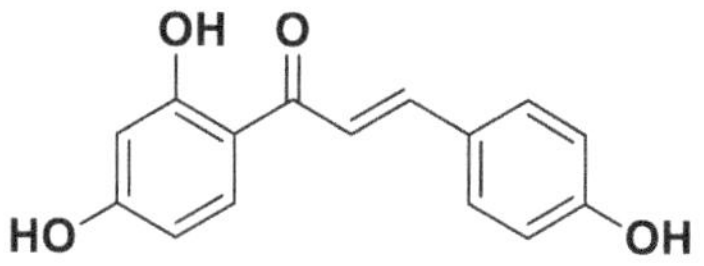

FIGURE 11.4 Chalcones as VEGFR-2 inhibitors.

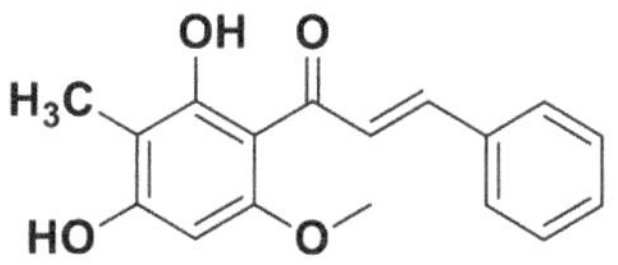

11.4 INHIBITION OF MATRIX METALLOPROTEINASES-2/9 BY CHALCONES

Matrix metalloproteinases (MMPs) are the enzyme class responsible for the degradation of small molecules present in the extracellular matrix (Genís et al., 2006). Tissue inhibitors of metalloproteinases are known to inhibit MMP activity. MMP-2 and MMP-9 are the prime components responsible for the angiogenesis (Sang, 1998). MMP-9 has been seen to degrade the type-IV collagens present in the basement membrane, which leads to the clearance of matrix neighboring the endothelial cells. When the cryptic regulatory sequence gets exposed, it leads to proliferation and migration, which consequently leads to triggering of the angiogenic switch (Rundhaug, 2005). As the matrix-bound VEGF gets released by the cleavage process and the circulating endothelial precursor stem cells are released from bone marrow, the angiogenesis process initiates (Figure 11.5) (Stetler-Stevenson, 1999). A handful of molecules bearing chalcone pharmacophore have been preclinically perceived to block MMP-2 and MMP-9 very selectively and may be rationally designed in future for the complete blockade of angiogenic switching.

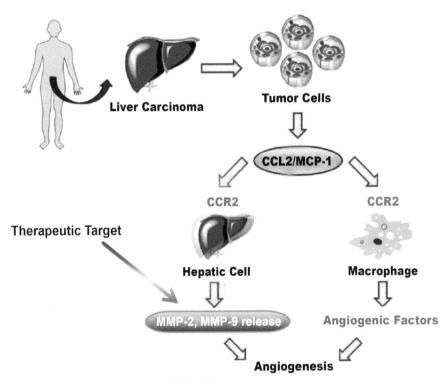

FIGURE 11.5   Illustration of the MMP-2/9 pathway.

Novel prenylated chalcones **(19-** and **20)** obtained from the twigs of *Dorstenia turbinate* presented remarkable and complete inhibition of MMP-2 in brain tumor-derived glioblastoma cells. The authors also highlighted the importance of two prenylated groups in respective positions that play a dominant role in interaction with the MMP-2. The presence of free hydroxyl groups perceptibly improved the inhibition of MMP-2 (Ngameni et al., 2006).

The wonder chalcone molecule isoliquiretigenin (**18**) has been seen to modulate a large number of biochemical targets. The compound of natural origin considerably reduces the VEGF expressions, MMP-2/9 expressions, and HIF-1α levels in cancer cells and demonstrates antimigratory activities, which are the potential biomarkers for the antiangiogenic process. In addition to the above activities, the chalcone also inhibits phosphatidylinositol-3 kinase (PI3K) expression and p38 and Akt kinase phosphorylation and the suppression of NF-kβ deoxyribonucleic acid (DNA)-binding activity (Wang et al., 2013a).

The antiangiogenic perspective of 4-maleamide peptidyl chalcones (**21** and **22**) in human prostate cancer cells (PC-3 and LNCaP) was studied by Rodrigues and co-workers where they observed alterations in the cell cycle, cellular invasion, changes in migration, anticlonogenic activity, and antiangiogenic potentials and marked MMP-9 inhibition in micromolar concentrations (Rodrigues et al., 2011).

The MMP inhibitory ability and TM4SF5 antagonistic activity of 4-(*p*-toluenesulfonylamino)-4′-hydroxy chalcone (TSAHC) (**23**) (Figure 11.6), a chalcone derivative, were profoundly studied, and its antiangiogenic behavior was established. The compound has been found to express antitumor activity in rats with no significant toxicity at a dose of 50 mg/kg (Park et al., 2012).

The antiangiogenic, antiproliferative, antimetastatic, and proapoptotic perspectives of garcinol (**24**), a chalcone-like compound isolated from *Garcinia indica* fruits, have been reported recently on pancreatic cancer cells. The polyisoprenylated benzophenone produces dose- and time-dependent induction of apoptosis (G_0–G_1 phase cell cycle arrest) in BxPC-3 and Panc-1 cell lines at 40 μM concentration. The compound expresses the activity by targeting transcription factor NF-kB, metastatic factors (PGE2, MMP-9, IL-8, and VEGF), and apoptotic factors (caspase-3/9, X-IAP, PARP, and cIAP), thereby preventing chemoresistance and neovascularization (Parasramka and Gupta, 2011). In addition to it, the molecule exhibit dose-dependent inhibition of cell invasion against pancreatic and prostate cancer cell lines by modulating a signal transducer and activator of the transcription-3 (STAT-3) signaling pathway (Ahmad et al., 2012).

The MMP inhibitory attributes (both collagenase and gelatinase subfamily) of 4′-hydroxy chalcone (**6**) and its derivatives as potential anticancer and chemopreventive agents remained the center of attention of Ptent WO2003037315A1. The antiangiogenic perspective of this compound continues to be the sole heart of this invention (Angiolab Inc., Assignee 2003).

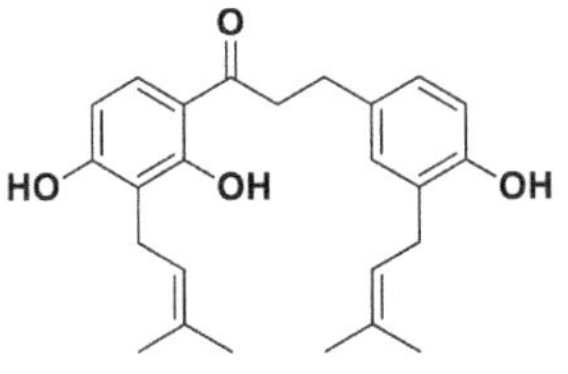

(19)

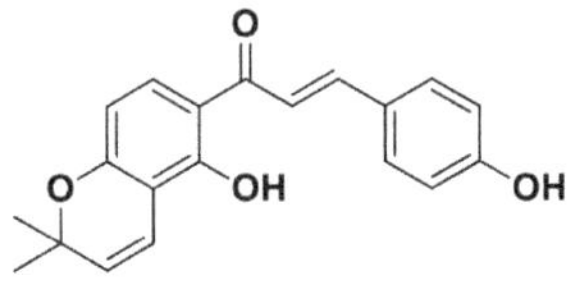

(20)

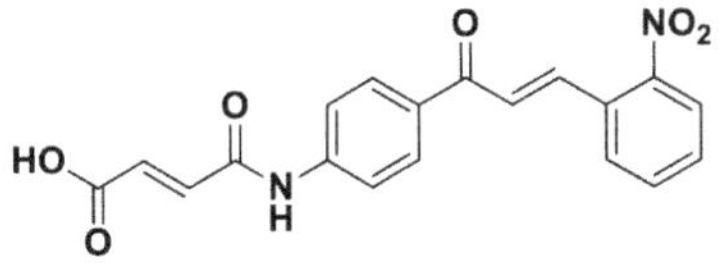

(21)

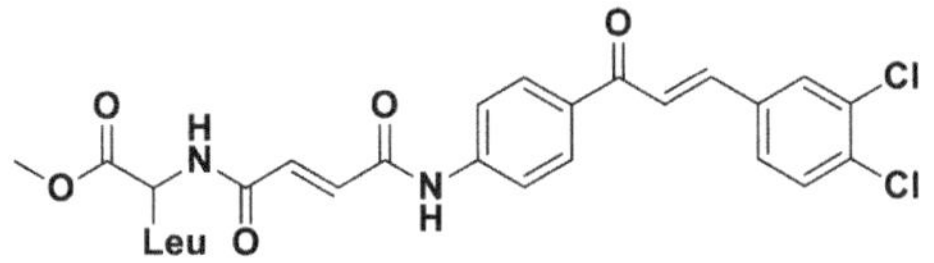

(22)

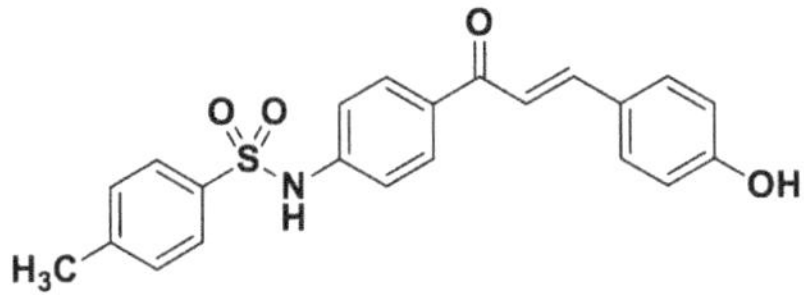

(23)

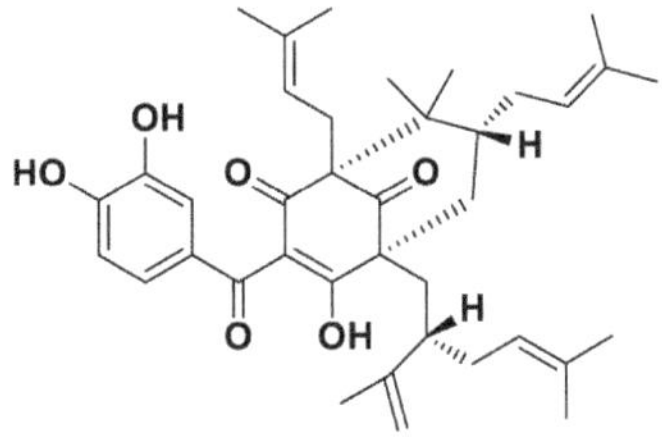

(24)

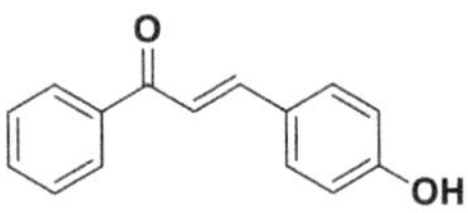

(6)

FIGURE 11.6 Chalcones as MMP-2/9 inhibitors.

11.5 CONCLUSION

This chapter focuses on the budding perspective of chalcone-based natural inhibitors (isoliquiretigenin, butein, garcinol, hydroxysafflor yellow A, broussochalcone A, 2,4-dihydroxy-6-methoxy-3,5-dimethylchalcone, 4′-hydroxy chalcone, and Parasiticin-A, -B, and -C) and synthetic inhibitors (4-(*p*-toluenesulfonylamino)-4′-hydroxy chalcone, 4-maleamide peptidyl chalcone, and quinolyl-thienyl chalcone) that will prevent angiogenic switching (fibroblast growth factor angiogenin, TGF-β) by directly inhibiting the three vital therapeutic targets: VEGF,

VEGFR-2, and MMP-2/9, which will block neovascularization, vascular formation, and network formation. The highlighted studies will positively motivate young minds, medicinal chemists, future researchers, and allied scientists for developing or exploring potential angiogenic inhibitors.

KEYWORDS

- **anticancer**
- **angiogenesis**
- **chalcone**
- **natural products**
- **vascular endothelial growth factor**
- **vascular endothelial growth factor receptor-2**
- **matrix metalloproteinases**

REFERENCES

Ahmad, A., Sarkar, S. H., Aboukameel, A., Ali, S., Biersack, B., Seibt, S., Li, Y., Bao, B., Kong, D., Banerjee, S., & Schobert, R. (2012). Anticancer action of garcinol in vitro and in vivo is in part mediated through inhibition of STAT-3 signaling. *Carcinogenesis*, *33*(12), 2450–2456.

Angiolab Inc., Assignee (2003). Pharmaceutical composition containing chalcone or its derivatives for matrix metalloproteinase inhibitory activity. Patent WO2003037315A1.

Auerbach, W., & Auerbach, R. (1994). Angiogenesis inhibition: A review. *Pharmacology & Therapeutics*, *63*(3), 265–311.

Baka, S., Clamp, A. R., & Jayson, G. C. (2006). A review of the latest clinical compounds to inhibit VEGF in pathological angiogenesis. *Expert Opinion on Therapeutic Targets*, *10*(6), 867–876.

Battegay, E. J. (1995). Angiogenesis: Mechanistic insights, neovascular diseases, and therapeutic prospects. *Journal of Molecular Medicine*, *73*(7), 333–346.

Bertl, E., Becker, H., Eicher, T., Herhaus, C., Kapadia, G., Bartsch, H., & Gerhäuser, C. (2004). Inhibition of endothelial cell functions by novel potential cancer chemopreventive agents. *Biochemical and Biophysical Research Communications*, *325*(1), 287–295.

Breier, G. (2000). Angiogenesis in embryonic development—A review. *Placenta*, *21*, S11-S15.

Chung, C. H., Chang, C. H., Chen, S. S., Wang, H. H., Yen, J. Y., Hsiao, C. J., Wu, N. L., Chen, Y. L., Huang, T. F., Wang, P. C., & Yeh, H. I. (2013). Butein inhibits angiogenesis of human endothelial progenitor cells via the translation dependent signaling pathway. *Evidence-Based Complementary and Alternative Medicine*, *2013*, 1–10.

D'Haene, N., Sauvage, S., Maris, C., Adanja, I., Le Mercier, M., Decaestecker, C., Baum, L., & Salmon, I. (2013). VEGFR1 and VEGFR2 involvement in extracellular galectin-1-and galectin-3-induced angiogenesis. *PLoS One*, *8*(6), e67029.

Ferrara, N. (2002). VEGF and the quest for tumour angiogenesis factors. *Nature Reviews Cancer*, *2*(10), 795.

Ferrara, N., Gerber, H. P., & LeCouter, J. (2003). The biology of VEGF and its receptors. *Nature Medicine*, *9*(6), 669.

Ferrara, N., & Kerbel, R. S. (2005). Angiogenesis as a therapeutic target. *Nature*, *438*(7070), 967.

Genís, L., Gálvez, B. G., Gonzalo, P., & Arroyo, A. G. (2006). MT1-MMP: Universal or particular player in angiogenesis? *Cancer and Metastasis Reviews*, *25*(1), 77–86.

Guo, F., Feng, L., Huang, C., Ding, H., Zhang, X., Wang, Z., & Li, Y. (2013). Prenylflavone derivatives from Broussonetia papyrifera, inhibit the growth of breast cancer cells in vitro and in vivo. *Phytochemistry Letters*, *6*(3), 331–336.

Lee, Y. S., Lim, S. S., Shin, K. H., Kim, Y. S., Ohuchi, K., & Jung, S. H. (2006). Anti-angiogenic and anti-tumor activities of 2'-hydroxy-4'-methoxychalcone. *Biological and Pharmaceutical Bulletin*, *29*(5), 1028–1031.

Li, J., Huang, S., Armstrong, E. A., Fowler, J. F., & Harari, P. M. (2005). Angiogenesis and radiation response modulation after vascular endothelial growth factor receptor-2 (VEGFR2) blockade. *International Journal of Radiation Oncology, Biology, Physics*, *62*(5), 1477–1485.

Mahapatra, D. K., Asati, V., & Bharti, S. K. (2015a). Chalcones and their therapeutic targets for the management of diabetes: Structural and pharmacological perspectives. *European Journal of Medicinal Chemistry*, *92*, 839–865.

Mahapatra, D. K., Asati, V., & Bharti, S. K. (2019a). An updated patent review of therapeutic applications of chalcone derivatives (2014-present). *Expert Opinion on Therapeutic Patents, 29*(5), 385–406.

Mahapatra, D. K., & Bharti, S. K. (2016). Therapeutic potential of chalcones as cardiovascular agents. *Life Sciences*, *148*, 154–172.

Mahapatra, D. K., Bharti, S. K., & Asati, V. (2015a). Chalcone scaffolds as anti-infective agents: Structural and molecular target perspectives. *European Journal of Medicinal Chemistry*, *101*, 496–524.

Mahapatra, D. K., Bharti, S. K., & Asati, V. (2015b). Anti-cancer chalcones: Structural and molecular target perspectives. *European Journal of Medicinal Chemistry*, *98*, 69–114.

Mahapatra, D. K., Bharti, S. K., & Asati, V. (2017). Chalcone derivatives: Anti-inflammatory potential and molecular targets perspectives. *Current Topics in Medicinal Chemistry*, *17*(28), 3146–3169.

Mahapatra, D. K., Bharti, S. K., Asati, V., & Singh, S. K. (2019b). Perspectives of medicinally privileged chalcone based metal coordination compounds for biomedical applications. *European Journal of Medicinal Chemistry, 174*, 142–158.

McColl, B. K., Stacker, S. A., & Achen, M. G. (2004). Molecular regulation of the VEGF family – Inducers of angiogenesis and lymphangiogenesis. *Apmis*, *112*(7–8), 463–480.

Nagy, J. A., Dvorak, A. M., & Dvorak, H. F. (2007). VEGF-A and the induction of pathological angiogenesis. *Annual Review of Pathology: Mechanisms of Disease 2*, 251–275.

Nelson, G., Alam, M. A., Atkinson, T., Gurrapu, S., Kumar, J. S., Bicknese, C., Johnson, J. L., & Williams, M. (2013). Synthesis and evaluation of p-N,N-dialkyl substituted chalcones as anti-cancer agents. *Medicinal Chemistry Research*, *22*(10), 4610–4614.

Ngameni, B., Touaibia, M., Patnam, R., Belkaid, A., Sonna, P., Ngadjui, B. T., Annabi, B., & Roy, R. (2006). Inhibition of MMP-2 secretion from brain tumor cells suggests chemopreventive properties of a furanocoumarin glycoside and of chalcones isolated from the twigs of Dorstenia turbinata. *Phytochemistry, 67*(23), 2573–2579.

Otrock, Z. K., Mahfouz, R. A., Makarem, J. A., & Shamseddine, A. I. (2007). Understanding the biology of angiogenesis: Review of the most important molecular mechanisms. *Blood Cells, Molecules, and Diseases, 39*(2), 212–220.

Parasramka, M. A., & Gupta, S. V. (2011). Garcinol inhibits cell proliferation and promotes apoptosis in pancreatic adenocarcinoma cells. *Nutrition and Cancer, 63*(3), 456–465.

Park, K.H. Lee, J.W., Ryu, Y.B., Ryu, H.W., & Lee, S.A. (2012). Method for screening anticancer compounds inhibiting function of TM4SF5 and anticancer composition containing chalcone compounds. US Patent 20120282619.

Przybylski, M. (2009). A review of the current research on the role of bFGF and VEGF in angiogenesis. *Journal of Wound Care, 18*(12), 516–519.

Ribatti, D. (2008). The discovery of the placental growth factor and its role in angiogenesis: A historical review. *Angiogenesis, 11*(3), 215–221.

Ribatti, D. (2009). Endogenous inhibitors of angiogenesis: A historical review. *Leukemia Research, 33*(5), 638–644.

Rizvi, S. U. F., Siddiqui, H. L., Nisar, M., Khan, N., & Khan, I. (2012). Discovery and molecular docking of quinolyl-thienyl chalcones as anti-angiogenic agents targeting VEGFR-2 tyrosine kinase. *Bioorganic & Medicinal Chemistry Letters, 22*(2), 942–944.

Rodrigues, J., Abramjuk, C., Vásquez, L., Gamboa, N., Domínguez, J., Nitzsche, B., Höpfner, M., Georgieva, R., Bäumler, H., Stephan, C., & Jung, K. (2011). New 4-maleamic acid and 4-maleamide peptidyl chalcones as potential multitarget drugs for human prostate cancer. *Pharmaceutical Research, 28*(4), 907–919.

Rundhaug, J. E. (2005). Matrix metalloproteinases and angiogenesis. *Journal of Cellular and Molecular Medicine, 9*(2), 267–285.

Sang, Q. X. A. (1998). Complex role of matrix metalloproteinases in angiogenesis. *Cell Research, 8*(3), 171.

Sawamiphak, S., Seidel, S., Essmann, C. L., Wilkinson, G. A., Pitulescu, M. E., Acker, T., & Acker-Palmer, A. (2010). Ephrin-B2 regulates VEGFR2 function in developmental and tumour angiogenesis. *Nature, 465*(7297), 487.

Shahi, P. K., & Pineda, I. F. (2008). Tumoral angiogenesis: Review of the literature. *Cancer Investigation, 26*(1), 104–108.

Shibuya, M. (2006). Vascular endothelial growth factor receptor-1 (VEGFR-1/Flt-1): A dual regulator for angiogenesis. *Angiogenesis, 9*(4), 225–230.

Sridhar, S. S., & Shepherd, F. A. (2003). Targeting angiogenesis: A review of angiogenesis inhibitors in the treatment of lung cancer. *Lung Cancer, 42*(2), 81–91.

Stetler-Stevenson, W. G. (1999). Matrix metalloproteinases in angiogenesis: A moving target for therapeutic intervention. *The Journal of Clinical Investigation, 103*(9), 1237–1241.

Sun Yat-Sen Cancer Center, South China Institute of Botany, Chinese Academy of Sciences, Assignee (2003). Use of 2',4'-dihydroxy-6'-methoxy-3',5'-dimethylchalcone for preparing anticancer medicine. Patent CN1454895 A, Nov 12, 2003.

Syam, S., Abdelwahab, S. I., Al-Mamary, M. A., & Mohan, S. (2012). Synthesis of chalcones with anticancer activities. *Molecules, 17*(6), 6179–6195.

Tahergorabi, Z., & Khazaei, M. (2012). A review on angiogenesis and its assays. *Iranian Journal of Basic Medical Sciences, 15*(6), 1110.

Varinska, L., Van Wijhe, M., Belleri, M., Mitola, S., Perjesi, P., Presta, M., Koolwijk, P., Ivanova, L., & Mojzis, J. (2012). Anti-angiogenic activity of the flavonoid precursor 4-hydroxychalcone. *European Journal of Pharmacology*, *691*(1-3), 125–133.

Wang, K. L., Hsia, S. M., Chan, C. J., Chang, F. Y., Huang, C. Y., Bau, D. T., & Wang, P. S. (2013a). Inhibitory effects of isoliquiritigenin on the migration and invasion of human breast cancer cells. *Expert Opinion on Therapeutic Targets*, *17*(4), 337–349.

Wang, L., Chen, G., Lu, X., Wang, S., Han, S., Li, Y., Ping, G., Jiang, X., Li, H., Yang, J., & Wu, C. (2015). Novel chalcone derivatives as hypoxia-inducible factor (HIF)-1 inhibitor: Synthesis, anti-invasive and anti-angiogenic properties. *European Journal of Medicinal Chemistry*, *89*, 88–97.

Wang, Z., Wang, N., Han, S., Wang, D., Mo, S., Yu, L., Huang, H., Tsui, K., Shen, J., & Chen, J. (2013b). Dietary compound isoliquiritigenin inhibits breast cancer neoangiogenesis via VEGF/VEGFR-2 signaling pathway. *PLoS One*, *8*(7), e68566.

Wei, H., Zhang, X., Wu, G., Yang, X., Pan, S., Wang, Y., & Ruan, J. (2013). Chalcone derivatives from the fern *Cyclosorus parasiticus* and their anti-proliferative activity. *Food and Chemical Toxicology*, *60*, 147–152.

Weidner, N. (1993, November). Tumor angiogenesis: Review of current applications in tumor prognostication. In *Seminars in Diagnostic Pathology*, Vol. 10, No. 4, pp. 302–313.

Wilting, J., & Christ, B. (1996). Embryonic angiogenesis: A review. *Naturwissenschaften*, *83*(4), 153–164.

Xi, S., Zhang, Q., Xie, H., Liu, L., Liu, C., Gao, X., Zhang, J., Wu, L., Qian, L., & Deng, X. (2009). Effects of hydroxy safflor yellow A on blood vessel and mRNA expression with VEGF and bFGF of transplantation tumor with gastric adenocarcinoma cell line BGC-823 in nude mice. *China Journal of Chinese Materia Medica*, *34*(5), 605–610.

Zhu, X. F., Xie, B. F., Zhou, J. M., Feng, G. K., Liu, Z. C., Wei, X. Y., Zhang, F. X., Liu, M. F., & Zeng, Y. X. (2005). Blockade of vascular endothelial growth factor receptor signal pathway and antitumor activity of ON-III (2′, 4′-dihydroxy-6′-methoxy-3′,5′-dimethylchalcone), a component from Chinese herbal medicine. *Molecular Pharmacology*, 67(5), 1444–1450.

CHAPTER 12

Direct Electrochemical Oxidation of Blood

V. A. RUDENOK

Izhevsk State Agricultural Academy, Izhevsk, Udmurt Republic, Russia, 426069

ABSTRACT

Protection of the human internal environment is an important aspect of the ecological problem. In a healthy human, this function is performed by the liver monooxigenase system that helps to remove hydrophobic toxic substances from the organism by their hydroxylating oxidation with molecular hydrogen catalyzed by special detoxicating ferment cytochrome P-450.

12.1 INTRODUCTION

The new area in efferent medicine—electrochemical oxidation of blood—is being successfully developed. Originally, the efficiency of direct current action on blood was found while working on fuel cells. Direct current was let to flow via various liquids, including blood. It was established that it destructively affects bacteria and their toxins.

In Russia, this method was developed in the works of the Academician Yu. M. Lopukhin. However, accidentally, the significant limitations of the method emerged. Under the electric field action, the blood corpuscles adsorbed, during the electrolysis in stagnant liquid in an electrolyzer, on the surface of electrodes and blocked them. The circuit current was interrupted. The attempt to recover the lost electrical conductivity by rigid mechanical actions upon the electrodes unblocked them and the electrolysis could continue as long as required, but, at the same time, a number of other problems emerged, which prevented the development in this area. The investigations followed the way of electrolysis of sodium chloride aqueous solutions with further introduction of electrolysis products into the blood.

The process development received a boost after the elaboration of the method of direct electrochemical oxidation of blood by its electrolysis directly in the blood vessel. This technology significantly simplifies the method and the process instrumentation. The test on animals demonstrated its efficiency in various diseases. At the same time, a new capability of the method, previously inaccessible, was discovered. As the electrolysis is conducted directly in the blood flow, it allows saturating the blood with elemental hydrogen. High hydrogen reducibility contributes to the reduction of nonequilibrium radicals in the course of cancer chemotherapy, facilitating the recovery process. This was the impulse to develop a variety of processes that allow selectively saturating blood either only with hypochlorite or only hydrogen, depending on the aim of the method application.

12.2   DIRECT ELECTROCHEMICAL OXIDATION OF BLOOD

Here, important is the issue of creating artificial systems modeling the liver detoxicating functions. This is conditioned by the fact that the existing methods of extracorporal detoxication (hemodialysis, hemosorption) are not efficient enough in removing hydrophobic toxins (Vasiliev et al., 1991).

Electrochemical methods can be used to clean blood and tissues from poisonous and ballast substances by their hydroxylating oxidation. The idea of modeling the liver detoxicating function with the help of electrochemical oxidation was formulated earlier (Yao and Wolfson, 1975). Despite the profound interest in the method of electrochemical oxidation, it did not find proper clinical application as, for the first time, only the principal capabilities of organism detoxication by direct electrooxidation of blood and other biological fluids were considered.

The systemic investigations were carried out to reveal the principal capabilities of electrooxidation of hydrophobic toxins in blood, lymphatic fluid, and plasma, to study kinetics and oxidation mechanism of typical toxins, to define the influence of electrochemical oxidation on different homeostasis indexes of intact animals, as well as to demonstrate how much effective is the electrochemical oxidation when used at the organism level (Vasiliev et al., 1991).

In this research, it was found out that electrooxidation of organic substances can proceed on different electrodes, but the majority of typical toxins are oxidized with a sufficient rate only on electrodes—catalysts of platinum group metals. A platinum electrode is the most suitable catalyst for modeling the hydroxylating function of cytochrome P-450 since practically all toxins are oxidized on it. The preliminary research studies, conducted by different authors and referenced in this paper, are dedicated to the investigation of electrooxidation of various endogenic toxins on the platinum anode: bilirubin, cholesterol, urea, glucose, xenobiotics: methanol, ethanol, phenol, formaldehyde, and barbituric acid derivatives.

The device for electrochemical oxidation of blood in the form of an electrochemical cell with a volume of up to 150 mL equipped with input and output nozzles was tested. The cell is equipped with two electrodes, one of them is platinum. It is noted that under the same current load the increased electrode actual surface leads to a sharp increase in the blood trauma due to the increased number of blood corpuscles getting into the field of a double electric layer. Besides, during the blood contact with the electrodes, the development of blood clots is observed, the electrode surface gets rapidly covered with proteins and decomposed blood corpuscles, which block the electrode surface, and the electrooxidation process is stopped. The current does not pass through the cell. Two methods are proposed to respond to this phenomenon: the method of vibrating electrodes (Sergienko et al., 1984) and method of oxidation in the boiling bed (Vasiliev et al., 1986). The electrode vibration with a frequency of 22 Hz and an amplitude of 2 mm provides a normal passage of current through the cell. The method of oxidation in the boiling bed means that the process proceeds at the current high densities when its main part is spent on water electrolysis with oxygen and hydrogen evolution. The evolved gas bubbles form a boiling bed near the electrodes preventing the formation of a condensed film of proteins and blood corpuscles on the electrode surface. It is pointed out that both methods allow conducting the blood electrolysis for a long time. It should have generated some optimism in further works; however, in the course of extended systematic research studies, a broad range of the method applicability opportunities was found (Petrosyan, 1991); in addition, the method of direct electrochemical oxidation was considered unpromising at that time due to the complexity of practical implementation and low reliability of operation under clinical conditions.

Therefore, Petrosyan (1991) worked out and produced the device for electrochemical detoxication with indirect electrochemical oxidation of blood. The electrolysis of physiological saline is carried out in the electrochemical cell, and the product is introduced into the blood system with the help of a dropper tube. The method has shortcomings; thus, the volume of the solution introduced should not exceed 0.1 of the total blood volume at a low concentration of the final product, decreasing its efficiency.

Rudenok et al. (2004), for the first time, were able to solve the problem of blood corpuscles sticking onto the electrode surface and eliminate the effect of spontaneous stoppage of current flowing through the system by conducting the electrolysis inside the blood vessel without applying rigid mechanical actions upon metal electrodes. The mechanism of this phenomenon is still not entirely clear. Probably, the defining role is played here by "flow potential" preventing the blood corpuscles to nonreversibly adsorb as a dielectric barrier on the electrode surface. The method of direct electrochemical oxidation of blood envisages the electrolysis directly in the blood vessel based on bipolar polarization of the extended electrode. The process modeling in the electrochemical cell demonstrated that with the bipolar inclusion of the extended wire platinum electrode, the potential is distributed along its length, and at one of its ends the polarization potential corresponds to the oxidation process of chloride ion, and at the second one (cathode) the polarization potential corresponds to the potential of hydrogen reduction (Figure 12.1). Here, the detoxicating action is made by hypochlorite being synthesized but not by the electrical current in the blood (Rudenok and Marasinskaya, 2005; Rudenok et al., 2008).

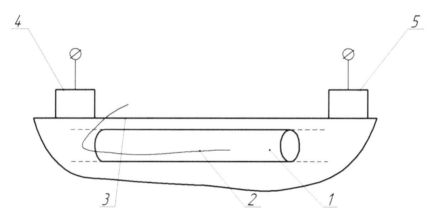

FIGURE 12.1   Scheme of direct electrochemical oxidation of blood. 1, blood vessel; 2, wire electrode; 3, skin covering; 4 and 5, applicator electrodes.

The hypochlorite action was positively tested on rabbits and calves with staphylococcal infection and pneumonia (Rudenok et al., 2004, 2011), and the method was introduced in one of the veterinarian clinics to cure skin conditions of dogs (Shabalina et al., 2006). More detailed investigations of the method using biochemical and other techniques were clinically made (Rudenok et al., 1999).

The experiments were conducted on 18 nonpedigree dogs of 1.5–2 years old, with the bodyweight from 20 up to 30 kg, selected by the principle of analogs. Three groups of six animals in each were formed: reference—clinically healthy, and the first and second test groups with the diagnosis of moist eczema on the back.

The reference animals underwent morphological and biochemical blood investigations, and the microelectrophoretic activity of their erythrocytes was determined with the help of the device "Cyto-Expert." The data served as a reference. The dogs in the first test group underwent medical treatment following the scheme described below, without the direct electrochemical oxidation (DECO) of blood. The dogs in the second test group underwent medical treatment in combination with blood DECO. The eczemas of dogs were medically treated following the traditional techniques by parenteral administration of glucocorticoids (4 mg/animal), gamavit (following the instruction), and spot treatment with antiseptics (terramycin).

The animals in the second test group additionally underwent blood DECO with the help of the device for organism detoxication to treat endogenic and exogenic intoxication conditioned by the accumulation of various toxic substances. For this, the conductor of platinum wire was introduced into the forearm cavity of the lateral subcutaneous vein. Two bipolar electrodes connected with the direct current source were applied to the skin covering at end parts, and 3 mA current was imposed within 15 min. The procedure was repeated 10 times every second day.

Before the experiment and then after 1, 3, 7, 14, 21, and 28 days, all sick animals underwent clinical checkup, morphological and biochemical blood investigations, and the microelectrophoretic activity of their erythrocytes was determined with the help of the device "Cyto-Expert" for expressing the diagnostics of endotoxemia and microelectrophoresis of cells. The changes in the electrophoretic activity of erythrocytes in the process of eczema treatment by various methods allowed the authors to speak about the intoxication decrease and (or) its elimination.

The blood investigation of the dogs in the reference group demonstrated that the main morphological and biochemical indexes were within the physiological norm. The following blood changes of the dogs in both test groups were found before the treatment: decrease in the number of erythrocytes and hemoglobin, ESR increase, leukocytosis, eosinophilia, neutropenia, and lymphocytosis. The conducted investigations of the blood erythrocyte content when treating animals with eczema applying blood DECO demonstrated their number increase at all investigation stages. By the end of treatment, they exceeded the reference data by 2.78%. In the dogs in the first test group, even by the end of therapy, the index was by 8.33% lower in comparison with that of the clinically healthy ones. Hemoglobin in the blood of the dogs in the first and second test groups before the experiment was 104.0 ± 1.26 and 102.0 ± 1.11 g/L, respectively, and 125.0 ± 3.6 and 135.0 ± 4.0 g/l, respectively, after the experiment. In the dogs in the first test group, it did not reach the reference level by 6.72%, in comparison with the second test group. The data obtained indicate a favorable influence of this treatment method on the dynamics of erythrocytes and hemoglobin.

It was found that the amount of leucocytes in the blood of the dogs in the first and second groups starts decreasing during the treatment, but this process is more intensive in the dogs additionally treated with blood DECO. As can be seen from Table 12.1, the amounts of eosinophils, stab neutrophils, and lymphocytes decrease and approach the indexes of the reference group in 28 days after starting the treatment. This indicates the arresting of allergic and inflammatory reactions. Neutropenia and lymphocytosis, indicating the antigenic stimulation of the organism, disappear by the end of treatment. However, in the dogs in the first test group, the amount of lymphocytes decreases considerably, possibly being the reaction to the administration of glucocorticoids, and the immune suppression, resulting in organism weakening and emergence of a new pathology of these animals in the future, can develop. The dogs in the second test group, additionally treated with blood DECO, do not have such immune suppression that can be recommended as the way of immunity improvement.

The positive dynamics of the total protein in the animals of both test groups is observed. However, the most vivid changes are seen in the dogs treated with medicals and blood DECO. The increased level of protein exchange also indicates the improved immunity response from the animal organism as the reaction to blood DECO.

Aspartate aminotransferase (AST) and alanine aminotransferase (ALT) are important indexes in the evaluation of the medical effect of the above eczema treatment schemes. Before the treatment, the amount of AST in the animals in the first and second test groups exceeded the indexes of clinically healthy animals in 2.8 and 3.3 times, respectively. However, the AST content in the blood serum of the dogs in the first test group exceeded the reference data by 86.67%, even by the end of treatment, and in the dogs in the second test group, in which blood DECO was applied, by only 13.33%.

TABLE 12.1 Leukogram of the Dogs During Eczema Treatment by Various Methods

Indexes	Animal Groups	Before Treatment	After 7 days	After 14 days	After 28 days
1	2	3	6	7	9
Eosinophils, %	Reference	$4.0 \pm 0.4^*$			
	1 test group	$7.5 \pm 0.47^*$	$5.8 \pm 0.44^*$	$4.8 \pm 0.34^*$	$4.2 \pm 0.23^*$
	2 test group	$7.5 \pm 0.73^*$	$5.5 \pm 0.47^*$	$4.2 \pm 0.44^*$	$4.0 \pm 0.40^*$
Stab neutrophils, %	Reference	$3.5 \pm 0.24^*$			
	1 test group	$5.2 \pm 0.18^*$	$4.3 \pm 0.37^*$	$4.0 \pm 0.40^*$	$3.7 \pm 0.37^*$
	2 test group	$5.3 \pm 0.23^*$	$4.8 \pm 0.18^*$	$4.3 \pm 0.37^*$	$3.3 \pm 0.24^*$
Segmented neutrophils, %	Reference	$62.7 \pm 0.83^*$			
	1 test group	$55.5 \pm 0.84^*$	$60.8 \pm 0.87^*$	$64.3 \pm 0.67^*$	$69.3 \pm 0.97^*$
	2 test group	$55.7 \pm 0.73^*$	$58.2 \pm 0.91^*$	$60.5 \pm 0.84^*$	$62.8 \pm 0.72^*$
Lymphocytes, %	Reference	$28.1 \pm 0.5^*$			
	1 test group	$30.0 \pm 0.80^*$	$27.3 \pm 0.61^*$	$25.2 \pm 0.71^*$	$20.5 \pm 0.4^*$
	2 test group	$29.8 \pm 0.72^*$	$29.3 \pm 0.83^*$	$28.8 \pm 0.66^*$	$27.6 \pm 0.2^*$
Monocytes, %	Reference	$1.7 \pm 0.37^{****}$			
	1 test group	$1.8 \pm 0.34^{***}$	$1.8 \pm 0.34^{***}$	$1.7 \pm 0.37^{****}$	$2.3 \pm 0.37^{**}$
	2 test group	$1.7 \pm 0.37^{****}$	$2.2 \pm 0.34^{**}$	$2.2 \pm 0.34^{**}$	$2.3 \pm 0.37^{**}$

Note: $^*P<0.001$, $^{**}P<0.002$, $^{***}P<0.005$, $^{****}P<0.01$.

Before the treatment, the ALT content in the blood of the dogs in the first and second groups differed from the indexes of clinically healthy animals by 2.2 and 2.4 times, respectively. Despite the fast drop in the ALT amount in the animals in both test groups, the content of this ferment in the dogs, who underwent only the medical treatment, did not reach the level of the clinically healthy ones and differed by 1.4 times. In the animals in the second test group, who were treated with blood DECO in combination with the basic traditional therapy, the ALT index reached the level of clinically healthy dogs.

Thus, blood DECO elicits the positive reliable dynamics of such laboratory indexes as AST and ALT, important during eczemas. The investigations carried out demonstrated that blood DECO inclusion into the treatment scheme during eczema treatment favorably influences the dynamics of alkaline phosphatase, total bilirubin, cholesterol, creatinine, and urea.

Since eczema is accompanied by intoxication, the authors applied the method of express diagnostics of endotoxemia based on microelectrophoresis with the help of the device "Cyto-Expert" (Table 12.2).

TABLE 12.2 Dynamics of Electrophoretic Activity of Erythrocytes in the Dogs During Eczema Treatment by Various Methods ($P < 0.001$)

Indexes	Animal Groups	Before Treatment	After 7 days	After 14 days	After 28 days
1	2	3	6	7	9
Oscillation amplitude of erythrocytes, mcm	Reference	13.69 ± 0.71			
	1 test group in % to the reference	5.49 ± 0.63	7.91 ± 0.32	9.1 ± 0.39	11.2 ± 0.61
		40.1	57.8	66.5	80.5
	2 test group in % to the reference	5.58 ± 0.53	8.99 ± 0.41	10.7 ± 0.38	13.4 ± 0.52
		40.8	65.7	78.2	98.0
Share of mobile erythrocytes, %	Reference	96.5 ± 0.5			
	1 test group	59.9 ± 0.6	71.8 ± 1.2	76.5 ± 0.6	83.7 ± 0.5
	2 test group	59.1 ± 1.0	74.9 ± 1.0	81.2 ± 0.7	96.0 ± 1.0

The investigations revealed the following. Before the treatment, the amplitude of erythrocyte oscillations in the animals in the first test group differed from the reference by 40.1% and in the animals in the second test group by 40.8%. By the end of treatment, this index maximally approached the data of the group of clinically healthy animals. The number of mobile erythrocytes in relation to their total number demonstrated the tendency to growth: from $59.9\% \pm 0.6$ up to $83.7\% \pm 0.5$ and from $59.1 \pm 1.0\%$ up to $96.0 \pm 1.0\%$ in the dogs in the first and second test groups, respectively. However, this process was more intensive in the animals treated with DECO. Thus, blood DECO, in combination with traditional eczema treatment scheme, results in a reliable stable increase in the number of mobile erythrocytes in the animals of the second test group in a shorter period of time in comparison with the animals in the first test group treated only with medications that should prove the DECO detoxication effect.

Thus, blood DECO, following the above technique, activates the production of erythrocytes, white blood cells, increases blood hemoglobin, and also provides vivid biostimulating and detoxicating actions. Based on the aforesaid, the authors make the conclusion that blood DECO inclusion into the intensive therapy complex during dog eczema treatment improves metabolic processes, increases the activity of the organism protective factors, and contributes to faster arresting of inflammatory reaction and also the healing of the affected skin areas.

As indicated before, the problems of blood corpuscles sticking onto the electrode surface in the course of direct electrolysis in blood resulted in the necessity to develop the methods of indirect electrochemical electrolysis. As the process was conducted in a stagnant solution in a remote cell, it contributed to the permanent loss of the second important product of electrolysis–elemental hydrogen. In the process of long-term accumulation of hypochlorite in the solution, hydrogen completely volatilized from the liquid and was not considered by the experimentalists.

At the same time, it is known that the introduction of hydrogen into the blood decreases the necrologic effect in the presence of chemotherapy medications, not decreasing the therapeutic effect (Nakashima-Kamimura et al., 2009). Moreover, since in the process of electrolysis, the equivalent amount of hydrogen is evolved in the blood vessels simultaneously with hypochlorite, the authors attempted to use the electrolysis process in the blood vessel to accumulate hydrogen in the blood (Rudenok et al., 1999). This allowed coming close to the possibility of experimental detection of the efficiency of hydrogen obtained during the electrolysis as the deoxidant of nonequilibrium radicals being the reason of the side effects of chemotherapy in oncology. In this regard, it was necessary to expand the functional capabilities of the device due to the selective synthesis of either only hydrogen or only hypochlorite in the blood flow. For that purpose, the electrolysis technique was changed; the device scheme is shown in Figure 12.2.

The device consists of wire electrode 1, one part of which is submerged into blood vessel 2 and the second part is submerged into the attachable vessel 3 with the solution of electrolyte 4. Porous diaphragm 5 permeable for solution 4 is mounted in the bottom part of vessel 3. Additional electrode 6 is submerged into solution 4; the second additional electrode 7 is applied to the surface of skin covering 8 near the end of wire electrode 1 submerged into the blood. Electrodes 6 and 7 are connected to the direct current source (not indicated) via contacts 9, and vessel 3 is applied to skin covering 8 from the side opposite to the place, where electrode 7 is applied, with the possibility of electrical contact of solution 4 with the skin covering via porous diaphragm 5. When the electric current flows between electrodes 6 and 7 along the organism tissues, wire electrode 1, being on the way of the current lines, is polarized so that the part of wire electrode 1 submerged into the blood in blood vessel 2 is polarized to the potential with the sign opposite to the sign of electrode potential of electrode 7. The second part of wire electrode 1, being above the surface of skin covering 8 and submerged in the solution of electrolyte 4 inside attachable vessel 3, is polarized to the potential with the sign opposite to the sign of electrode potential of electrode 6.

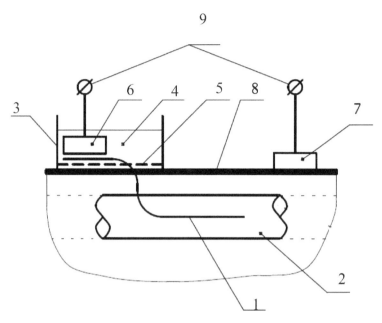

FIGURE 12.2 Scheme of direct electrochemical oxidation of blood, which allows synthesizing only either hypochlorite or hydrogen in the blood flow (Patent no. 2566199).

Thus, wire electrode 1 is polarized by the bipolar scheme in such a way that its one end is polarized cathodically and the second one anodically. At the same time, oxidation proceeds on the surface of one part of wire electrode 1 and reduction on the second one. Depending on the polarity of electrode 7, at the end of wire electrode 1, either chlorine ion in the blood plasma will be oxidized with hypochlorite ion formation or water molecules will be reduced with elemental hydrogen formation.

Accordingly, the opposite electrode processes will proceed in attachable vessel 3 on the surface of the part of electrode 1 submerged into solution 4. Consequently, the operation conditions of electrode 1 model the operation conditions of the bipolar electrode split by double diaphragms 5 and 8, excluding the penetration of the electrolysis products from one electrode space into another.

The literature review dedicated to the perspective area of blood electro-chemical oxidation directly in the blood vessel, in particular, revealed a great interest of researchers in this direction. However, mostly, the problems were detected, the solution of which is still far to be clear. The experimental and ethical difficulties do not give a chance to rapidly complete all the tasks put. However, the promising perspectives can become an incentive to the in-depth study of this problem.

12.3   CONCLUSION

The ways for developing the method of organism detoxication and treatment by an electrochemical synthesis of sodium hypochlorite and elemental hydrogen in the bloodstream immediately in the blood vessel from the components contained by the blood are discussed.

KEYWORDS

- **electrochemical oxidation**
- **blood**
- **detoxication**

REFERENCES

Nakashima-Kamimura N., Mori T., Ohsawa I., Asoh S., Ohta S., Molecular hydrogen alleviates nephrotoxicity induced by an anti-cancer drug cisplatin without compromising anti-tumor in mice. Cancer Chemotherapy and Pharmacology 2009, 64, 4, 753–761.

Petrosyan E.A. Pathogenetic principles and substantiation of suppurative surgical infection treatment by the method of indirect electrochemical oxidation: Extended abstract of dissertation of doctor of medical sciences. Leningrad, 1991.

Rudenok V.A., Alimov A.M., Zakomyrdin A.A., Milaev V.B. Electrochemical synthesis of hypochlorite and hydrogen in blood flow. In Proceedings of Kuban State Agrarian University. Kuban State Agrarian University Publishers, Krasnodar, 1999, pp. 181–182.

Rudenok V.A., Marasinskaya E.I., Zakomyrdin A.A. (applicant and patent holder). Organism detoxication method and device for the method implementation: Patent 2229300, Russian Federation, MPK 7A61K 33/14/, –No. 2002120848/14; applied on July 30, 2002, published on May 27, 2004, Bulletin No. 15, 5 p.

Rudenok V.A., Marasinskaya E.I. Gas emission during electrolysis in a blood vessel. Modern problems of agrarian science and ways of solving them. In *Proceedings of All-Russian Scientific Practical Conference*, Feb. 15–18, 2005, Izhevsk State Agricultural Academy, Izhevsk, 2005, pp. 210–212.

Rudenok V.A. Marasinskaya E.I., Zakomyrdin A.A. Organism detoxication by direct electrochemical oxidation of blood. Veterinary, 2008, 4, 41–44.

Rudenok V.A., Zakomyrdin A.A., Marasinskaya E.I. Synthesis of sodium hypochlorite medication by direct electrochemical oxidation of blood. In *Topical Problems of Veterinary Pharmacology, Toxicology and Pharmacy: Proceedings of III Session of Pharmacologists and Toxicologists of Russia*. Saint Petersburg Academy of Veterinary Medicine Publishers, Saint Petersburg, 2011, pp. 390–394.

Sergienko V.I., Martynov A.K., Khapilov N.A., Author's certificate 1074493 USSR. Bulletin, 1984, No. 7.

Shabalina E.V., Rudenok V.A., Milaev V.B., Kochurova N.V. Use of the method of direct electrochemical oxidation of blood in the complex treatment of dermhelminthiasis of dogs caused by *Demodex canis*. In Veterinary Medicine for Domestic Animals: Collection of Papers. Kazan Academy of Veterinary Medicine, Kazan, 2006, Vol. 3, pp. 139–141.

Vasiliev Yu. B., Grinberg V.A., Guseva E.K., Chechkov A.A. Electrochemistry, 1986, 22.

Vasiliev Yu. B., Sergienko V.I., Grinberg V.A., Martynov A.K. Removal of toxins from the organism with the help of electrochemical oxidation. Issues of Medical Chemistry, 1991, 37(2), 74–78.

Yao S.J., Wolfson S.K. US Patent No. 3878564, Apr. 22, 1975.

Index